Crisis Management of Chronic Pollution

Crisis Management of Chronic Pollution

Editor

Dalmiro Martell

scitus academics

Crisis Management of Chronic Pollution
Edited by **Dalmiro Martell**

ISBN: 978-1-68117-709-0
Library of Congress Control Number: 2016958729

SCITUS Academics LLC,
616, Corporate Way, Suite 2, 4766,
Valley Cottage, NY 10989

www.scitusacademics.com

Notice

Printed in United States of America on Acid Free Paper

Preface

Chronic pollution is permanent pollution caused either by repeated or possibly continuous emissions of pollutants or by the presence of certain pollutants. Today, we all recognize the importance of pollutants in the air or in water, which contribute to poor health. Measures of air quality are usually reported in conjunction with our daily weather, in addition to the impacts of a lack of access to safe drinking water or, possibly, pollution from industry into lakes and rivers. In many cases, clear links between types and levels of specific contaminants in the air and in water and their health effects have been established. Consequently, the effects of pollution on human health must be assessed for both indoor and outdoor environments. Since the 19th century industrial revolution, the health risks associated with the release of heavy metals and toxic chemicals from manufacturing plants have also become a much more serious health problem. The effects of soil pollution on our health have, however, been considerably lower until recently. While studies to know the real costs are in their infancy, soil contaminants can be responsible for millions of dollars in health effects. Cancers (arsenic, asbestos, dioxins), neurological damage and lower IQ (lead, arsenic), kidney disease (lead, mercury, cadmium) and skeletal and bone diseases (lead, fluoride, cadmium) are serious problems. Soil pollution can have harmful effects on human health. The harmful effects of soil pollution can be caused by direct contact with contaminated soil or by contact with other resources, such as water, which can be directly exposed to contaminated soil. It is particularly important to have very heavy metals and persistent organic chemicals. Human activity introduces heavy metals (such as cadmium, arsenic and mercury) through mining, smelting, industry, agriculture and fossil fuel burning into our soils. Our disposal of materials such as paint, electronic waste and sewage that contain heavy metals also contributes to the concern of heavy metal contamination.

This book is a compilation of all works carried out over the last decade or more and links them to the actions of decision-makers and the scenarios of stakeholders. This text bridges a gap in chronic pollution literature.

The intention of the text is to bring together academic and industry researchers and practitioners to focus on the development of advanced technologies to overcome chronic pollution. It highlights emerging technologies and strategies that facilitate a safe environment for future generations.

Table of Contents

1

CHAPTER

The Investigation and Assessment on Groundwater Organic Pollution

Hongqi Wang[1], Shuyuan Liu[1] and Shasha Du[1]

[1] College of Water Sciences, Beijing Normal University, Key Laboratory of Water and Sediment Sciences, Ministry of Education.Beijing, China

1. INTRODUCTION

Groundwater is an important part of the water resource. It plays an irreplaceable role in supporting the national economy and social development. In China, more than 1/3 of the total water resources are utilized. As surveys shown, over 400 cities of all exploit groundwater. More seriously, many of them use groundwater as the only source of water supply.

A series of problems emerge gradually with the utilization of groundwater. Just as river waters have been over-used and polluted in many parts of the world, so have groundwater. The organic solvents and dioxins pollution of Love Canal occurred in 1978 is one of the most widely known examples, which contributes high rates of cancer and an alarming number of birth defects. Similar things occur frequently in recent decades. Governance of groundwater is so urgent a major matter of peace and prosperity. After years of researches, the nature and pollution mechanism of the contaminants in the groundwater have already got comprehended.

General scope of the organic contamination in groundwater is reviewed in this chapter. We will detail account the types of groundwater organic contamination, the pollution source of groundwater. and the fate and transport of chemicals in groundwater. Also a detailed description of the investigation and assessment method in this chapter. At last, we give some comments and suggestion on the groundwater investigation and assessment.

The figure 1 described some source of groundwater contamination, and the transport of chemicals in groundwater. We can see the landfills, leaking sewers, oil storage tanks, pesticides and fertilizer, and septic tank in the picture, all of these could be the pollution source of groundwater. We also can know the groundwater transport and flow in the unsaturated zone and saturated zone.

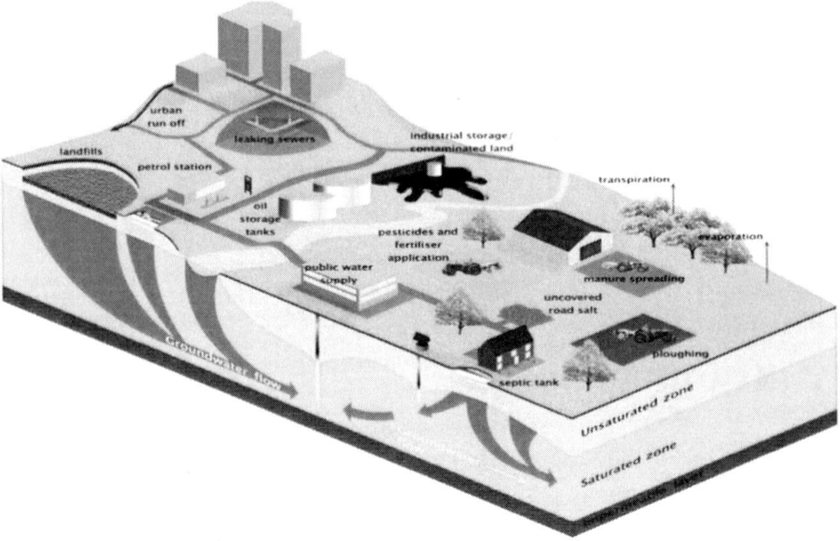

FIGURE 1. Diagram of the Groundwater Pollution [1].

2. A REVIEW OF ORGANIC COMPOUNDS IN GROUNDWATER

Many studies have been conducted since 1970 to characterize concentrations of organic compounds in groundwater. In 1977, 16 drinking water wells have been closed in Gray town of Maine state because of there were at least 8 synthetic organics that were detected in drinking water wells. And in 1986, there were at least 33 organics that were detected in drinking water wells in USA [2]. It has been reported trace organic pollutants to be detected in all of 50 states. The U.S. Geological Survey(USGS) collected and assorted the test data the 1926 drinking water wells in the nation's rural areas from 1986 to 1999. And at least one VOCs were detected from 232 wells and the positive rate was 12%, with the highest positive rate were Chloroform, tetrachloroethylene and so on [3].

Similar conditions are to be found in other countries. In the 80s of last century, based on an inventory of the presence of halogenated substances in raw water of 232 groundwater pumping stations in The Netherlands a compilation of more than 100 organic substances identified in contaminated groundwater, the detection rate of trichloroethylene up to 67% [4]. The organic pollutants could be detected in groundwater in Britain. Flordward studied on 209 water supply wells in Britain shown that the main pollutant in the groundwater are the trichloroethylene and tetrachloroethylene. Beginning in 1974, Environment Agency of Japan conducted a nationwide comprehensive survey of chemical environmental safety. The trichloroethylene in groundwater was reported for the first time. The European Union is the largest pesticide consumer in the world, more than 600 pesticides were applied. Six of the top 10 were European countries in the pesticide application. Atrazine exists in groundwater all over the Europe, and the content always beyond the European Union drinking water standard(0.1μg/l) 10-100 times.

The research on organic pollutions of China is at starting stage, but there has serious organic contamination events in some areas. Based on a study of water pollution in sewage system in the Gaobeidian Prefecture of Beijing, 1988, organic substances identified in shallow wells and deep wells in South-east agriculture districts in Beijing. And 32 organic substances identified in deep wells, and 52 in shallow wells. Most of that are carcinogens (e.g. Chloroform and benzene) [5]. Analysis of the years groundwater monitoring data, and it is shown that the quality of groundwater is gradually worse.

A study on groundwater organic pollution in region of Beijing, Tianjin and Tangshan conducted by Institute of Chemistry of Chinese Academy of Sciences shows that the type of organic pollutants up to 133 [6]. The researchers Chen Honghan, He Jiangtao and others [7] have summed up the characteristics of organic contamination of shallow groundwater in a study area of the Taihu Lake basin. The results show that the detection probabilities of compounds in groundwater are higher but the concentrations of the compounds are lower. The concentrations of all the components of BTEX and halocarbons are lower than the standards set by the U.S. Environmental Protection Agency (EPA) for drinking water except for benzene in a few sampling sites.

3. TYPES OF GROUNDWATER ORGANIC CONTAMINATION

Different types of groundwater contamination sources can pose different threats to human health and different problems in health risk assessment (table 1).

3.1. Volatile Organic Compounds (Vocs)

Volatile organic compounds (VOCs) are organic compounds with chemical and physical properties that allow the compounds to move freely between water and air. VOCs have been used extensively in industry, commerce, and households in the United States since the 1940's. Many products contain VOCs including fuels, solvents, paints, glues, adhesives, deodorizers, refrigerants, and fumigants. In general, these compounds have low molecular weights, high vapor pressures, and low-to-medium water solubilities [8]. Many of these compounds show evidence of animal or human carcinogenicity, mutagenicity, or teratogenicity. And these compounds are quite persistent in groundwater, because of their relatively low biological and chemical reactivity. This persistence is assisted by low temperatures, absence of light and contact with the atmosphere, and comparatively low microbial concentrations typical of groundwater environments. By comparison with other organic compounds, VOCs may be transported for relatively long distances in groundwater, as a result of their relatively weak sorption affinity and their resistance to degradation. Because of human-health concerns, many VOCs have been the focus of national regulations, monitoring, and research during the past 10 to 20 years.

3.2. The Pesticides

Pesticides consist of a large group of chemicals that are used in agriculture and residential settings to control plant and animal infestation. Pesticides are commonly applied on farms, fruit orchards, golf courses, and residential lawns and gardens. There are several different types of pesticides: Herbicides, Insecticides, Nematocides, Fungicides. Some pesticides do not break down easily in water and can remain in the groundwater for a long period of time. Likewise, the insecticide DDT, though banned for nearly twenty years, can still be found at trace levels in some groundwater. After prolonged exposure to high doses, some pesticides can cause cancer; some can also result in birth defects and damage to the nervous system. The use of pesticides and herbicides is one of the main ways of organic pollution of groundwater. Many water wells and irrigation wells have been closed for the byproducts from pesticides and herbicides be detected in shallow water in Colorado.

TABLE 1. 50 organic pollutants most commonly detected in groundwater [11]

Ordering	Component	CASRN	Types
1	Trichloromethane	67-66-3	VOCs
2	Tetrachloroethylene	127-18-4	VOCs
3	1,1,1-Trichloroethane	71-55-6	VOCs
4	Trichloroethylene	79-01-6	VOCs
5	1,1-Dichloroethene	75-35-4	VOCs
6	Methyl tert-butyl ether	1634-04-4	VOCs
6*	cis-1,2-Dichloroethylene	156-59-2	VOCs
8	1, 2, 4-Trimethylbenzene	95-63-6	VOCs
9	Toluene	108-88-3	VOCs
10	Prometon	1610-18-0	Pesticides

11	1,1-Dichloroethane	75-34-3	VOCs
12	Bromacil	314-40-9	Pesticides
13	Tebuthiuron	34014-18-1	Pesticides
14	1, 3-Dichlorobenzene	541-73-1	VOCs
15	1,2-Dichloropropane	78-87-5	VOCs
16	Carbon disulfide	75-15-0	VOCs
17	Deethylatrazine	6190-65-4	Pesticides
17*	1,4-Dichlorobenzene	106-46-7	VOCs
19	Sulfamethoxazole	723-46-6	Medicine
20*	1,2-Dichlorobenzene	95-50-1	VOCs
20	2-Hydroxyatrazine	2163-68-0	Pesticides
22*	Trichlorofluoromethane	75-69-4	VOCs
22	Bentazon	25057-89-0	Pesticides
24	Atrazine	1912-24-	Pesticides

		9	
25	Picloram	1918-2-1	Pesticides
26	Diuron	330-54-1	Pesticides
27*	Benzene	71-43-2	VOCs
27*	Tetrachloromethane	56-23-5	VOCs
29	Chlorobenzene	108-90-7	VOCs
30*	2-Butanone	78-93-3	VOCs
30	Acetone	67-64-1	VOCs
32*	m- + p-Xylene	106-42-3	VOCs
32*	trans-1,2-Dichloro- ethylen	156-60-5	VOCs
32*	1,2-Dibromoethane	106-93-4	VOCs
35	Ethylbenzene	100-41-4	VOCs
36	caffeine	58-08-2	Medicine
37	Isopropylbenzene	98-82-8	VOCs
38	o-Xylene	95-47-6	VOCs

38*	1,1,2-Trichloroethane	79-00-5	VOCs
38*	Bromodichloromethane	75-27-4	VOCs
38*	1,1,1,2-Tetrachloroethane	630-20-6	VOCs
38*	n-Propylbenzene	103-65-1	VOCs
43	Chloromethane	74-87-3	VOCs
44	1,1,2-Trichloro-1,2,2-trifluoroethane	76-13-1	VOCs
45	Dichlorodifluoromethane	75-71-8	VOCs
46	Metolachlor	51218-45-2	Pesticides
46*	Simazine	122-34-9	Pesticides
48	Bromoform	75-25-2	VOCs
48*	Imidacloprid	138261-41-3	Pesticides
48*	1,3,5-Trimethylbenzene	108-67-8	VOCs

3.3. The Other Organic Contamination

Tens of thousands of manmade chemicals are used in today's society with all having the potential to enter our water resources. There are a variety of pathways by which these organic contaminants can make their way into the aquatic environment [9]. If the groundwater is the drinking water sources, there will be potentially dangerous on human health. Pharmaceuticals and other organic contaminants are a set of compounds that are receiving an increasing amount of public and scientific attention. Water samples were collected from a network of 47 groundwater sites across 18 states in 2000 [10]. All samples collected were analyzed for 65 organic contaminants representing a wide variety of uses and origins. Thus, sites sampled were not necessarily used as a source of drinking water but provide a variety of geohydrologic environments with potential sources of organic contaminants. organic contaminants were detected in 81% of the sites sampled, with 35 of the 65 organic contaminants being found at least once. The most frequently detected compounds include N,N-diethyltoluamide (35%, insect repellant), bisphenol A (30%, plasticizer), tri(2-chloroethyl) phosphate (30%, fire retardant), sulfamethoxazole (23%, veterinary and human antibiotic), and 4-octylphenol monoethoxylate (19%, detergent metabolite).

4. SOURCES OF GROUNDWATER CONTAMINATION

Organic contamination includes all of natural and synthetic that could cause adverse effect on human health or ecology environment.

4.1. Natural Pollution Sources

Naturally formed waters such as ocean water and connate brines can be sources of groundwater contamination under certain circumstances. Changes in pumping rates can cause fresh-water aquifers to be contaminated by intrusion of seawater. Similarly, changes in the groundwater flow field or leakage through imperfectly sealed wells can cause contamination of groundwater supply by naturally occurring brines or other poor-quality waters. Generally, trace amount of natural organic compounds existence in groundwater in most of regions. The major is humic acid, especially in forest and grassland. Although itself could not impair the groundwater quality, it could be enhance the heavy metal and other organic matters activities in groundwater.

4.2. Organic Contamination Come From Human Activities

As the human population grows, groundwater pollution from human activity also increases. There are a number of possible sources that could lead to groundwater contamination. Such as crude oil leakage in oil production, organic waste discharge, spills and leaks from underground storage tank and so on.

4.2.1. City and industrial wastewater

The treatment and disposal of sewage present health risks in both developed and undeveloped countries. In undeveloped countries, sewage may be directly applied to the land surface. In more developed areas, sewage is generally transported to municipal treatment plants or disposed of in septic tanks and cesspools. Groundwater contamination can result in all these cases. Sewage provides a source of pathogens, nitrates, and a variety of organic chemicals to groundwater. Land application of sewage can provide a direct contaminant source via infiltration. Treatment plants can act as contaminant sources in several ways. Leaks may occur in sewer lines and infiltration may occur from the ponds and lagoons within the treatment plants. In addition, the sewage sludge that is a product of sewage treatment processes is often disposed on land in conjunction with agricultural activity. Depending on the characteristics of the sludge, the soil characteristics, and the application process, such land application can act as a large non-point source of groundwater contamination. Land disposal of treated waste water can pose comparable risks. Depending on hydrogeologic conditions, septic tanks and cesspools may allow untreated sewage to enter the groundwater flow system. In addition, use of solvents to clean out the systems can cause groundwater contamination by synthetic organic compounds. The material cleaned out from septic tanks must eventually be disposed of, often by land application.

Industrial Wastewaters are applied to land in ponds or lagoons that are either designed to percolate the liquid into the soil or to store and/or evaporate the liquid above ground. In either case, such facilities act as potential groundwater contamination sources. Facilities designed to intentionally infiltrate into the ground include cooling ponds for power generation and for other industrial processes. The liquids in such facilities may contain potentially hazardous materials. Storage and evaporation ponds are often lined to prevent infiltration, but are likely to act as groundwater contamination sources under some circumstances, depending on surface runoff characteristics, the integrity and permeable of the liner(s), and the groundwater flow system. Poorly designed evaporation ponds may, in many cases, function as infiltration ponds.

In the United States, the big city and small town are commonly found in contaminated groundwater. An test on 39 groundwater supply in small towns conducted by the U.S. EPA, it reported that 11 VOCs could be detected in treated or untreated groundwater [12].

4.2.2. Land disposal of municipal and industrial waste

Land disposal of solid waste is the groundwater contamination source of most current concern to the general public in many developed countries and of most current regulatory interest.

Solid waste can be disposed in landfills, facilities engineered to safely contain the waste. While landfills may often prevent exposure of solid waste at the land surface, many landfills provide a direct connection with groundwater. In the past, landfill siting was based on the availability of inexpensive, undeveloped land requiring little modification for waste disposal, rather than on hydrogeologic suitability. Disposed materials often are very susceptible to leaching into groundwater.

Landfills may be grouped according to the type of materials they contain. Municipal landfills accept only non-hazardous materials, but are still likely to contain materials which pose potential health risks. Industrial landfills may contain either "hazardous" or "non-hazardous" materials. Until recently, little was known about how they were operated or what they contained. Open dumps and abandoned disposal sites generally have no engineering design. Their connection with the groundwater system and the type of materials present is often unknown. It is often in abandoned disposal sites that large volumes of highly toxic materials are found. The most hazardous solid waste disposal generally results from industrial and manufacturing activities as well as some governmental energy and defense activities. Populations of both developed and developing countries, where there is current or historical industrial activity, face potential health risks from solid waste disposal. It is reported that there will be the highest content and most types of organic contaminants in groundwater which is near the landfills. If there has 1 kilometers distance it still exist in the groundwater [13].

4.2.3. Petrochemical pollution

In recent years, there has been increasing awareness of the large number of potentially leaking underground storage gasoline tanks. For much of the twentieth century, underground storage tanks were constructed of unprotected carbon steel. Corrosion causes leaks in such tanks over some period of time, ranging from a few years to tens of years. Although the leakage from individual tanks is often small, it is often enough to contaminate a large volume of groundwater. In addition, the large number of buried tanks-several million in the United States-makes them a potentially significant groundwater contamination source. Above ground storage tanks pose less of a threat than underground tanks. Leak detection and maintenance is easier and the connection with the groundwater system is less direct. However leaks from such tanks may still act as groundwater contamination sources.

4.2.4. Agricultural Activities

Numerous agricultural activities can result in non-point sources of groundwater contamination. Fertilizers, pesticides, and herbicides are applied as part of common agricultural practice throughout the world. These applications can act as sources of contamination to groundwater supplies serving large populations. Whether or not fertilizers, pesticides, and herbicides become sources of groundwater contamination depends on changing hydrogeologic conditions, application methods, and biochemical processes in the soil. In developing countries, animal and/or human waste is used for fertilizer. This is an example of the land application of sewage discussed earlier. There are the same concerns with pathogens and nitrates. The manufactured inorganic fertilizers widely used in developed countries, and finding increasing usage in all countries, also pose the threat of nitrate contamination of groundwater systems. Pesticide and herbicide application provides a source of numerous toxic organic chemicals to groundwater supplies.

Even without the introduction of fertilizers, pesticides, and herbicides, irrigation activities can lead to groundwater contamination. Naturally occurring minerals in the soil can be leached at higher rates leading to hazardous concentration levels in the groundwater. Evaporation of irrigation water can cause evaporative concentration of certain chemicals in the root zone. Flushing of these chemicals can then lead to hazardous concentration levels in groundwater.

Agricultural activities related to animals also can be groundwater contamination sources. These include the feeding of animals and the storage and disposal of their waste. Animal wastes and feedlot runoff are commonly collected in some sort of pit or tank creating the contamination threat described earlier for sewage disposal.

More than 300 pesticides were applied in Asia. The Japan is a country with the largest amount of pesticide on unit area cultivated land. Indonesia, Korea, India and China are the major consumers. But, there did not have pesticides routine monitoring in the developing countries in Asia [14].

4.2.5. Surface Water And Atmospheric Contaminants

Groundwater is but one component of the hydrologie cycle. Groundwater quality is very much influenced by surface-water conditions and vice versa. Contamination of any surface water bodies that recharge the groundwater system is a source of groundwater contamination. This includes "natural" recharge sources such as lakes and rivers as well as "man-made" recharge sources such as artificial recharge ponds/injection wells and infiltration of urban runoff. More generally, it is important to consider the interaction of all environmental sources and pathways of pollution. Environmental contaminant sources cannot be divided into separate, isolated compartments. For example, atmospheric pollution can lead to deposition of hazardous fallout to surface waters and to soils, and eventually lead to groundwater contamination.

5. THE FATE AND TRANSPORT OF CHEMICALS IN GROUNDWATER

5.1. Volatilization

Volatilization occurs in whether the vadose zone or saturated zone when the dissolved contaminants and non-aqueous phase contaminants exposed to gas. The factors affecting volatilization include solubility of the compound, molecular weight and water-saturated state of the geological media. The evaporation rate must be measured fundamentally in order to determine pollutions transporting into the atmosphere, changes of the pollution load in the vadose zone and groundwater. The process that the contaminants of deep soil volatilize to the atmosphere can be assumed as one-dimensional diffusion, which can be described with Fick's second law. Volatilization of the water-soluble organic matter, such as benzene dissolved in water is generally described by Henry's Law [15].

5.2. Adsorption

Adsorption in Soil and sediment makes an important influence on the behavior of organic pollutants. The mobility and biological toxicity reduced as organics are detained in the soil and sediment. Generally, adsorption is affected by sediments and soil properties, such as organic percentage, the type and quantity of clay minerals, cation exchange, pH and the physical and chemical properties of the contaminants. During the adsorption, the organic contaminants in the water adsorbed on the surface of the soil particles by the simultaneous distribution role of both water and solid, the driving force is mainly based on principle of "like dissolves like" and electrostatic adsorption of the polar group, and the following formulas is established [15]:

$$C_{sa} = K_a C_{wa} \quad (1)$$

The equation (1) is existed when the adsorption systems reach equilibrium. Where, Csa is the amount of organic pollutants adsorbed per unit weight of soil particles; Cwa is concentration of organic pollutants;Ka is the total sorption coefficient.

The adsorption of organic contaminants in soil or sediment usually described by Ka (soil absorption coefficient) or Koc (organic carbon absorption coefficient). The former refers to the ratio of the concentration of organic matter in the soil or sediment and its aqueous phase concentration. As well, the latter factor represents the ratio of the concentration of organic matter adsorbed by organic carbon in the soil or sediment and its aqueous phase concentration.

5.3. Biochemical Processes

Microorganisms may play an important role in contamination transformations within groundwater and on the soil. They can act as catalysts for many types of reactions. When modeling biochemical reactions in groundwater, additional processes must be considered. These include the changes in the availability of substrate for the microorganisms to utilize and reactions on the particles that the microorganisms are attached to. When microbial reactions are significant, there is a possibility of clogging of pores due to precipitation reactions or to biomass accumulation [16].

Microorganisms not only influence chemical reactions, but may be contaminants themselves. There is much current uncertainty about the fate and survival time of viruses, bacteria, and larger enteric organisms in groundwater [17-18]. Distribution of microorganisms will vary greatly with depth. Potential outbreaks of waterborne diseases due to biologic pollutants are of particular concern where there is land disposal of human waste (often via septic tanks) and animal waste. The potential for transmittal of waterborne diseases in groundwater is particularly high in areas of rapid velocities such as karst regions.

Biodegradation mainly depends on two factors [19], the intrinsic characteristics of the pollutants (the structure of organics, physical and chemical properties) and microorganism (the activity of microbial populations), and the environmental factors controlling the reaction rate (temperature, pH, humidity, dissolved oxygen). As the U.S. Environmental Protection Agency researched [20], soil microbial degradation of organic pollutants can be expressed as a one-order response equation:

$$\frac{dc}{dt} = -KXC = -k_T c \tag{2}$$

Where, C is the mass fraction of soil organic matter[mg/g]; X is the number of active microbial in the organic matter of soil degradation[10^6 /g]; t is degradation time[d]; K is the one-order biodegradation rate constant [g/(d 10^6)]; k_r is substrate removal constant[d^{-1}].

From the above equation,

$$\frac{C}{C_0} = \exp\left(-KX_t\right) = \exp(-k_T t) \tag{3}$$

Substituted into with half-life formula,

$$t_{\frac{1}{2}} = ln2 / k_T \tag{4}$$

The half-life of degradation of residual contamination is determined.

5.4. Fate And Transport In Unsaturated Zone [21]

In many cases, the receptor medium for release of a contaminant will be the unsaturated zone. In contrast to the saturated zone, pores in the unsaturated zone are not completely saturated with liquid. This fundamentally affects the processes governing flow and chemical transport. A number of processes will affect the contaminant within the unsaturated zone before it enters the saturated groundwater system and potentially is tapped by supply wells. The uncertainties in characterizing releases just described lead to uncertainties in defining the source terms and initial and boundary conditions for modeling unsaturated transport. Analogously, uncertainties in characterizing unsaturated transport processes lead to uncertainties in defining the source terms and initial and boundary conditions for modeling saturated transport.

For the most part, computer simulation of contaminant transport has focused on movement in the saturated zone. Assumptions are made regarding the time required for movement through the unsaturated zone. Often some sort of lag between source release and entry of chemicals into the saturated flow system is introduced into source terms. It is important to be aware of the unsaturated processes that are actually occurring, the uncertainty associated with these processes, and the role of monitoring in reducing these uncertainties.

5.5. Saturated Transport [21]

Once a chemical has been released into the ground and has either moved through the unsaturated zone or directly entered the saturated zone, saturated transport processes will determine if, how fast, and at what concentration a chemical reaches a supply well. A great deal of research has been carried out on understanding and modeling these processes. There is increasing recognition that chemical transport must be viewed as a stochastic process.

The same elements of uncertainty are present for saturated transport as for unsaturated transport. The important differences are that in saturated transport, water content equals porosity, hydraulic conductivity is no longer a function of water content or head, gravity rather than suction head is the driving force, and the scale of concern may be much larger

6. INVESTIGATIONS FOR GROUNDWATER ORGANIC POLLUTION

6.1. Current Situation Investigation

The current situation investigation main contents are as following:

Pollution source investigation: In groundwater polluted areas, investigate the non-point-source, line-source and point-source, and the type, pollution intensity, spatial distribution of natural source.

Investigation of unsaturated zone vulnerability: Investigate the unsaturated zone of thickness, lithological composition, composition, water permeability, the capability of degradation contaminations and so on.

Investigation of the pollution condition at the groundwater: Make sure the category, quantity or concentration of the pollutants, ascertain the pollution range, variation trend and the factors relation. All of these need samples collection in filed and laboratory test.

6.2. Pollution Source Investigation

With the developed, groundwater pollution attracted wide attention. In view of existing situation, we launched the survey of pollution sources, including the following aspects: industrial pollution sources, domestic pollution sources, agricultural pollution sources and surface polluted waters.

6.2.1. Industrial Pollution Sources

According to industrial pollution sources, we must investigate the situation as: the company name, position, sewage, waste residue (tailings) emissions, discharge, scale, pathways and outfall location, types of pollutants, quantity, composition and hazards, and the abandoned site of major polluting enterprises, abandoned wells, oil and survey of solvents and other underground storage facilities.

6.2.2. Domestic Pollution Sources

The survey include the distribution of dumps, scale, waste disposal methods and effects, the generation of dump leaching filtrate and components, geological structure of storage site; amount of sewage generated, treatment and disposal of the way, main pollutants and their concentration and hazards

6.2.3. Agricultural Pollution Sources

Agricultural pollution sources investigation mainly include land use history and current situation, the varieties, numbers, operations, time of farmland application of chemical fertilizers and pesticides, range of sewage irrigation, main pollutants and concentration, the number of sewage irrigation and sewage irrigation amount. The scale of farms and so on.

6.2.4. Surface polluted waters

The surface polluted waters mainly about rivers, lakes, ponds, reservoirs and drains. We survey the distribution of polluted waters, the scales, the utilizations and water quality.

The coastal areas have to survey the situation of seawater invasion and saline water distribution.

7. THE ASSESSMENT ON GROUNDWATER ORGANIC POLLUTION

7.1. The Methods Of Groundwater Organic Pollution Assessment

7.1.1. The four steps of nas

The four steps of NAS was proposed by National Academy of Sciences, United States(NAS), was an assessment method on human health risk that led by the accident, air, water, soil and other medium. The method mainly in the following aspects: the hazard identification (qualitative evaluation the degree of hazards of the chemical substances on the human health and ecological); dose-response assessment (quantitative assessment the toxicity of chemical substances, established a relationship between the dose of chemical substances and the human health hazard); exposure assessment (quantitative or qualitative estimate or calculate the exposure, exposure frequency, exposure duration and exposure mode); exposure attribute (using the data to estimate the strength of the health hazards in the different conditions or the probability of the certain health effects). This method can qualitative analysis or quantitative analysis of groundwater contamination, or combine them, the results could be quantify and analysis, and provide more detailed information to the decision-makers.

7.1.2. The Four Steps Of Epa

In 1989, U.S. Environmental Protection Agency (EPA) promulgated the "risk assessment guidance for superfund: Human health evaluation manual", there was a similar assessment method to NAS method [22]. The steps following as data collection, exposure assessment, toxicity assessment, risk characterization. Contrast the two methods, NAS is more common methods, the use range wider, suitable for a variety of health risk assessment; the EPA method is more specific, it emphasis on the various parameters of the collection of contaminated sites, for the evaluation of contaminated sites, it more operational.

7.1.3. The mmsoils model

The MMSOILS model is multi-media model which describe the groundwater, surface water, soil and air in the migration of chemicals, exposure and food chain accumulation [23]. Contaminate sites is multi-phase, multi-media complex. The model including the migration and transformation of pollutants module and human exposure module. Migration and transformation module include: (1)atmospheric transport pathway; (2)soil erosion; (3)groundwater migration pathway; (4)surface water pathway; (5)food chain bioaccumulation. Human exposure are: (1)adopt from drinking water, animals and plants and soil; (2)atmospheric volatiles and particulate inhalation; (3)soil, surface water and groundwater contact with skin. The model could be simulate a comprehensive migration pathway and widely used in foreign countries, and the parameters could be analysis the uncertainty.

7.1.4. The Drastic Method

The DRASTIC method is a national standards system that developed by US EPA to evaluation aquifer vulnerability. It including: Depth to Water(D); Net Recharge(R); Aquifer media(A); Soil Media(S); Topography(T); Impact of the Unsaturated Zone Media(I); Conductivity of Aquifer Hydraulic(C). Assignment of each element from 1 to 10, and them proportional to the degree of vulnerability of groundwater. At the same time, each element is assigned a weight, the weight should be reflect the sensitivity of groundwater. The model can objectively assess the groundwater vulnerability of different areas, and its assumption that all regions of the aquifer has a uniform trend. But all the geological, hydrogeological and other conditions are different, and the model calculations defect, the DRASTIC method has some limitations.

7.2. Health–Based Risk Assessment

7.2.1. Estimating population exposure levels

An important step in health risk assessment is the quantification of actual human exposure. Exposure can be expressed as either the total quantity of a substance that comes in contact with the human system or the rate at which a quantity of material comes in contact with the human system (mass per time or mass per time per unit body weight. The exposure assessment evaluates the type and magnitude of exposures to chemicals of potential concern at a site. The exposure assessment considers the source from which a chemical is released to the environment, the pathways by which chemicals are transported through the environmental medium, and the routes by which individuals are exposed. Parameters necessary to quantitatively evaluate dermal exposures, such as permeability coefficients, soil absorption factors, body surface area exposed, and soil adherence factors are developed in the exposure assessment. Exposure to chemicals in water can occur via direct ingestion, inhalation of vapors, or dermal absorption. Ingestion includes drinking of fluids as well as using water for rinsing and cooking of foods. Dermal absorption includes swimming and bathing.

Determination of average exposure levels for a particular population is quite difficult. This is due to difficulties in acquiring sufficient water-quality data, in identifying the exposed individuals, and in quantifying the concentrations in the different exposure pathways. For a given groundwater contamination problem, the U.S. Environmental Protection Agency stresses the importance of identifying both the currently affected population as well as possible changes in future land use. Subpopulations that may be especially sensitive to exposure should also be identified [24].

When attempting to estimate exposure to larger population entire countries, for example other concerns arise. Cothern [25] computed the average population exposure to volatile organic compounds in the United States, based on data from several thousand ground- and surface-water supplies. National exposure was estimated as a straight extrapolation of the concentration intervals from the original data. Best- and worst-case assumptions were applied for handling the "below detectable" category. Crouch *et al.* [26] applied an alternative approach to estimate population exposure levels. Rather than estimating a distribution for exposure, they made the worst-case assumption that individuals are exposed to water at the maximum measured concentration for their water supply.

7.2.2. Health risk calculations

According to the Risk Assessment Guidance for Superfund Volume I: Human Health Evaluation Manual (Part E, Supplemental Guidance for Dermal Risk Assessment) (U.S. EPA) [27], we calculation the dermal absorbed dose (DAD) and ingestion absorbed dose (IAD) [28].

$$\mathrm{DAD} = \frac{DA_{event} \times EV \times ED \times EF \times SA}{BW \times AT} \qquad (4)$$

Where:

DAD=Dermally Absorbed Dose (mg/kg-day),

DA_{event}=Absorbed dose per event (mg/cm^2-event),

SA=Skin surface area available for contact (cm^2),

EV=Event frequency (events/day),

EF=Exposure frequency (days/year),

ED=Exposure duration (years),

BW=Body weight (kg),

AT=Averaging time (days).

$$IAD = \frac{\rho \times U \times EF \times ED}{BW \times AT}$$

(6)

Where:

IAD= Ingestion absorbed dose (mg/kg-day),

ρ= Pollutant concentration in groundwater (mg/L),

U=Drinking amount per days (L/d),

EF=Exposure frequency (days/year),

ED=Exposure duration (years),

BW=Body weight (kg),

AT=Averaging time (days).

The DAD and IAD can be represent with continuous ingestion dose (CDI).

Based on the carcinogenesis of contamination, the risk could be classified into cancer risk and noncancer hazard.

1. Noncancer hazard: Generally, the reaction of the body to non-carcinogenic substance has a dose threshold.

Lower than the threshold, they could not affect our health adversely. The non-carcinogenic risk to represent with hazard index (HI). It is defined as a ratio that continuous ingestion dose with reference dose [28].

$$\mathrm{HI} = \mathrm{CDI} / \mathrm{RfD}$$

(7)

Where: CDI= continuous ingestion dose (mg/kg-days), RfD= reference dose (mg/kg-days).

1. Cancer risk: There does not have dose threshold for the carcinogenic. Once it exist in environments, it will affect human health adversely. Cancer risk will be represent with risk. It is defined as a product of continuous ingestion dose with carcinogenesis slope factor.

$$\mathrm{CDI} \times \mathrm{SF}$$

(8

$$1 - \exp(-\mathrm{CDI} \times \mathrm{SF})$$

(9))

(If the low dose exposure risk>0.01)

Where: SF= carcinogenesis slope factor (mg^{-1}•kg•d)

When calculating the risk of a variety of substances in a variety of ways, figure out all non-cancer risk and cancer risk respectively, then add all risks together. Regardless of synergistic effect and antagonistic effect.

8. THE COUNTERMEASURES AND SUGGESTIONS

8.1. The Countermeasures For Groundwater Pollutions

Groundwater treatment technologies are mainly as follows: pump and treat, air sparging, in-situ groundwater bioremediation and in-situ reactive walls.

8.1.1. Pump and treat technology

Pump and treat is the most common form of groundwater remediation. It is often associated with treatment technologies such as Air Stripping and Liquid-phase Granular Activated Charcoal.

Pump and treat involves pumping out contaminated groundwater with the use of a submersible or vacuum pump, and allowing the extracted groundwater to be purified by slowly proceeding through a series of vessels that contain materials designed to adsorb the contaminants from the groundwater. For petroleum-contaminated sites this material is usually activated carbon in granular form. Chemical reagents such as flocculants followed by sand filters may also be used to decrease the contamination of groundwater. Air stripping is a method that can be effective for volatile pollutants such as BTEX compounds found in gasoline.

For most biodegradable materials like BTEX, MTBE and most hydrocarbons, bioreactors can be used to clean the contaminated water to non-detectable levels. With fluidized bed bioreactors it is possible to achieve very low discharge concentrations which will meet or exceed discharge standards for most pollutants.

Depending on geology and soil type, pump and treat may be a good method to quickly reduce high concentrations of pollutants. It is more difficult to reach sufficiently low concentrations to satisfy remediation standards, due to the equilibrium of absorption (chemistry)/desorption processes in the soil.

At the figure 2, we can know how does pump and treat technology work. This system usually consists of one or more wells equipped with pumps.

When the pumps are turned on, they pull the polluted groundwater into the wells and up to the surface. At the surface, the water goes into a holding tank and then on to a treatment system, where it is cleaned [29].

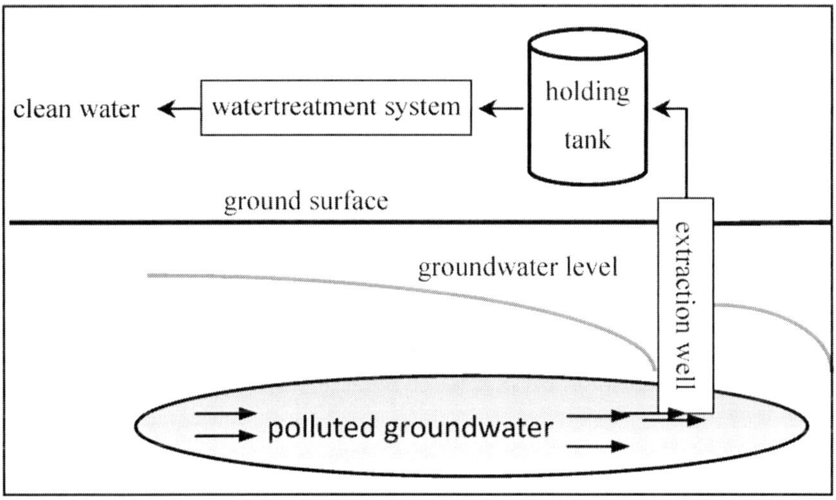

FIGURE 2. Pump and Treat Technology [29]

8.1.2. Air sparging [30]

Air sparging is an in situ groundwater remediation technology that involves the injection of a gas (usually air/oxygen) under pressure into a well installed into the saturated zone. Air sparging technology extends the applicability of soil vapor extraction to saturated soils and groundwater through physical removal of volatilized groundwater contaminants and enhanced biodegradation in the saturated and unsaturated zones. Oxygen injected below the water table volatilizes contaminants that are dissolved in groundwater, existing as a separate aqueous phase, and/or sobbed onto saturated soil particles. The volatilized contaminants migrate upward in the vadose zone, where they are removed, and generally using soil vapor extraction techniques. This process of moving dissolved and non-aqueous volatile organic compounds (VOCs), originally located below the water table, into the unsaturated zone has been likened to an in situ, saturated zone, air stripping system. In addition to this air stripping process, air sparging also promotes biodegradation by increasing oxygen concentrations in the subsurface, stimulating aerobic biodegradation in the saturated and unsaturated zones(figure 3).

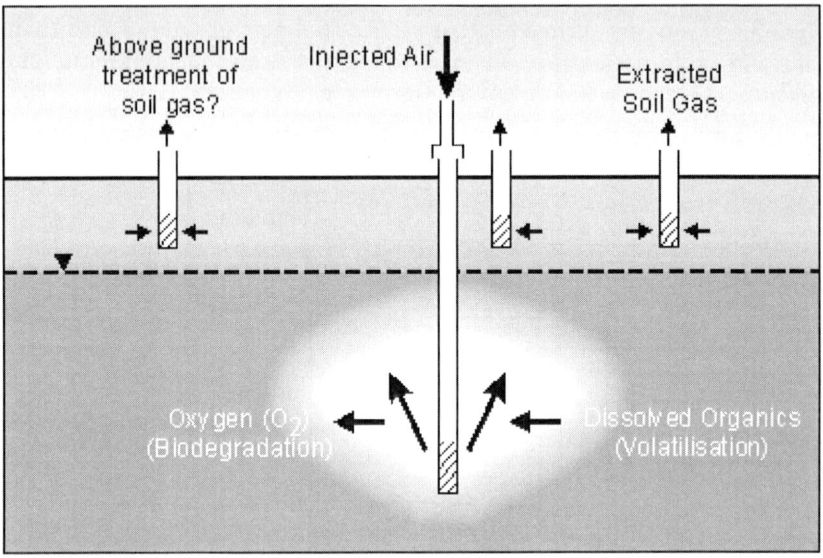

FIGURE 3. Air Sparging [31]

8.1.3. In-situ groundwater bioremediation [32]

In-situ groundwater bioremediation is a technology that encourages growth and reproduction of indigenous microorganisms to enhance biodegradation of organic constituents in the saturated zone. In-situ groundwater bioremediation can effectively degrade organic constituents which are dissolved in groundwater and adsorbed onto the aquifer matrix.

In-situ groundwater bioremediation can be effective for the full range of petroleum hydrocarbons. While there are some notable exceptions (e.g., MTBE) the short-chain, low-molecular-weight, more water soluble constituents are degraded more rapidly and to lower residual levels than are long-chain, high-molecular-weight, less soluble constituents. Recoverable free product should be removed from the subsurface prior to operation of the in-situ groundwater bioremediation system. This will mitigate the major source of contaminants as well as reduce the potential for smearing or spreading high concentrations of contaminants.

In-situ bioremediation of groundwater can be combined with other saturated zone remedial technologies (e.g., air sparging) and unsaturated zone remedial operations (e.g., soil vapor extraction, bioventing).

Bioremediation generally requires a mechanism for stimulating and maintaining the activity of these microorganisms. This mechanism is usually a delivery system for providing one or more of the following: An electron acceptor (oxygen, nitrate); nutrients (nitrogen, phosphorus); and an energy source (carbon). Generally, electron acceptors and nutrients are the two most critical components of any delivery system.

In a typical in-situ bioremediation system, groundwater is extracted using one or more wells and, if necessary, treated to remove residual dissolved constituents. The treated groundwater is then mixed with an electron acceptor and nutrients, and other constituents if required, and re-injected upgradient of or within the contaminant source. Infiltration galleries or injection wells may be used to re-inject treated water. In an ideal configuration, a "closed loop" system would be established. All water extracted would be re-injected without treatment and all remediation would occur in situ. This ideal system would continually recirculate the water until cleanup levels had been achieved. If your state does not allow re-injection of extracted groundwater, it may be feasible to mix the electron acceptor and nutrients with fresh water instead. Extracted water that is not re-injected must be discharged, typically to surface water or to publicly owned treatment works (POTW).

8.1.4. In–situ reactive walls [33]

In-situ reactive walls are an emerging technology that have been evaluated, developed, and implemented only within the last few years. This technology is gaining widespread attention due to the increasing recognition of the limitations of pump and treat systems, and the ability to implement various treatment processes that have historically only been used in above-ground systems in an in situ environment. This technology is also known in the remediation Industry as "funnel and gate systems" or "treatment walls".

The concept of in-situ reactive walls involves the installation of impermeable barriers downgrading of the contaminated groundwater plume and hydraulic manipulation of impacted groundwater to be directed through porous reactive gates installed within the impermeable barrier. Treatment processes designed specifically to treat the target contaminants can be implemented in these reactive or treatment gates. Treated groundwater follows its natural course after exiting the treatment gates. The flow through the treatment gates is driven by natural groundwater gradients, and hence these systems are often referred to as passive treatment walls. If a groundwater plume is relatively narrow, a permeable reactive trench can be installed across the full width of the plume, and thus preclude the necessity for installation of impermeable barriers.

In-situ reactive walls eliminate or at least minimize the need for mechanical systems, thereby reducing the long-term operation and maintenance costs that so often drive up the life cycle costs of many remediation projects. In addition, groundwater monitoring and system compliance issues can be streamlined for even greater cost savings.

Bioventing, also a modification of vapor extraction technology, is briefly contrasted with air sparging. With bioventing, extraction or injection of air into the vadose zone increases subsurface oxygen concentration, promoting bioremediation of unsaturated soil contaminants. This technique is applicable to all biodegradable contaminants, but has been applied most frequently and reportedly most successfully to sites with petroleum hydrocarbon contamination

8.2. The Suggestions For Groundwater Pollutions

The past 40 years, groundwater subjected to pollution, it cannot be ignored that there has a serious threat to human health and ecological security problems. The research on groundwater pollution risk assessment will help understand the relationship between the soil conditions and groundwater pollution, identify the high-risk regions of groundwater pollution, provide a powerful tools for the land use and groundwater resource management, and help the policy maker and managers to develop effective management strategies and protection measures on groundwater. So we can offer some suggestions as following:

1. Continue to strengthen the research on the fate and transport in hydrogeological conditions. Hydrogeological conditions of the contaminated sites have a vital role in organic pollution of groundwater. We should pay attention to the impact that the thickness of the unsaturated zone, the aquifer lithology of unsaturated zone and groundwater, the groundwater runoff conditions on the organic pollution investigation and contaminated aquifer restoration. Unsaturated zone is the only avenue for the organic pollutant into the groundwater system. In the protection of groundwater quality, we should take impact of the physical, chemical, and biological characteristics of the unsaturated zone soil on the transport and degradation of organic pollutants into consideration.

2. In the future research, the natural attenuation of typical organic contamination in groundwater should be reinforce research, especially the organic degradation mechanism of microbes.

3. In recent years, the environmental hormone pollution research and prevention has begun to attract the attention of the world. Environmental hormone research has become the forefront and hot topic of environmental science research. But the mechanism of environmental hormone is not clearly, we should take more attention on these.

4. The research on groundwater pollution risk assessment to be carried out on the typical regions. To provide practical experience on established an reasonable and feasible groundwater pollution risk assessment system.

5. Exerting governmental function adequately and improving the laws, regulations and norms on groundwater quality monitor and assessment. Strengthening the cross-disciplinary exchanges and studies and establishing the groundwater pollution monitor network and the chemical toxicological database.

With entering a new era of environmental protection, the research of groundwater pollution risk assessment is bound to make new contributions to human survival and to protect and improve the natural environment, and to advance the theory research of environmental science.

9. ACKNOWLEDGEMENTS

This study is granted by the Specific Research on Public Service of Environmental Protection in China (No. 201009009). The authors appreciate the tutor and classmates for help.

REFERENCES

1. Foundation for Water Research. FWR: Groundwater. http://www.euwfd.com/html/groundwater.html (accessed 18 September 2012)

2. Barbash, J., P. V. Roberts. Volatile Organic Chemical Contamination of Groundwater Resources in the U.S. Water Pollution Control Federation 1986;58(5) 343-348.

3. Moran, Michael J. Occurrence and Status of Volatile Organic Compounds In Ground Water From Rural, Untreated, Self-supplied Domestic Wells In the United States, 1986-99. Rapid City, SD: U.S. Department of the Interior, U.S. Geological Survey; 2002.

4. Zoftman B C J. Persistence of organic contaminants in groundwater. In: Duivenboodne W V (ed.). Quality of Groundwater. Netherlands: Elsevier Scientific Publishing Company; 1981, p464-480.

5. Xu Z F, Lei S H, et al. A Preliminary Study on the Analysis of Organic Pollutants in a Suburban Sewage System of Beijing. Environmental Chemistry 1988;7(1) 69-80.

6. Liu Z C, Nie Y F, et al. The Pollution Control on Groundwater System. Beijing: China Environmental Science Press; 1991.

7. Chen H H, He J T, Liu F, et al. Organic contamination characteristics of shallow groundwater in a study area of Taihu Lake basin, Jiangsu, China. Geological Bulletin of China 2005;24(8) 735-739

8. Rathbun, R.E.. Transport, behavior, and fate of volatile organic compounds in streams: U.S. Geological Survey Professional Paper; 1998.

9. Heberer T, Furhmann B, Schmidt-Baumler K, Tsipi D, Koutsouba V, Hiskia A. Occurrence of pharmaceutical residues in sewage, river, ground, and drinking water in Greece and Berlin (Germany). In: Daughton CG, Jones-Lepp TL, editors. American Chemical Society Symposium Series 791. Pharmaceuticals and personal care products in the environment. Scientific and regulatory issues; 2001. p396.

10. Kimberlee K. B, Dana W. K, Edward T. F, et al. A national reconnaissance of pharmaceuticals and other organic wastewater contaminants in the United States -I) Groundwater. Science of the total Environment 2008;402 192-200.

11. Liu F, Wang S M, Chen H H. Progress of investigation and evaluation on groundwater organic contaminants in western countries. Geological Bulletin of China 2010;29(6) 907-917

12. Patrick, R.; Ford, E.; Quarles, J. Groundwater Contamination in the U.S.A.;2ed.;University of Pennsylvania Press: Philadephia; 1987. p538

13. Zhang L.Y, Han J. L, An S.J, et al. Detection and removal of organic pollutants in refuse leaching. China Environmental Science 1998;18(2) 184-188.

14. Hirata,T., Nakasugi, O., Yoshioka, M. et al. Groundwater pollution by volatile organochlorines in Japan and related phenomena in the subsurface environment. Water Science and Technology 1992;25(11) 9-16

15. Li C, Wu Q. Research Advances of Groundwater Organic Contamination. Geotechnical Investigation & Surveying 2007;1 27-30.

16. National Research Council. Groundwater Models, Scientific and Regulatory Applications. Washington, D.C., National Academy Press;1990.

17. Keswick, B.H., Gerba, C.P. Secor, S.L., and Cech, I. Survival of enteric viruses and indicator bacteria in groundwater. Journal Environmental Science and Health 1982;17(6) 903-912.

18. Bitton, G., Farrah, S.R., Rushkin, R.H., Butner, J. and Chou, Y.J. Survival of pathogenic and indicator organisms in groundwater. Groundwater 1983;21(4) 405-410.

19. Zhang, W. H, Li, G. H, Shao, H.G, et al. Study on Biodegradation of Oil Pollutants on the Polluted Soils of the Aeration Zone. Research of Environmental Science 2005;15(2) 60-62.

20. Michael, P. Bioremediation of petroleum waste from the fefining of lubricant oils. Environmental Progress 1993;12(1) 5-11.

21. Eric R., Carl C., et al. Groundwater Contamination Risk Assessment: A Guide to Understanding and Managing Uncertainties. IAHS Press; 1990. p22-26.

22. U.S.EPA. Risk Assessment Guidance for Superfund, Volume: Human Health Evaluation Manual (Part A). Interim final. EPA/540/l-89/002,1989.

23. Chen, H., Liu, Z. Q, Li, G.H. A study the establishment and evaluation method of the soil risk guideline of contaminated sites. Hydrogeology & Engineering Geology 2006;21(2):84-88.

24. U.S. EPA. Risk Assessment Guidance for Superfund, Volume 1, Human Health Evaluation Manual (Part A), Interim Final. Office of Emergency and Remedial Response, Washington, D.C.; 1989.

25. Cothern, R.C. Techniques for the assessment of carcinogenic risk due to drinking water contaminants. CRC Critical Review of Environmental Control 1986;16(4): 357-399.

26. Crouch, E.A.C., Wilson, R., and Zeise, L. The risks of drinking water. Water Resources Research 1983;19(6): 1359-1375.

27. U.S.EPA. Risk Assessment Guidance for Superfund Volume I: Human Health Evaluation Manual (Part E, Supplemental Guidance for Dermal Risk Assessment) Final. EPA/540/R/99/005, 2004.

28. Han B., He J.T, Chen H.H. Primary study of health based risk assessment of organic pollution in groundwater. Earth Science Frontiers 2006;13(1) 224-229.

29. U.S.EPA. A Citizen's Guide to Pump and Treat. Office of Solid Waste and Emergency Response. EPA 542-F-01-025, 2001.

30. Ralinda R. Miller, P.G. Air Sparging. Technology Overview Report, GWRTAC SERIES; 1996.

31. Commonwealth Scientific and Industrial Research Organisation. CSIRO: Land and Water. Soil and Groundwater Remediation. http://www.clw.csiro.au/research/urban/protection/remediation/projects_2.html. (accessed 18 September 2012)

32. U.S.EPA. Chapter X: In-Situ Groundwater Bioremediation. (ed.) How To Evaluate Alternative Cleanup Technologies For Underground Storage Tank Sites: A Guide For Corrective Action Plan Reviewers. EPA 510-R-04-002; 1995.

33. Suthersan, S.S. IN SITU REACTIVE WALLS, Remediation engineering: design concepts, Boca Raton: CRC Press LLC; 1999.

2

CHAPTER

Mercury in Marine and Oceanic Waters—A Review

Barbara Gworek, Olga Bemowska-Kałabun, Marta Kijeńska, Justyna Wrzosek- Jakubowska

Institute of Environmental Protection–National Research Institute Warsaw Poland

ABSTRACT

Mercury contamination in water has been an issue to the environment and human health. In this article, mercury in marine and oceanic waters has been reviewed. In the aquatic environment, mercury occurs in many forms, which depend on the oxidation-reduction conditions. These forms have been briefly described in this article. Mercury concentrations in marine waters in the different parts of the world have been presented. In the relevant literature, two models describing the fate and behavior of mercury in saltwater reservoirs have been presented, a conceptual model which treats all the oceans as one ocean and the "ocean margin" model, providing that the ocean margins manifested themselves as the convergence of continents and oceans, covering such geological features, such as estuaries, inland seas, and the continental shelf. These two conceptual models have been summarized in the text. The mercury content in benthic sediments usually reflects is level in the water reservoir, particularly in reservoirs situated in contaminated areas (mines, metallurgical plants, chemically protected crops). The concentrations of mercury and its compounds determined in the sediments in surface waters in the different parts of the world have been presented. Due to the fact that the pollution caused by mercury is a serious threat for the marine environment, the short paragraph about mercury bioaccumulation in aquatic organisms has been included. The cited data demonstrated a large scatter of mercury contents both between the fish species and the water areas. Mathematical models, valuable tools which provide information about the possible responses of ecosystems, developed to simulate mercury emissions, both at a small scale, for local water reservoirs, and at a global scale, as well as to model mercury bioaccumulation in the chain web of aquatic systems have been described.

Keywords: Mercury, Marine waters, Benthic se diments, Pollution, Environment

1 INTRODUCTION

Mercury is an element the emissions of which from its natural sources exceed its anthropogenic emissions. Therefore, it is difficult to estimate mercury emission levels from its natural sources. This is related, e.g., to the absence of sufficient data on this issue and the complexity of processes affecting natural mercury emissions (to mention only geological processes which are characterized by enormous spatial and temporal variabilities). In general, natural mercury emissions into the environment can be divided into original emissions, e.g., volcanic emissions, and secondary reemissions of natural origin. Natural mercury emissions can also be divided into emissions from oceans which are estimated at 2680 Mg/year and those of terrestrial origin, estimated at 1850 Mg/year [Mason 2009; Pirrone et al. 2010; Pirrone and Cinnirella 2012) with volcanic emissions to the atmosphere estimated as 94–112 Mg/year (Nriagu and Becker 2003; Nriagu and Becker 2004). Natural mercury emission processes mainly include

- Degassing from mercury deposits;
- Degassing from aquatic and terrestrial systems (through the reduction of Hg^{2+} to Hg^{0});
- Geological activity—volcanic and geothermal processes (underwater exhalations from geothermal vents);
- Biomass burning—e.g., fires of forests and steppes;
- Erosion of mercury-containing minerals; and
- Plant growth.

Mercury emissions into the air from sources of natural origin are an important element of global mercury flows, as it is the largest element of the global mercury cycle. Many research teams have estimated these emissions, using different numerical models in addition to direct measurements. Their balance is hampered by the share of reemissions from anthropogenic sources which is difficult to estimate (Mason 2009; Pirrone et al. 2010; Pirrone and Cinnirella 2012).

In the aquatic environment, mercury occurs in many forms which depend on the oxidation-reduction conditions. The forms $HgCl_4^{2-}$ and $HgOH^-$ dominate in the good oxidation conditions, whereas sulfur-related forms (HgS^{2-} and CH_3HgS^-) prevail in the reduction conditions.

In the intermediate conditions, the alkyl forms of mercury, MeHgCl and EtHgCl, can most often be found (Kabata-Pendias and Mukherjee 2007; Kabata-Pendias 2011; Tyler 1992). Soluble forms of mercury, such as $[HgOH]^+$, $[HgCl]^+$, $[HgCl_2]$, $[HgCl_3]^-$, $[HgCl]^{2-}$, and $[HgS_2]^{2-}$, can often be encountered. Higher concentrations of Cl^- ions, which form stable complexes with mercury, such as $HgCl_3^-$, $HgCl_2^-$, $HgCl_4^{2-}$, or $HgBrCl^-$, lead to increased dissolution of solid phases of mercury (Grassia and Nettib 2000).

Moreover, mercury can occur in soluble non-ionic organic compounds and other organic and inorganic compounds. Examples of sparingly soluble forms of mercury include CH_3Hg^+ or $Hg(CN)_2$. There are many paths of mercury transport into the aquatic environment. Inorganic forms, such as Hg(II) and methylmercury (MeHg), can be directly introduced into reservoirs through (wet and dry) depositions from the atmosphere (Lina et al. 2006). On the other hand, Hg(II) and MeHg are transported into water reservoirs through surface runoff as well as through leaching from the upper levels of a soil profile to groundwater and, subsequently, to surface waters (Fitzgerald et al. 2007; Morel et al. 1998; US EPA 1997).

Wet and dry atmospheric depositions are the most frequent paths of transport of Hg (THg) into the surface waters of the Arctic and the Antarctic. It is estimated that the total annual quantity of mercury penetrating the atmosphere from natural and anthropogenic mercury emission sources is 5000–6500 t (Berg et al. 2006; Bone et al. 2007; Bookman et al. 2008; Camargo1993; Cheng and Hu 2010; Jiang et al. 2006; Gbor et al. 2007; Gray and Hines 2006; Lamborg et al. 2002; Mason et al. 1994; Streets et al. 2009), with the emissions from anthropogenic sources estimated at about 2200 t Hg/year (Bergan et al. 1999; Gbor et al. 2007; Pacyna et al. 2006). China is the country with the highest mercury emissions (Wang et al. 2016). Every year, thousands of tons of mercury from air deposition are transported into aquatic ecosystems. Mercury-containing flowing and standing surface waters, such as rivers, streams, and estuaries, are the main paths of mercury transport into marine ecosystems.

In the aquatic environment, mercury undergoes many different chemical and biochemical processes which condition its speciation and transport between the solid and aqueous phases (Fitzgerald et al. 2007; Morel et al. 1998). In the aquatic environment (water, sediments, aquatic fauna, and flora), most mercury occurs in organic and inorganic forms of divalent mercury and Hg(0), as a form of mercury dissolved in the aqueous phase (Ullrich et al. 2001). Mercury adsorption and desorption processes in the aquatic environment play a dominating role in the distribution of different forms of mercury in the particular elements in the aquatic environment. In waters, these processes are also responsible for the course of mercury transport, transformation, and uptake by living organisms, and they also condition the toxicity of this element.

In the aquatic environment, mercury can be adsorbed on sediment particles, thus constituting a substantial mercury resource. In the sediments in water reservoirs, both as a result of chemical reactions and under the impact of biological factors, e.g., those related to the activity of microorganisms, methylmercury, $CH3Hg^{+}$, and dimethylmercury, $(CH_3)_2Hg$, emerge (Benoit et al. 2003; St. Louis et al. 1994).

MeHg is the most common form of organic mercury in the environment. Methylmercury is a neurodevelopmental toxicant (Obi et al. 2015), and it is also the most toxic form of mercury (Henriques et al. 2015). MeHg and dioxin-like compounds are considered as the most important toxic compounds in the case of large-scale consumers of seafood (Szefer 2013).

Methylation is a result of abiotic and biotic processes, which are affected by such factors as pH, temperature, the presence of sulfates, and the availability of organic carbon.

The impact of mercury in the environment on human health was found for the first time in relation to the Minamata disease in the 1950s, which caused mass-scale poisoning by methylmercury. It had accumulated in aquatic organisms which were subsequently eaten by humans. A similar case of poisoning by mercury accumulated in fish also took place in Sweden (Zaib et al. 2015).

Apart from spectacular cases of poisoning, the presence of mercury in the environment also affects the human population in a more concealed manner. Every year, Trasande et al. (Trasande et al. 2005) found mercury concentrations exceeding 5.8 μg/L—a level related to IQ loss—in blood samples taken from 316,588–637,233 children. Humans are mainly exposed to methylmercury as a result of their consumption of oceanic fish (Drevnick et al. 2015).

One of the key stages of the biogeochemical cycling of mercury in the environment is its biomobilization, which is mediated by microorganisms. Microorganisms mediate in the following four types of mercury transformation: in the reduction of Hg(II) to Hg(0), in the degradation of $CH_3Hg(I)$ and other organic mercury compounds, in the methylation of Hg(II), and in the oxidation of Hg(0) to Hg(II). The transformation processes listed above also unfold in the environment without any involvement of microorganisms. Table 1 shows a summary list of known mercury transformation mechanisms.

Table 1 Mechanisms of mercury transformation processes

Mechanisms	Reduction of Hg(II)	Degradation/demethylation of $CH_3Hg(I)$	Methylation	Oxidation of Hg(0)
Biotically	Enzymatically—mercury reductase	Enzymatically—organomercurial lyase (OL)	Transfer of methyl groups by corrinoid coenzyme (bacterial methyltransferase)	Hydroperoxidase (e.g., catalase)
	Indirectly—reduced metabolites	Oxidative demethylation—anaerobic bacteria	Disturbances in methionine synthesis pathway in fungi	–
Abiotically	Free radicals related to humic substances	Photodegradation	Photochemically—induced by humic and fulvic acids	In the atmosphere via H_2O_2 under low pH conditions
	Disproportionation of Hg(I)	–	Transformation of $CH_3Hg(I)$ to $(CH_3)Hg$ in the presence of H_2S	–

Source: Barkay (2000)

Biological mercury transformations, along with the processes of its chemical transformations, provide the basis for mercury cycling in the biosphere; furthermore, the role played by the biological transformations themselves is not fully known and probably depends on mercury concentrations and the conditions occurring in the environment.

The reduction of Hg^{2+} and the degradation of organic mercury compounds are processes constituting natural detoxification mechanisms, which unfold in bacteria and enable their growth and development in the presence of these toxic compounds. Bacteria that are resilient to the harmful impact of Hg^{+2} are capable of producing mercury, which catalyses the reaction with the involvement of nicotinamide adenine dinucleotide phosphate (NADPH). Due to its low water solubility and high vapor pressure, Hg^0 which emerges as a result quickly escapes into the atmosphere. Some bacteria which are resilient to Hg^{2+} due to the presence of organomercurial lyase (OL) also demonstrate resistance to the toxic impact of organic mercury compounds (Walsh et al. 1988). The resistance to the harmful impact of organic forms of mercury is related to the action of both enzymes, lyase and reductase, while all the bacteria which are not vulnerable to the impact of organic mercury compounds are at the same time resilient to the toxic effect of Hg^{+2} (a wide range of resilience). However, a large part of bacteria do not show the capacity to produce organomercurial lyase; as a result, they are vulnerable to the presence of this form of mercury in the environment (a narrow range of resilience).

Whereas the reactions of the reduction of Hg^{2+} to Hg^0 may unfold without any involvement of microorganisms, in contrast, the process of the degradation of organic forms of mercury without any involvement of microorganisms unfolds extremely slowly and is of no greater significance in the processes of mercury cycling in nature (Barkay et al. 1992). The chemical reactions in which Hg^0 escapes into the atmosphere include the reduction of Hg^{2+} to $2Hg^+$, with the subsequent disproportionation to Hg^0 and Hg^{2+} (Baltisberger et al. 1979), and the reduction through the interaction of Hg^{2+} with free-radical electrons of humic acids (Alberts et al. 1974).

As shown in Table 1, abiotic sources of methylmercury include transformations of Hg^{2+} as a result of its reactions with fulvic and humic acids (Nagase 1984) and the transfer of methyl groups from the soluble fraction of organic matter to Hg^{2+}. These reactions are initiated phototochemically in the presence of sulfur (Akagi et al. 1974).

The intensity of methylation grows as the temperature and the content of Hg(II) in the environment increase, while the optimum level is reached at pH of less than 5.

The methylation reaction is also stimulated by the presence of other metals which play the role of catalysts. Photochemical methylation takes place in the presence of chemical donors of methyl groups, such as methanol or ethanol; moreover, the greatest efficiency is achieved in the presence of acetic acid. The anaerobic conditions and the presence of colored compounds as photosensitizers also enhance the efficiency of this process (Barkay 2000).

2 MERCURY IN MARINE AND OCEANIC WATERS

In the biogeochemical mercury cycling, a substantial part of it reaches seas and oceans and its important source is its atmospheric deposition (Bindler 2003; Wang et al. 2004).

In oceanic waters, mercury mainly occurs in the forms of Hg^0, Hg^{2+}, MeHg, and diMeHg and in colloidal form (Morel et al. 1998). In marine waters, mercury forms compounds with chlorine ($HgCl_3^-$ and $HgCl_4^{2-}$) to a greater extent than oxides, as is the case in freshwaters (Mason and Fitzgerald 1993). It has been demonstrated that in saltwaters, Hg^{2+} forms complexes with halides, such as those of chlorine, and these complexes do not undergo the reduction and methylation processes (Gardfeldt et al. 2003; Whalin et al. 2007) as quickly as the other Hg^{2+}compounds do. Dissolved gaseous mercury (DGM) arises as a result of the transformations of both Hg^0 and Me_2Hg (Lamborg et al. 1999; Mason et al. 1995), with the form Hg^0 dominating in the upper layer of the ocean (Table 2) (Gardfeldt et al. 2003; Laurier et al. 2003).

Table 2 Mercury concentrations in marine waters in the different parts of the world, in ng/L

Survey site	THg[a] Seawaters	HgCR[a]	DGM[a]	diMeHg[a]	HgR[a]	MeHg[a]	HgC[a]	Source
Minamata Bay (Japan)	1,600–3,600	···						Hamhara et al. (1961)
Atlantic Ocean	400–1,600	···	···	···	···	···	···	Aidan'yan and Belavskaya (1963)
Red Sea	700–2,000	···	···	···	···	···	···	Aidin'yan (1962)
Pacific Ocean, Ramapo Deep	80–150	···	···	···	···	···	···	Hamaguchi et al. (1961)
Baltic Sea	0.6 ± 0.2	···	···	···	···	~20 % THg	···	Pempkowiak et al. (1998)
Gulf of Gdańsk	277–630	···	···	···	···	···	···	BMEPC (1987)
Gulf of Finland	10–140	···	···	···	···	···	···	
Bothnian Bay	2–40	···	···	···	···	···	···	
Gulf of Riga	10–40	···	···	···	···	···	···	
Matsalu Bay	5–130	···	···	···	···	···	···	
North Sea	0.56 ± 0.24	···	···	···	···	···	···	Coquery and Cossa (1995)
North Sea	0.5–200	···	···	···	···	···	···	Schmidt (1991)
Mediterranean Sea (Tunisia)	990–27,060	···	···	···	···	···	···	Nasfi (1995)
Atlantic Ocean (N-E)	2.7							Cossa et al. (1996)
Sepetiba Bay (Brazil)	···	0.12–0.67	0.032–0.092	···	0.11–0.36	<0.02–0.55	···	Marins et al. (2000)
Sepetiba Bay (Brazil)	···	···	<2.6	···	0.2–32.2	0–246	···	Marins et al. (2000)
South China Sea	0.8–2.3	···	···	···	···	0.05–0.22	···	Fu et al. (2010)
Andaman Sea (Thailand)	<10–540	···	···	···	···	···	···	Thongra-Ra and Parkpian (2002)
Yellow Sea	1.34–5.5	···	···	···	···	···	···	Ci et al. (2011)
Wuli Estuary	39–430	···	···	···	···	0.046–0.28	···	Wang et al. (2009a)
Pacific Ocean (N)	···	···				173 ± 118[b]	···	Sunderland et al. (2009)
Pacific Ocean (equatorial part)	···	···	···	<10–670[b]	···	<50–500[b]	···	Mason and Fitzgerald (1990, 1993)
Atlantic Ocean (N)	···	···	···	81 ± 68[b]	···	ND[c]	···	Mason et al. 1998
Atlantic Ocean (S and equatorial part)	···	···	···	<10–110[b]	···	25–200[b]	···	Mason and Sullivan (1999)
Mediterranean Sea	···	···	···	3 ± 3[b]	···	280 ± 120[b]	···	Kotnik (2007)
	···	···	···	3 ± 2[b]	···	290 ± 26[b]	···	Horvat et al. (2003)
	···	···	···	<20–290[b]	···	<150[b]	···	Cossa et al. (1997)

··· no data or not applicable

[a] *THg* total Hg, *HgCR* total dissolved Hg, *DGM* dissolved gaseous Hg, *HgR* reactive Hg, *diMeHg* dimethylmercury, *MeHg* monomethylmercury, *HgC* Hg in elemental form

[b] Value given in fM

[c] ND below the detection limit

The impact of anthropogenic sources on the contents of mercury and its forms can be seen to a greater extent in the surface layers of the oceans (i.e., in the upper layer of the ocean waters down to a depth of about of 100 m) (Strode et al. 2007). In the near-surface layer, i.e., in the layer between 100 and 1000 m, which is the water-mixing layer, the displacement of mercury of anthropogenic origin is hampered. It depends on many factors, e.g., the local and regional movements of water masses, on the production formation and degradation, as well as the accumulation of intermediate substances (Mason and Sheu 2002).

The particular oceans are different in their total mercury concentrations, with the average concentration of about 1.5 picomoles (pM) (Lamborg et al. 2002). Higher concentrations were recorded in the Mediterranean Sea—2.5 pM (Cossa et al. 1997), and the northern part of the Atlantic—2.0 pM (Mason et al. 1998), whereas a lower concentration was found in the Antarctic Ocean—0.8 pM (Laurier et al. 2004; Sunderland and Mason 2007). Simulations carried out using mathematical models have indicated that the mercury concentrations in most oceans are not balanced with its atmospheric deposit and that they will also systematically grow over the next several dozen years (Sunderland and Mason 2007).

The mercury exchange with the ocean water currents unfolds most quickly in the Atlantic Ocean (17.99–24 Sv), while in the other intervals, it has similar values, falling within the range of 6 to 12.4 Sv.

It is believed that the mercury exchange at the interface between the ocean surface and the atmosphere unfolds relatively quickly. Just as in the freshwater systems, Hg transformation processes in saltwaters are quite complex, too. Hg^{2+} reaches the ocean through its dry and wet depositions, while Hg^0 does so through its dry deposition; in addition, Hg2+ subsequently undergoes biological or photochemical reduction to Hg^0 and absorption on the solid particles of the suspended sediment or methylation (Strode et al. 2007).

It is difficult to determine the intensity of the methylation process in saltwaters, since old data are not reliable given the analytical detection limits of measurement equipment; in consequence, difficulties emerge in respect of the MeHg measurements in these waters. The average levels of MeHg contents determined in oceanic waters fell in the range of 2–35 % (Sunderland et al. 2009). In saltwater systems, mercury methylation may unfold in sediments in the continental shelf areas (Hammerschmidt and Fitzgerald 2006), in estuaries (Gill et al. 1999; Heyes et al. 2006), in a seawater column (Monperrus et al. 2007), or in hydrothermal vents situated in deep ocean layers (Kraepiel et al. 2003).

In the relevant literature, the following two models describing the fate and behavior of mercury in saltwater reservoirs have been presented: a conceptual model which treats all the oceans as one ocean, based on the data and modeling presented by Rolfus and Fitzgerald (1995), and the "ocean margin" model developed by Cossa et al. (1996), providing that the ocean margins manifested themselves as the convergence of continents and oceans, covering such geological features, such as estuaries, inland seas, and the continental shelf. However, it should be borne in mind that these models are very general and only reflect basic knowledge of the mercury paths and transformations in the marine environment.

2.1 The Fate of Mercury in Saltwaters (a Conceptual Model of One Ocean)

Rolfus and Fitzgerald (1995) created a simple ocean model, designed to help determine the fate of mercury in saltwaters and its accumulation in fish and mostly based on the data which these authors had collected earlier. The model provided that the ocean consisted of the following three compartments: the coastal zone, the upwelling zone, and the open ocean. The open ocean zone represents more than 90 % of the total surface area of the oceans, but it is characterized by low fish productivity. The coastal and upwelling zones represent, respectively, about 10 and 0.1 % of the total surface area and generate almost all the fish production (in the model, each such area accounts for almost 50 % of the total fish production). The mercury sources adopted in the model include atmospheric deposition from the atmosphere, the flux from river systems, and the flux from deep ocean waters. On the basis of the dependence observed in the North Atlantic between the increased concentration of reactive mercury in the surface ocean waters and the predicted rate of atmospheric deposition, it is also assumed that its point of entry is the mixed layer of the waters. From this layer, reactive mercury is transported in the form of complexes with suspended particles to the regions and layers of the ocean where its methylation occurs (these areas are naturally poorer in oxygen) and mercury is released from the complexes as they fall into lower regions of the reservoir. This model also assumes that monoMeHg forms in regions which are poor in oxygen, under the thermocline of the open ocean zone and the upwelling zone, from where it is subsequently transported to the mixed layer, to a depth of less than 100 m, where it is incorporated into the lower levels of the trophic chain (Mason and Fitzgerald 1990, 1993, 1996). After its transport from the mixed layer, in the subthermocline water layer, most of reactive mercury is methylated to dimethylmercury (diMeHg), although monomethylmercury can also form directly from reactive mercury. In marine waters, dimethylmercury is unstable and most of it quickly decomposes to form MeHg (Mason and Fitzgerald 1996). Some of MeHg is then converted to Hg^0, which is transported to the surface zones of the ocean. In this layer, as a result of its supersaturation, elemental Hg can be released into the atmosphere. This is the main mechanism of removal of mercury from saltwaters, reaching a level of 1 % per day in the open ocean zone. It is presumed that the transport of reactive mercury from the mixed layer of the ocean is a factor which determines the rate of methylmercury formation (Mason and Fitzgerald 1996).

The reduction processes are related to both abiotic and biotic factors of the marine environment (Mason et al. 1995); specifically, abiotic factors are responsible for about 10–30 % of the reduction. It was demonstrated that the abiotic reduction might unfold under the impact of sunlight. In an experimental aquatic system, they demonstrated that in the conditions of the combination of fulvic and humic acids with a synthetic light source, the reduction of dissolved divalent forms of Hg occurred. In turn, Weber (1993) suggested that the process of abiotic mercury reduction was mediated by methyltin compounds (e.g., $MeSnCl_3$, Me_2SnCl_2, Me_3SnCl) and humic acids. However, for the most part, the reduction of reactive mercury results from the action of microorganisms, e.g., bacteria and cyanobacteria in the mixed layer of the ocean (Mason et al. 1995).

The model described here also assumed that mercury methylation in the coastal compartments (including the estuary regions) occurred in both the sediments and in the water column, in the vicinity of the oxycline (Rolfus and Fitzgerald 1995), from where methylmercury was transported to the mixed layer and incorporated into the lower trophic levels of the marine food web. The total deposition of mercury to marine reservoirs was estimated at 10 Mmol in a year, while the input from rivers and their estuaries was estimated at about 10 % of this value (about 1 Mmol/year); in turn, the input to the upwelling layer from deeper and colder waters was estimated at about 0.5 Mmol/year. Several authors, including Rolfus and Fitzgerald (1995), found that the enhanced mercury deposition resulting from higher anthropogenic emissions caused greater bioaccumulation in the food chain and, primarily, higher mercury contents in marine fish.

2.2 The Fate of Mercury in Saltwaters (the *Ocean Margin* Model)

Estuaries and coastal regions are particularly vulnerable to anthropogenic mercury contamination, since contaminants are both transported to them in river waters and discharged directly to the oceanic waters. In these regions, there is a higher level of reactive forms of mercury and particle-bound mercury released into the atmosphere from local sources of anthropogenic origin. It is well known that reactive Hg deposits faster than elemental Hg and that higher concentrations of oxidants may occur in the atmospheric layer over the coastal waters. On the basis of research results, it has been found that in certain mercury-contaminated estuaries, the main source of their contamination is the direct discharge of mercury into water rather than its atmospheric deposition. On the basis of data collected for the regions mentioned above, Cossa et al. (1996) developed a mass balance model for mercury for these areas and the quantity of THg estimated on the basis of the model was about 3.3 Mmol. In this model, it was assumed that the mercury flux from the river systems to the oceanic margins represented the highest share of the total mercury input to the coastal waters and that annually, about 4.8 Mmol of mercury reached oceanic waters in this way. Cossa et al. (1996) also noted that the concentrations of total mercury were very variable and that, moreover, the highest concentration was found in the waters of rivers flowing through urbanized and industrial areas. More than 90 % of mercury are transported in particle-bound form. Moreover, a substantial part of this form of mercury seems to be unreactive and is deposited in sediments.

The authors estimated that the dry deposition of mercury in coastal waters amounted to about 2 Mmol/year and found that a substantial part of mercury was chemically reactive and participated in reactions unfolding in the marine environment.

Atmospheric deposition in remote ocean areas and then its transport via upwelling to these regions (2.5–3.5 Mmol/year) were also considered to be a significant source of mercury in the waters of marine coastal regions.

The impact of anthropogenic sources on the contents of mercury and its forms is more conspicuous in the surface layer of the oceans (i.e., in the upper layer of the ocean waters down to a depth of about of 100 m) (Strode et al. 2007). In the near-surface layer, i.e., in the layer between 100 and 1000 m, which is the water-mixing layer, the displacement of mercury of anthropogenic origin is hampered. It depends on many factors, e.g., the local and regional movements of water masses, on the formation and degradation, and the accumulation of intermediate substances (Mason and Sheu 2002).

It is believed that the process of mercury exchange at the interface between the ocean surface and the atmosphere unfolds relatively quickly. Just as in freshwater systems, Hg transformation processes in saltwaters are also quite complex. Hg^{2+} reaches the ocean through its dry and wet depositions, while Hg^0 does so through its dry deposition; in addition, Hg^{2+} subsequently undergoes biological or photochemical reduction to Hg^0 and absorption on the solid particles of the suspended sediment or methylation (Strode et al. 2007)

It is difficult to determine the intensity of the methylation process in saltwaters, since old data are not reliable given the analytical detection limits of measurement equipment; in consequence, difficulties emerge in respect of the MeHg measurements in these waters. The average levels of MeHg contents determined in oceanic waters fell in the range of 2–35 % (Sunderland et al. 2009).

In saltwater systems, mercury methylation may unfold in sediments in the continental shelf areas (Hammerschmidt and Fitzgerald 2006), in estuaries (Gill et al. 1999; Heyes et al. 2006), in a seawater column (Monperrus et al. 2007), or in hydrothermal vents situated in deep ocean layers (Kraepiel et al. 2003).

The mercury flux from sediments to the waters in coastal waters was regarded as its less significant source.

Taking into account the assumptions listed above and enumerating them in approximate order of significance, it is found that the flux of total mercury in the area of coastal waters includes, *inter alia*,

- The sedimentation of particles, with mercury adsorbed on their surface originating from surface flow in riverine and deep regions;
- The transport from the open ocean waters; and
- The emissions into the atmosphere which are, on a global scale, usually offset by atmospheric deposition, with a regional imbalance resulting from the domination of deposition processes over releases in the northern latitudes and a converse situation in the southern latitudes.

The model in question was also used to describe the fate and behavior of methylmercury in oceanic coastal waters (Cossa et al. 1996). It follows from the assumptions of the model that more than half of MeHg present in oceanic coastal waters originates from the waters of the bottom of the ocean, whereas the other part consists of the product of methylation of reactive mercury in coastal waters and MeHg from other sources. Mercuric ions in the anoxic layers of sediments in marine water reservoirs are transformed to monomethylmercury, primarily through microbial transformations driven by sulfate-reducing bacteria, such as *Desulfovibrio desulfuricans* (Compeau and Bartha 1985), and the organic matter content has a significant impact on the course of the process (Choi and Bartha 1994). Research has demonstrated that the bioavailability of MeHg bound to sediment particles is substantially greater than the bioavailability of divalent inorganic mercury bound to sediment particles. Dissolved species of both methylmercury and divalent inorganic mercury demonstrate greater bioavailability than particle-bound forms (Choi and Bartha 1994).

3 MERCURY IN BENTHIC SEDIMENTS

Both in soil and benthic sediments, mercury is mainly related to the presence of organic matter. In sediments, methylmercury which has formed through methylation represents not more than 1.5 % of the total quantity of mercury; this quantity is determined as an equilibrium level between the formation and removal processes. In each aquatic environment, a certain state of equilibrium is established between the reactions of methylation and demethylation, depending on the stability of the particular forms of mercury. Dimethylmercury is unstable in sediments but retains relatively high stability under the impact of such factors as its high sulfate content, salinity, anaerobic conditions, or continuous access to methane (Weber et al. 1998).

When comparing the results of the measurements of mercury contents in samples taken from sediments, it should be taken into consideration that slight changes in the conditions of the storage of the collected samples (temperature, redox conditions, and oxygenation) have a very large effect on changes in the contents of the particular forms of mercury (Horvat and Gibicar2005).

The mercury content in benthic sediments usually reflects its level in the water reservoir, particularly in reservoirs situated in contaminated areas (mines, metallurgical plants, chemically protected crops). Table 3 shows the concentrations of mercury and its compounds determined in the sediments in surface waters in the different parts of the world. There are substantial differences in mercury concentrations among the sediments originating from different water reservoirs and different regions (Table 3).

Table 3 Mercury concentrations in marine sediments, in mg/kg d.s.

Location	HgT total mercury	MeHg	Source
Marine sediments			
Baltic Sea	2–340	–	Pempkowiak et al. (1998)
Baltic Sea Proper	100 ± 50	–	Borg and Jonsson (1996)
Baltic Sea (Aland Sea)	180 ± 60	–	
Baltic Sea (Bothnian Sea)	100 ± 30	–	
Bothnian Bay	400 ± 240	–	
Gulf of Puck	0.74–5.7	–	Falandysz et al. (1993)
Gulf of Gdańsk	3.5–160	–	
Gulf of Gdańsk	0.25	–	Szumiło-Pilarska et al. (2016)
Gulf of Puck	2.8–180	–	Boszke et al. (2002)
Denmark Strait	60–220	–	Brzezińska et al. (1984)
Baltic Sea (Bosex Area)	140–190	–	
South Baltic Sea	30 ± 10	–	
Baltic Sea Proper	20–360	–	
Gulf of Gdańsk	310 ± 310	–	
Gulf of Riga	30–790	–	Ojaver (1995)
Mediterranean Sea (Israel)	10–900	–	Herut et al. (1994)
Mediterranean Sea (Italy)	100–5,330	–	Barghigiani and Ristori (1995)
South China Sea (Malaysia)	20–127	0.01–0.053	Kannan and Falandysz (1998)
Victoria Harbour, Hong Kong	47–855	0.1–1.5	Shi et al. (2007)
East China Sea (China)	<0.0005–0.0798	–	Shi et al. (2005)
	0.0041–0.0476	–	Fang and Chen (2010)
	0.042–0.072	–	Fang et al. (2004)
Coastal sediments			
Bay of Fundy, (USA/Canada)	25–514[a]	0.5–7.38	Sunderland et al. (2006)
Hainan Coast	0.02–0.1	–	Qiu et al. (2011)
Southeast China Coast	0.0023–0.9036	–	Ding et al. (2009)
Andaman Sea (Thailand)	0.047–2.135	–	Thongraa-Ra and Parkpian (2002)
Bohai Sea	0.8–25	–	Wang et al. (2009b)

[a] pmol/g d.s.

It is believed that natural THg concentrations fall within the interval from 10 to 200 ng/g d.s. (Fergusson 1990), while the quotient of the THg concentration and methylmercury can be used as a marker to describe the state of contamination of a reservoir (Kannan and Falandysz 1998); moreover, in the uncontaminated regions, this indicator takes values of less than 1. In the benthic sediments of the Gulf of Gdańsk, the determined average concentration of methylmercury was 0.65 ng/g d.s., representing about 0.5 % THg, while the indicator in question fell within the interval from 0.02 to 2.27. The direction of mercury transformations in the anoxic conditions of the lower layers of the water column and benthic sediments is determined by a high affinity of mercuric ions with sulfides.

In the anoxic conditions, the dominant forms of mercury include HgS, HgS_2H_2, HgS_2H^-, and HgS_2^{2-}, as well as CH_3HgS (the most important form of mercury in these conditions). Mercuric sulfide HgS, which is sparingly soluble in the aquatic environment, is mainly deposited in benthic sediments and determines the solubility of Hg(II) in waters with low oxygen content or without oxygen; moreover, it is believed that the deposition of HgS in sediments is correlated with the contents of organic matter and iron oxide ions (Boszke et al. 2002). The solubility of HgS increases with the growing content of sulfide ions, which enable the formation of soluble complexes with mercury. This is of large importance in the anoxic areas of aquatic ecosystems, where there are high concentrations of dissolved mercury. Although mercuric sulfide HgS is strongly bound to sediments, part of it may occur in dissolved form as a result of microbial transformations or under the impact of the aerobic conditions (Gagnon et al. 1997). The redeposition of mercury from benthic sediments to the water column occurs as a result of diffusion or the activity of benthic organisms. Moreover, in the latter case, it is greater by a factor of 2 to 10 than the redeposition as a result of diffusion (Rutgers van der Loeff et al. 1984). It is also believed that the existence of oxidizing conditions in the surface layer of sediments constitutes a geochemical barrier to the diffusion of methylmercury from the near-surface layers of sediments with oxidizing conditions to the water column. The mercury input from deeper sediment layers to the near-surface layers of sediments as a result of diffusion was estimated at 3 % of the total mercury input (Gagnon et al. 1997).

Research carried out in the water-sediment system demonstrated that Hg(II) underwent strong sorption on clay particles and biotic elements, which was much weaker than that on sand particles. It was demonstrated that it also underwent strong complexation on organic matter. Following their absorption on solid elements, Hg(II) and methylmercury settle and accumulate on the surface of the sediment layer, which undergoes dynamic transformations on the boundary of the water column, resuspension and bioturbation. The content of methylmercury in water and sediments is a function of its transformations in the methylation and demethylation processes. In the aquatic environment, methylmercury undergoes degradation under the impact of two microbial transformations and under the impact of light (Barkay et al. 2003).

Equilibrium research demonstrated that the in situ formation of mercury was often its main source in water reservoirs and benthic sediments (Benoit et al. 2003; Gilmour et al. 1998). Despite the contamination of aquatic ecosystems with mercury from anthropogenic sources, its direct increase can be observed in aquatic organisms rather than in these ecosystems. This primarily results from the presence of regulatory mechanisms, such as the activity of methylating microorganisms and the contents of organic matter and sulfur (Grieb et al. 1990). The quantity of accumulated mercury in aquatic organisms is affected by such factors as the level in the trophic chain (the higher level in the trophic chain it is, the more mercury is stored in it), the age of organisms, the microbial activity, the mercury concentration in the sediment layer, the content of the soluble fraction of organic matter, salinity, pH, and the redox potential (Kamman et al. 2005). The benthic sediments of oceans are regarded as the ultimate "recipient," which cumulates highly insoluble forms of HgS (Whalin et al. 2007; Wiener et al. 2003). The first stage related to mercury bioaccumulation in aquatic ecosystems is the transformation from its inorganic form to methylmercury, which occurs in both the aquatic and sedimentation phases. Although the mercury bioaccumulation process prevail over its demethylation, the mechanism of methylmercury synthesis has not been completely identified (Horvat 1996).

Cores of sediments from distant, uncontaminated areas are used to study mercury deposition in historically remote times. The overall background THg level demonstrated by Fuji (1976) is about 50 μg/kg in river sediments, 100–300 μg/kg in lake sediments, and 50–80 μg/kg in marine sediments. The MeHg content in the pool of total mercury (THg) reaches about 10 % (US EPA 1997).

Mercury is a strongly dispersed element. Its contents in sedimentary and magmatic rocks are similar, i.e., about 10 μg/kg. Higher mercury contents can be found in rocks which are rich in organic matter, i.e., in clay rocks. Natural mercury contents in sediments do not exceed 0.05 mg/kg. Due to the poor solubility of mercuric carbonates, phosphates, and sulfides and also due to the formation of bonds and connections with organic compounds, mercury undergoes ionic adsorption by organic and inorganic substances. Mercury adsorption is conditioned by many factors, depending, e.g., on the chemical form of mercury, the quantities and chemical properties of organic and inorganic colloids in sediments, the type of cations in sorption complexes, and the redox potential or the reaction of sediments. An increase in the mercury content in surface sediments is usually related to human activities in a given area, but it is rarely affected by geological factors, such as the presence of underlying mercury-containing rocks. The major mercury sources in sediments are the same as the mercury sources in soils (e.g., atmospheric deposition, pesticides, the application of sewage sludge and municipal waste for soil fertilization, coal combustion, metallurgy, and so-called artisanal and small-scale gold mining (ASGM)) and surface waters from which Hg migrates to sediments (Lacerda and Salomons 1999).

4 MERCURY BIOACCUMULATION IN AQUATIC ORGANISMS

As a result of the intensive fixation of mercury by bottom sediments, there is the risk of its accumulation in aquatic organisms (Wilken and Hintelmann 1991). Surveys on different species of seawater and freshwater fish have indicated that mercury concentrations in their tissues grow as their body mass and age increase (Squadrone et al. 2013).

The cycle of mercury transformations in the aqueous environment is a complex one. Biomethylation, the products of which can easily move between the different elements of the environment, may play an important role in the transfer of mercury. Inorganic compounds of mercury (II) in the aqueous environment undergo biochemical transformations effected by microorganisms in the anaerobic conditions, as a result of which highly toxic methylmercury compounds emerge. Both metallic mercury and the mercury contained in different compounds undergo biomethylation in hardly soluble sediments on the bottom of water reservoirs (Boening2000; He et al. 2005; Squadrone et al. 2013; Stein et al. 1996; Wang et al. 2004; Winfrey and Rudd 1990).

It is important to investigate the methylmercury absorption and accumulation pathways in the food chain. For this purpose, research is carried out on mercury concentrations in animal tissue and determines the position which these organisms occupy in the food chain—by using the stable nitrogen isotope ^{15}N (the content of which in tissues grows with each successive trophic level occupied by the consumer) (Post 2002; Roach et al. 2013). Short-chain alkyl mercury compounds, including those of metylmercury, demonstrate strong bioaccumulation properties. As a lithophilous compound, methylmercury is also capable of easily penetrating through mucous membranes.

A percentage share of mercury in the form of methylmercury in organisms grows for a higher trophic level. Mostly because of their small body size, short lifetime, and low position in the food chain, aquatic invertebrates accumulate smaller amounts of methylmercury in their organisms compared with those in fish. Wong et al. (1997) demonstrated that an average mercury concentration in invertebrates fell within the range from 0.07 to 0.18 μg/g. Surveys to determine the mercury contents in aquatic organisms have also been carried out on crayfish. They are considered to be scavengers which feed on the lowest trophic levels; as a result of this, the methylmercury contents in their bodies are slow compared with its contents in organisms situated at higher trophic levels.

The pollution caused by mercury is a serious threat for the marine environment. This is both a hygienic and ecotoxic problem. When this element occurs at levels exceeding its natural level in the seas and oceans, it poses a large threat for aquatic organisms.

Mercury concentrations detected in fish tissues vary, depending, e.g., on the fish species or the sea area from which a given fish originates. This is important when consumers choose a product. It is important to compare the place of origin of the fish contained in a fish product with the list of sea areas which are generally recognized to be highly vulnerable to mercury pollution.

Mercury contents in the organisms of sea fish depend not only on their species. The surveys carried out by Monteiro et al. (1996) on eight North Atlantic fish species demonstrated that mercury concentrations in the fish varied depending on the depth where they lived. The average mercury concentrations in the fish species covered by the surveys varied between 57 and 377 ppb in fresh tissue and were four times higher in mesopelagic species (depth of more than 300 m) than those in epipelagic (depth of less than 200 m). The fate of organic Hg species in marine ecosystems is dependent also upon surface water temperature, nutrient supply, and on the abundance of phytoplankton and its species composition (Szefer 2002).

In addition to the depth where sea fish feed, the mercury contents in their bodies are also affected by the fact whether they live in open waters or those situated closer to the coast. The coastal regions of seas and oceans are particularly vulnerable to anthropogenic mercury contamination, since both contaminants from rivers and those discharged directly into the oceans are transported to those areas. An additional source of mercury in these sea areas is the mercury exchange at the interface between the ocean surface and the atmosphere. Table 4 lists the data on mercury contents in fish coming from different sea areas in the world. These data demonstrate a large scatter of mercury contents both between the fish species and the water areas.

Table 4 Mercury contents in various sea fish from different sea areas in the world

Origin of sample	Concentration	Unit	Source
Caspian Sea	<0.05–0.79	mg/kg DM	Anan et al. (2005)
Pacific Ocean, Alaska	0.19–0.40	mg/kg DM	Meador et al. (2005)
Pacific Ocean, California	0.24–0.73	mg/kg DM	
Barents Sea, Greenland	0.19–1.10	mg/kg FM	Julshamn et al. (2006)
Indian Ocean, Mozambique	0.21–3.97	mg/kg DM	Kojadinovic et al. (2007)
Atlantic Ocean, Ghana	0.004–0.122	mg/kg FM	Voegborlo and Akagi (2007)
Atlantic Ocean, Azores	0.19–1.44	mg/kg FM	Afonso et al. (2007)
Black Sea, Turkey	0.025–0.084	mg/kg FM	Tuzen (2009)
Baltic Sea, Poland	0.018–0.118	mg/kg FM	Polak-Juszczak (2009)
Adriatic Sea, Croatia	0.001–0.52	mg/kg FM	Bilandžić et al. (2011)

DM dry mass, *FM* fresh mass

5 MATHEMATICAL MODELS

The management strategies intended to control the emissions of mercury of anthropogenic origin require the identification of the response of ecosystems to changes in the atmospheric deposition of this element. Mathematical models are valuable tools which provide information about the possible responses of ecosystems, since the measured data themselves are often insufficient and cannot be extrapolated in order to investigate the environmental impacts in other systems (Knightes et al. 2009).

Environmental models are an important decision-making tool for authorities. The spatial and temporal scales linking the tasks related to the control of the environment and the protection of its quality usually do not enable a comprehensive approach to the problem, and thus, they do not make it possible to understand the relation between economic activity and environmental quality. Moreover, environmental models facilitate the understanding of the key research needs and priorities of the data collection in the future (NRC 2007).

There are many models which are developed to simulate mercury emissions, both at a small scale, for local water reservoirs, and at a global scale, as well as to model mercury bioaccumulation in the chain web of aquatic systems. These models are used, e.g., to investigate the time scale required for the bioaccumulation of a specific mercury content level in the tissues of predators, in response to the differentiated sources of this element in the reservoir. These models are based on data on the mercury concentrations in entire aquatic ecosystems for which simulations are carried out and in the hydrodynamic processes (surface runoff, seepage, leaching, evaporation/evapotranspiration, etc.).

As the appropriate model is designed and developed, its important component are the data on which the model will be based, in particular the data on the dynamics of the processes unfolding in sediments, in the water column, and at the interface between them (Knightes et al. 2009). Table 5 shows the selected parameters used to develop the mathematical models and programs for mercury cycling in the environment. Table 6 presents a summary comparison of certain of the available models used to simulate the fate and behavior of mercury in the aquatic environment.

Table 5 The parameters used in the models of mercury recycling in the environment

tbcolw100ptChemical properties	Elemental Hg		Inorganic Hg^{2+}		$MeHg^{+}$	
Air/water partition coefficient K_{AW}	0.117	Estimated by Clapeyron for Henry's law constant at 25 °C	NA[a]		9.00×10^{-8}	Recommended literature value
Fraction absorbed on the skin	NR[b]	–	0	EA (2009)	0.1	EA (2009)
Diffusion coefficient in the air, m^2/s	6.34×10^{-6}	Estimated by the method according to Hemsoln and Cimbala (2003)	NA	–	8.61×10^{-6}	Estimated by the FSG method
Diffusion coefficient in water, m^2/s	2×10^{-9}	Estimated by the method according to Hayduk and Laudie (1974)	NA	–	1.7	Recommended literature value
log Kow – octanol/water partition coefficient (log), dimensionless	0.62	Recommended literature value	NA	–	1.9	Estimated using a non-hydrophobic relation with the octanol/wate partition coefficient log Kow
log Koc/w – organic carbon/water partition coefficient (log), cm^3/g	4.16[d]	–	NA	–	251.1	Recommended literature value
Relative molar mass, g/mol	200.59	Recommended literature value	NA		NA	–
Soil/water partition coefficient, cm^3/g	NA	–	500	Recommended literature value	1.13 (25 °C)	Recommended literature value
Vapor pressure, Pa	[illegible]	Recommended literature value	NA	–	100 (21 °C)	Recommended literature value
Water solubility, mg/L	0.056 (25 °C)	Recommended literature value	74,000 (20 °C)	Recommended literature value	0.5	EA (2009)
Soil-to-dust transport factor	0.5	EA 2009	0.5	EA 2009	1	EA (2009)

[a] *NA* not applicable; the models usually do not require these values for organic/inorganic compounds

[b] *NR* not relevant

[c] The experimentally determined value of K_{AW} for the Cl^- ionic concentration of 0.2×10^{-3} mol

[d] The value estimated from the soil/water partition coefficient for the inorganic components of mercury with the organic carbon content of 0.0348 (equivalent to 6 % of SOM)

Table 6 Models simulating mercury transport and transformation in aquatic system

Model	Properties	Survey site	Investigated media	Assessed forms of mercury
Coastal system				
ECoS (the Estuarine Contaminant Simulator model)	Estimation of contamination with Hg	Ria de Aveiro Coastal Lagoon (Portugal)	Suspended sediments, stagnant water	THg
2D STATRIM (the 2D STAtionary TRIeste gulf Mercur, a model with two submodels, 2D MIKE21MT, a model of transport in sediments, and PCFLOW2D-HD, a hydrodynamic model)	Simulations of Hg transport and transformation	The Gulf of Trieste	Suspended sediments, stagnant water, plankton	Hg^{2+}, Hg^{0}, MeHg, THg
Modified PCFLOW 3D (a hydrodynamic model containing a module for transport in sediments)	Simulations of Hg transport and transformation	The Gulf of Trieste	Suspended sediments, stagnant water	Hg^{2+}, Hg^{0}, MeHg, THg
Other				
EMMMA Environmental Mercury Mapping, Modeling, and Analysis	Determination of the characteristics of contamination with Hg, simulations of Hg transport and transformation, simulations of bioconcentration in fish	–	Suspended sediments, water, fish	Hg^{2+}, Hg^{0}, MeHg, THg

Source: Wang et al. (2004)

6 CONCLUSION

The surveys on the mercury contents in oceanic and marine waters carried out in the second half of the twentieth century showed that its levels were comparable to those that now occur in the waters of the global ocean. This phenomenon relates to the fact that the mercury level in the global ocean has not yet reached the phase of a dynamic equilibrium with its level in the atmosphere. It follows from the following three major premises:

- The average residence time of mercury in oceanic waters is 20–30 years, whereas that in the atmosphere varies between 0.8 and 2 years; thus, the mercury discharged into the ocean is removed from there much more slowly than the mercury emitted into the atmosphere;
- Oceanic waters concentrate much larger masses of mercury of natural (geogenic) origin rather than those of anthropogenic origin, which are close to the masses originating from the preindustrial period; and
- A large part of the mercury which is vertically transported from the oceanic depths to the surface layers is transported back to the mixed layer, which is more active biologically.

The premises listed above indicate that an increase in the mercury concentration level in oceanic waters will be very slow and may take hundreds of years, even if an increase in the mercury concentration level in the atmospheric air is not observed (Selin et al. 2010; Mason et al. 2012). In 1990–1996, there was a significant reduction in the quantities of mercury and other dangerous substances discharged into oceanic waters. Since 1997, the quantities of mercury discharged into the oceans have remained at a relatively constant level.

REFERENCES

1. Afonso, C., Lourenço, H. M., Nunes, M. L., & Castro, M. (2007). Contaminant metals in black scabbard fish (*Aphanopus carbo*) caught off Madeira and the Azores. *Food Chemistry, 101*, 120–125.
2. Aidin'yan, N. (1962). Content of mercury in some natural waters. *Chemical Abstracts, 57*, 16336.
3. Aidin'yan, N., & Belavskaya, G. A. (1963). The problem of supergene migration of mercury.*Chemical Abstracts, 59*, 8471–1963.

4. Akagi, H., Takabatake, E., & Fujita, Y. (1974). Photochemical methylation of inorganic mercury in the presence of solid sulfur. *Chemistry Letters, 3*, 761–764.

5. Alberts, J., Schindler, J., Miller, R., & Nutter, D. (1974). Elemental mercury evolution mediated by humic acid. *Science, 184*, 895–897.

6. Anan, Y., Kunito, T., Tanabe, S., Mitrofanov, I., & Aubrey, D. G. (2005). Trace element accumulation in fishes collected from coastal waters of the Caspian Sea. *Marine Pollution Bulletin, 51*, 882–888.

7. Baltisberger, R., Hildebrand, D., Grieble, D., & Ballintine, T. (1979). A study of the disproportionation of mercury (I) induced by gas sparging in acidic aqueous solutions for cold vapor atomic absorption spectrometry. *Analytica Chimica Acta, 111*, 111–122.

8. Barghigiani, C., & Ristori, T. (1995). Preliminary results on the role of rivers in total Hg concentrations in marine sediments and benthic organisms of coastal area of Italy. *Water, Air, and Soil Pollution, 80*, 1017–1020.

9. Barkay, T. (2000). Mercury cycle. In J. Lederberg (Ed.), *Encylopedia of microbiology* (pp. 171–181). San Diego: Academic Press.

10. Barkay, T., Turner, R., Saouter, E., & Horn, J. (1992). Mercury biotransformations and their potential for remediation of mercury contamination. *Biodegradation, 3*, 147–159.

11. Barkay, T., Miller, S. M., & Summers, A. O. (2003). Bacterial mercury resistance from atoms to ecosystems. *FEMS Microbiology Review, 27*, 355–384.

12. Benoit, J. M., Gilmour, C. C., Heyes, A., Mason, R. P., & Miller, C. L. (2003). Geochemical and biological controls over methylmercury production and degradation in aquatic systems. In Y. Cai & O. C. Braids (Eds.), *Biochemistry of environmentally important trace elements* (pp. 262e–297e). Washington, DC: American Chemical Society.

13. Berg, T., Fjeld, E., & Steinnes, E. (2006). Atmospheric mercury in Norway: contributions from different sources. *Science of the Total Environment, 368*, 3–9.

14. Bergan, T., Gallardo, L., & Rodhe, H. (1999). Mercury in the global troposphere: a three-dimensional model study. *Atmospheric Environment, 33*, 1575–1585.

15. Bilandžić, N., Dokić, M., & Sedak, M. (2011). Metal content determination in four fish species from the Adriatic Sea. *Food Chemistry, 124*, 1005–1010.

16. Bindler, G. R. (2003). Estimating the natural background atmospheric deposition rate of mercury utilizing ombotrophic bogs in southern Sweden. *Environmental Science and Technology, 37*, 40–46.

17. BMEPC—Baltic Marine Environment Protection Commission. (1987). *Progress reports on cadmium, mercury, cooper and zinc.* Baltic Sea Environment Proceedings, 24. Helsinki Commission.

18. Boening, D. W. (2000). Ecological effects, transport, and fate of mercury: a general review.*Chemosphere, 40*, 1335–1351.

19. Bone, S. E., Charette, M. A., Lamborg, C. H., & Gonneea, M. E. (2007). Has submarine groundwater discharge been overlooked as a source of mercury to coastal waters.*Environmental Science and Technology, 41*, 3090–3095.

20. Bookman, R., Driscoll, C. T., Engstrom, D. R., & Effler, S. W. (2008). Local to regional emission sources affecting mercury fluxes to New York lakes. *Atmospheric Environment, 42*, 6088–6097.

21. Borg, H., & Jonsson, P. (1996). Large-scale metal distribution in Baltic Sea sediments. *Marine Pollution Bulletin, 32*, 8–21.

22. Boszke, L., Głosińska, G., & Siepak, J. (2002). Some aspects of speciation of mercury in a water environment. *Polish Journal of Environmental Studies, 11*(4), 285–298.

23. Brzezińska, A., Trzosińska, A., Żmijewska, W., & Wodkiewicz, L. (1984). Trace metals in suspended matter and surficial bottom sediments from the Southern Baltic. *Oceanologia, 18*, 79–94.

24. Camargo, J. A. (1993). Which source of mercury pollution. *Nature, 365*, 302.

25. Cheng, H., & Hu, Y. (2010). China needs to control mercury emissions from municipal solid waste (MSW) incineration. *Environmental Science and Technology, 44*, 7994–7995.

26. Choi, S. C., & Bartha, R. (1994). Environmental factors affecting mercury methylation in estuarine sediments. *Bulletin of Environmental Contamination and Toxicology, 53*, 805–812.

27. Ci, Z. J., Zhang, X. S., Wang, Z. W., & Niu, Z. C. (2011). Phase speciation of mercury (Hg) in coastal water of the Yellow Sea, China. *Marine Chemistry, 126*, 250–255.

28. Compeau, G. C., & Bartha, R. (1985). Sulfate reducing bacteria: principal methylators of mercury in anoxic estuarine sediment. *Applied and Environmental Microbiology, 50*, 489–502.

29. Coquery, M., & Cossa, D. (1995). Mercury speciation in surface waters of the North Sea.*Netherlands Journal of Sea Research, 34*, 245–257.

30. Cossa, D., Coquery, M., Gobeil, C., & Martin, J. (1996). Mercury fluxes at the ocean margins. In R. Baeyens, W. Ebinghaus, & O. Vasiliev (Eds.), *Global and regional mercury cycles: sources, fluxes and mass balances* (pp. 229–248). The Netherlands: Kluwer Academic Publishers.

31. Cossa, D., Martin, J. M., Takayanagi, K., & Sanjuan, J. (1997). The distribution and cycling of mercury species in the western Mediterranean. *Deep-Sea Research Part II: Topical Studies in Oceanography, 44*, 721–740.

32. Ding, Z. H., Liu, J. L., Li, L. Q., Lin, H. N., Wu, H., & Hu, Z. Z. (2009). Distribution and speciation of mercury in surficial sediments from main mangrove wetlands in China. *Marine Pollution Bulletin, 58*, 1319–1325.

33. Drevnick, P. E., Lamborg, C. H., & Horgan, M. J. (2015). Increase in mercury in Pacific yellowfin tuna. *Environmental Toxicology and Chemistry, 34*(4), 931–934.

34. EA Environment Agency. (2009). *Updated technical background to the CLEA model. Science Report SC050021/SR3*. Bristol: Environment Agency.

35. Falandysz, J., Dubrawski, R., & Bolalek, J. (1993). Zawartosc rtęci w osadach dennych z Zatoki Puckiej i Gdanskiej (Mercury content in the benthic sediments in the Gulfs of Puck and Gdańsk—in Polish). *Bromatologia i Chemia Toksykologiczna, 26*, 29–32.

36. Fang, T., & Chen, R. (2010). Mercury contamination and accumulation in sediments of the East China Sea. *Journal of Environmental Sciences (China), 22*, 1164–1170.

37. Fang, J., Wang, K. X., Tang, J. L., Wang, Y. M., Ren, S. J., Wu, H. Y., & Wang, J. (2004). Copper, lead, zinc, cadmium, mercury, and arsenic in marine products of commerce from Zhejiang coastal area, China,

May 1998. *Bulletin of Environmental Contamination and Toxicology, 73*, 583–590.

38. Fergusson, J. E. (1990). *The heavy elements—chemistry, environmental impact and health effects*. Oxford: Pergamon Press.

39. Fitzgerald, W. F., Lamborg, C. H., & Hammerschmidt, C. R. (2007). Marine biogeochemical cycling of mercury. *Chemical Reviews, 107*, 641–662.

40. Fuji, M. (1976). Mercury. In S. Kitamura, M. Kondo, Y. Takizawa, M. Fujii, & M. Fujiki (Eds.),*Mercury distribution in lithosphere and atmosphere*. Tokyo: Kodansha Scientific.

41. Fu, X. W., Feng, X. B., Zhang, G., Xu, W. H., Li, X. D., Yao, H., Liang, P., Li, J., Sommar, J., Yin, R. S., & Liu, N. (2010). Mercury in the marine boundary layer and seawater of the South China Sea: concentrations, sea/air flux, and implication for land outflow. *Journal of Geophysical Research-Atmospheres, 115*, 1–11.

42. Gagnon, C., Pelletier, E., & Mucci, A. (1997). Behavior of anthropogenic mercury in coastal marine sediments. *Marine Chemistry, 59*, 159–176.

43. Gardfeldt, K., Sommar, J., Ferrara, R., Ceccarini, C., Lanzillotta, E., Munthe, J., Wangberg, I., Lindqvist, O., Pirrone, N., Sprovieri, F., Pesenti, E., & Stromberg, D. (2003). Evasion of mercury from coastal and open waters of the Atlantic Ocean and the Mediterranean Sea. *Atmospheric Environment, 37*, S73–S84.

44. Gbor, P. K., Wen, D., Meng, F., Yang, F., & Sloan, J. J. (2007). Modeling of mercury emission, transport and deposition in North America. *Atmospheric Environment, 41*, 1135–1149.

45. Gill, G. A., Bloom, N. S., Cappellino, S., Driscoll, C. T., Dobbs, C., et al. (1999). Sediment-water fluxes of mercury in Lavaca Bay, Texas. *Environmental Science and Technology, 33*, 663–669.

46. Gilmour, C. C., Reidel, G. S., Ederington, M. C., Bell, J. T., Benoit, J. M., Gill, G. A., & Stordall, M. C. (1998). Methylmercury concentrations and production rates across a trophic gradient in the northern Everglades. *Biogeochemistry, 40*, 327–345.

47. Grassia, S., & Nettib, R. (2000). Sea water intrusion and mercury pollution of some coastal aquifers in the province of Grosseto (southern Tuscany, Italy). *Journal of Hydrology, 237*, 198–211.

48. Gray, J. E., & Hines, M. E. (2006). Mercury: distribution, transport, and geochemical and microbial transformations from natural and anthropogenic sources. *Applied Geochemistry, 21*, 1819–1820.

49. Grieb, T. M., Driscoll, C. T., Gloss, S. P., Schofield, C. L., Bowie, G. L., & Porcella, D. B. (1990). Factors affecting mercury accumulation in fish in the Upper Michigan Peninsula.*Environmental Toxicology and Chemistry, 9*, 919–930.

50. Hamaguchi, H., Kuroda, R., & Hosohara. (1961). Photometric determination of traces of mercury in seawater. *Chemistry Abstracts, 55*, 15222.

51. Hammerschmidt, C. R., & Fitzgerald, W. F. (2006). Methylmercury cycling in sediments on the continental shelf of southern New England. *Geochimica et Cosmochimica Acta, 70*, 918–930.

52. Hayduk, W., & Laudie, H. (1974). Prediction of diffusion coefficients for non-electrolysis in dilute aqueous solutions.*Journal of the American Institute of Chemical Engineers, 20*, 611–615.

53. He, Z. L., Yang, X. E., & Stoffella, P. J. (2005). Trace elements in agroecosystems and impacts on the environment.*Journal of Trace Elements in Medicine and Biology, 19*, 125–140.

54. Heinsohn, R. J., & Cimbala, J. M. (2003).*Indoor air quality engineering: environmental health and control of indoor air pollutants.* New York: Marcel Dekker.

55. Henriques, B., Rocha, L. S., Lopes, C. B., Figueira, P., Monteiro, R. J. R., Duarte, A. C., Pardal, M. A., & Pereira, E. (2015). Study on bioaccumulation and biosorption of mercury by living marine macroalgae: prospecting for a new remediation biotechnology applied to saline waters.*Chemical Engineering Journal, 281*, 759–770.

56. Herut, B., Hornung, H., & Kress, N. (1994). Mercury, lead, copper, zinc and iron in shallow sediments of Haifa Bay, Israel. *Fresenius Environmental Bulletin, 3*, 147–151.

57. Heyes, A., Mason, R. P., Kim, E. H., & Sunderland, E. (2006). Mercury methylation in estuaries: insights from using measuring rates using stable mercury isotopes. *Marine Chemistry, 102*, 134–147.

58. Horvat, M. (1996). *Global and regional mercury cycles: sources, fluxes and mass balances.* Netherlands: Kluwer Academic Publishers.

59. Horvat, M., & Gibicar, D. (2005). Speciation of mercury: environment, food, clinical, and occupational health. In R. Cornelis, J. Caruso, H. Crews, & K. Heumann (Eds.), *Handbook of elemental speciation II—species in the environment, food, medicine and*

occupational health(pp. 281–305). Chichester: John Wiley & Sons Ltd.

60. Horvat, M., Kotnik, J., Logar, M., Fajon, V., Zvonaric, T., & Pirrone, N. (2003). Speciation of mercury in surface and deep-sea waters in the Mediterranean Sea. *Atmospheric Environment, 37*(S1), 93–108.

61. Hosohara, K., Kozuma, H., Kawasaki, K.-H., & Tsuruta, T. (1961). Total mercury content in sea water. *Chemical Abstracts, 56*, 5766.

62. Jiang, G.-B., Shi, J.-B., & Feng, X.-B. (2006). Mercury pollution in China. *Environmental Science and Technology, 40*, 3672–3678.

63. Julshamn, K., Grøsvik, B. E., Nedreaas, K., & Maage, A. (2006). Mercury concentration in fillets of Greenland halibut (*Reinhardtius hippoglossoides*) caught in the Barents Sea in January 2006. *Science of the Total Environment, 372*, 345–349.

64. Kabata-Pendias, A. (2011). *Trace elements in soils and plants* (4th ed.). Boca Raton: CRC Press - Taylor & Francis Group.

65. Kabata-Pendias, A., & Mukherjee, A. B. (2007). *Trace elements from soil to human*. New York: Springer-Verlag Berlin Heidelberg.

66. Kamman, N., Chalmers, A., Clair, T., Major, A., Moore, R., Norton, S., & Shanley, J. (2005). Factors influencing mercury in freshwater surface sediments of northeastern North America.*Ecotoxicology, 14*, 101–111.

67. Kannan, K., & Falandysz, J. (1998). Speciation and concentrations of mercury in certain coastal marine sediments. *Water, Air, and Soil Pollution, 103*, 129–136.

68. Knightes, C. D., Sunderland, E. M., Barber, M. C., Johnston, J. M., & Ambrose, R. B. (2009). Application of ecosystem-scale fate and bioaccumulation models to predict fish mercury response times to changes in atmospheric deposition. *Environmental Toxicology and Chemistry, 28*(4), 881–893.

69. Kojadinovic, J., Potier, M., LE Corre, M., Cosson, R. P., & Bustamante, P. (2007). Bioaccumulation of trace elements in pelagic fish from the Western Indian Ocean.*Environmental Pollution, 146*, 548–566.

70. Kotnik, J. (2007). Mercury speciation in surface and deep waters of the Mediterranean Sea.*Marine Chemistry, 107*, 13–30.

71. Kraepiel, A. M. L., Keller, K., Chin, H. B., Malcolm, E. G., & Morel, F. M. M. (2003). Sources and variations of mercury in tuna. *Environmental Science and Technology, 37*, 5551–5558.

72. Lacerda, L., Salomons, W. (1999). Mercury contamination from New World gold and silver mine tailings. In: *Mercury contaminated sites.* Springer-Verlag Berlin, 131–142.

73. Lamborg CH, Rolfhus KR, Fitzgerald WF, Kim G (1999) The atmospheric cycling and air-sea exchange of mercury species in the South and equatorial Atlantic Ocean. *Deep-Sea Research Part II: Topical Studies in Oceanography 46*(5), 957–977

74. Lamborg, C. H., Fitzgerald, W. F., Damman, A. W. II., Benoit, J. M., Balcom, P. H., & Engstrom, D. R. (2002). Modern and historic atmospheric mercury fluxes in both hemispheres: global and regional mercury cycling implications. *Global Biogeochemical Cycles, 16,* 1104–1114.

75. Laurier, F. J. G., Mason, R. P., Whalin, L., & Kato, S. (2003). Reactive gaseous mercury formation in the North Pacific Ocean's marine boundary layer: a potential role of halogen chemistry. *Journal of Geophysical Research-Atmospheres, 108*(17), ACH 3-1–ACH 3-12.

76. Laurier, F. J. G., Mason, R. P., Gill, G. A., & Whalin, L. (2004). Mercury distributions in the North Pacific Ocean—20 years of observations. *Marine Chemistry, 90,* 3–19.

77. Lina, C.-J., Pongprueksaa, P., Lindberg, S. E., Pehkoned, S. O., Byune, D., & Jang, C. (2006). Scientific uncertainties in atmospheric mercury models I: model science evaluation.*Atmospheric Environment, 40,* 2911–2928.

78. Marins, R. V., Lacerda, L. D., & Paraquetti, H. H. M. (2000). Dissolved reactive mercury speciation in water profiles from Sepetiba Bay, Brazil. In J. O. Nriagu (Ed.), *11th Annual International Conference on Heavy Metals in the Environment.* Ann Arbor: University of Michigan, School of Public Health.

79. Mason, R. P. (2009). Mercury emissions from natural processes and their importance in the global mercury cycle. In N. I. Pirrone & R. Mason (Eds.), *Mercury fate and transports in the global atmosphere* (pp. 173–191). New York: Springer.

80. Mason, R. P., & Fitzgerald, W. F. (1990). Alkylmercury species in the equatorial Pacific. *Nature, 347,* 457–459.

81. Mason, R. P., & Fitzgerald, W. F. (1993). The distribution and biogeochemical cycling of mercury in the equatorial Pacific Ocean. *Deep-Sea Res Part I, Oceanography Research Papers, 40,* 1897–1924.

82. Mason, R., & Fitzgerald, W. (1996). Sources, sinks and biogeochemical cycling of mercury in the ocean. In W. Baeyens, R. Ebinghaus, & O. Vasiliev (Eds.), *Global and regional mercury cycles: sources, fluxes and mass balances* (pp. 249–272). the Netherlands: Kluwer Academic Publishers.

83. Mason, R. P., & Sheu, G. R. (2002). Role of the ocean in the global mercury cycle. *Global Biogeochemical Cycles, 16*(4), 40-1–40-14.

84. Mason, R. P., & Sullivan, K. A. (1999). The distribution and speciation of mercury in the south and equatorial Atlantic. *Deep Sea Researches, Part II, 46*, 937–956.

85. Mason, R. P., Fitzgerald, W. F., & Morell, F. M. M. (1994). The biogeochemical cycling of mercury: anthropogenic influences. *Geochimica et Cosmochimica Acta, 58*(15), 3194–3198.

86. Mason, R., Morel, F., & Egmond, H. (1995). The role of microorganisms in elemental mercury formation in natural waters. *Water, Air, and Soil Pollution, 80*, 775–787.

87. Mason, R. P., Rolfhus, K., & Fitzgerald, W. F. (1998). Mercury in the North Atlantic. *Marine Chemistry, 61*, 37–53.

88. Mason, R. P., Choi, A. L., Fitzgerald, W. F., Hammerschmidt, C. R., Lamborg, C. H., Soerensen, A. L., & Sunderland, E. M. (2012). Mercury biogeochemical cycling in the ocean and policy implications. *Environmental Research, 119*, 101–117.

89. Meador, J. P., Ernest, D. W., & Kagley, A. N. (2005). A comparison of the non-essential elements cadmium, mercury, and lead found in fish and sediment from Alaska and California.*Science of the Total Environment, 339*, 189–205.

90. Monperrus, M., Tessier, E., Point, D., Vidimova, K., Amouroux, D., et al. (2007). The biogeochemistry of mercury at the sediment–water interface in the Thau Lagoon. 2. Evaluation of mercury methylation potential in both surface sediment and the water column. *Estuarine, Coastal and Shelf Science, 72*, 485–496.

91. Monteiro, L. R., Costa, V., Furness, R. W., & Santos, R. S. (1996). Mercury concentrations in prey fish indicate enhanced bioaccumulation in mesopelagic environments. *Marine Ecology Progress Series, 141*, 21–25.

92. Morel, F., Kraepiel, A. M. L., & Amyot, M. (1998). The chemical cycle and bioaccumulation of mercury. *Annual Review of Ecological Systems, 29*, 543–566.

93. National Research Council (NRC) 2007. Models in environmental regulatory decision making. National Academies Press, Washington.

94. Nagase, H. (1984). Mercury methylation by compounds in humic material. *Science of the Total Environment, 32*, 147–156.

95. Nasfi, F. H. (1995). Total mercury content of sea water on the Tunisian shore. *Fresenius Environmental Bulletin, 4*, 161–168.

96. Nriagu, J., & Becker, C. (2003). Volcanic emissions of mercury to the atmosphere: global and regional inventories. *Science of the Total Environment, 304*, 3–12.

97. Nriagu, J., & Becker, C. (2004). Volcanic emissions of mercury to the atmosphere: global and regional inventories. *Science of the Total Environment, 327*(1–3), 331–333.

98. Obi, E., Okafor, C., Igwebe, A., Ebenebe, J., Johnson Afonne, O., Ifediata, F., Orisakwe, O. E., Nriagu, J. O., & Basu, N. (2015). Elevated prenatal methylmercury exposure in Nigeria: evidence from maternal and cord blood. *Chemosphere, 119*, 485–489.

99. OJAVER, E. *(1995). Mercury content in the ecosystem of the Gulf of Riga. (w:) Ecosystem of the Gulf of Riga between 1920 and 1990*, 80–103, Tallinn.

100. Pacyna, E. G., Pacyna, J. M., Steenhuisen, F., & Wilson, S. (2006). Global anthropogenic mercury emission inventory for 2000. *Atmospheric Environment, 40*, 4048–4063.

101. Pempkowiak, J., Cossa, D., Sikora, A., & Sanjuan, J. (1998). Mercury in water and sediments of the southern Baltic Sea. *Sciences of the Total Environment, 213*, 18192.

102. Pirrone N., Cinnirella S. (2012). Mercury fluxes and emissions from natural processes. In: Pirrone N. (ed.) *Atmospheric pathways, transport and fate of mercury in the global environment, UNEP mercury air transport and fate research partnership area*, F&T: 11–14.

103. Pirrone, N., Cinnirella, S., Feng, X., Finkelman, R. B., Friedli, H. R., Leaner, J., Mason, R., Mukherjee, A. B., Stracher, G. B., Streets, D. G., & Telmer, K. (2010). Global mercury emissions to the atmosphere from anthropogenic and natural sources. *Atmospheric Chemistry and Physics, 10*, 4719–4752.

104. Polak-Juszczak, L. (2009). Temporal trends in the bioaccumulation of trace metals in herring, sprat, and cod from the southern Baltic Sea in the 1994–2003 period. *Chemosphere, 76*, 1334–1339.

105. Post, D. M. (2002). Using stable isotopes to estimate trophic position: models, methods, and assumptions. *Ecology, 83*(3), 703–718.

106. Qiu, Y. W., Yu, K. F., Zhang, G., & Wang, W. X. (2011). Accumulation and partitioning of seven trace metals in mangroves and sediment cores from three estuarine wetlands of Hainan Island, China. *Journal of Hazardous Materials, 190*, 631–638.

107. Roach, K. A., Jacobsen, N. F., Fiorello, C. V., Stronza, A., & Winemiller, K. O. (2013). Gold mining and mercury bioaccumulation in a floodplain lake and main channel of the Tambopata River, Peru. *Journal of Environmental Protection, 4*, 51–60.

108. Rolfus, K., & Fitzgerald, W. (1995). Linkages between atmospheric mercury deposition and the methylmercury content of marine fish. *Water Air and Soil Pollution, 80*, 291–297.

109. Rutgers Van Der Loeff, M. M., Anderson, L. G., Hall, P. O. J., Iverfeldt, A., Josefson, A. B., Sunby, B., & Westerlund, S. F. G. (1984). The asphyxiation technique: an approach to distinguishing between molecular diffusion and biologically mediated transport at the sediment-water interface. *Limnology and Oceanography, 29*, 675–686.

110. Schmidt, D. (1991). Mercury in Baltic and North Sea waters. *Water, Air, and Soil Pollution, 62*, 43–55.

111. Selin, N. E., Sunderland, E. M., Knightes, C. D., & Mason, R. P. (2010). Sources of mercury exposure for US seafood consumers: implications for policy. *Environmental Health Perspectives, 118*, 137–143.

112. Shi, J. B., Liang, L. N., Yuan, C. G., He, B., & Jiang, G. B. (2005). Methylmercury and total mercury in sediments collected from the East China Sea. *Bulletin of Environmental Contamination and Toxicology, 74*, 980–987.

113. Shi, J.-B., Carman, C. M., Tang, C. W. Y., Zhang, G., Rudolf, S. S., & Dong, X. (2007). Spatial and temporal variations of mercury in sediments from Victoria Harbour, Hong Kong. *Marine Pollution Bulletin, 54*, 464–488.

114. Squadrone, S., Prearo, M., Brizio, P., Gavinelli, S., Pellegrino, M., Scanzio, T., Guarise, S., Benedetto, A., & Abete, M. C. (2013). Heavy metals distribution in muscle, liver, kidney and gill of European catfish (*Silurus glanis*) from Italian Rivers. *Chemosphere, 90*, 358–365.

115. St. Louis, V. L., Rudd, J. W. M., Kelly, C. A., Beaty, K. G., Bloom, N. S., & Flett, R. J. (1994). Importance of wetlands as sources of methyl mercury to boreal forest ecosystems. *Canadian Journal of Fisheries and Aquatic Sciences, 51*, 1065e1076.

116. Stein, E. D., Cohen, Y., & Winer, A. M. (1996). Environmental distribution and transformation of mercury compounds. *Critical Reviews in Environmental Science and Technology, 26*, 1–43.

117. Streets, D. G., Zhang, Q., & Wu, Y. (2009). Projections of global mercury emissions in 2050.*Environmental Science and Technology, 43*, 2983–2988.

118. Strode, S., Jaegle, L., Selin, N., Jacob, D., Park, R., Yantosca, R., Mason, R., & Slemr, F. (2007). Air-sea exchange in the global mercury cycle. *Global Biogeochemal Cycles, 21*, GB1017.

119. Sunderland, E. M., & Mason, R. P. (2007). Human impacts on open ocean mercury concentrations. *Global Biogeochemical Cycles, 21*(4), GB4022.

120. Sunderland, E. M., Gobas, F., Branfireun, B. A., & Heyes, A. (2006). Environmental controls on the speciation and distribution of mercury in coastal sediments. *Marine Chemistry, 102*, 111–123.

121. Sunderland, E. M., Krabbenhoft, D. P., Moreau, J. W., Strode, S. A., & Landing, W. M. (2009). Mercury sources, distribution, and bioavailability in the North Pacific Ocean: insights from data and models. *Global Biogeochemical Cycles, 23*, GB2010.

122. Szefer, P. (2002). Metal pollutants and radionuclides in the Baltic Sea—an overview.*Oceanologia, 44*(2), 129–178.

123. Szefer, P. (2013). Safety assessment of seafood with respect to chemical pollutants in European Seas. *International Journal of Oceanography and Hydrobiology, 42*(1), 110–118.

124. Szumiło-Pilarska, E., Grajewska, A., Falkowska, L., Hajdrycha, J., Meissner, W., Frączek, T., Bełdowska, M., & Bzoma, S. (2016). Species differences in total mercury concentration in gulls from the Gulf of Gdansk (Southern Baltic).*Journal of Trace Elements in Medicine and Biology, 33*, 100–109.

125. Thongraa-Ra, W., & Parkpian, P. (2002). Total mercury concentrations in coastal areas of Thailand: a review. *Science Asia, 28*, 301–312.

126. Trasande, L., Landrigan, P. J., & Schechter, C. (2005). Public health and economic consequences of methyl mercury toxicity to the developing brain. *Environmental Health Perspectives, 113*, 590–596.

127. Tuzen, M. (2009). Toxic and essential trace elemental contents in fish species from the Black Sea, Turkey. *Food and Chemical Toxicology, 47*, 1785–1790.

128. Tyler, G. (1992). *Mercury in soil—distribution, speciation and biological effect.* Copenhagen: Nordic Council of Ministers.

129. Ullrich SM, Tanton TW, Abdrashitova SA (2001) Mercury in the aquatic environment: A review of factors affecting methylation (Review). *Critical Reviews in Environmental Science and Technology 31*(3), 241–293

130. US Environment Protection Agency. (1997). *Mercury Study Report to Congress. EPA-452/R-97*. Washington, DC: Office of Research and Development.

131. Voegborlo, R. B., & Akagi, H. (2007). Determination of mercury in fish by cold vapour atomic absorption spectrometry using an automatic mercury analyser. *Food Chemistry, 100*, 853–858.

132. Walsh, C., Distefano, M., Moore, M., Shewchuk, L., & Vetdine, G. (1988). Molecular basis of bacterial resistance to organomercurial and inorganic mercuric salts. *FASEB Journal, 2*, 124–130.

133. Wang, Q., Kim, D., Dionysiou, D. D., Soriala, G. A., & Timberlake, D. (2004). Sources and remediation for mercury contamination in aquatic systems—a literature review. *Environmental Pollution, 131*, 323–336.

134. Wang, Z. W., Zhang, X. S., Xiao, J. S., Zhijia, C., & Yu, P. Z. (2009a). Mercury fluxes and pools in three subtropical forested catchments, southwest China. *Environmental Pollution, 157*, 801–808.

135. Wang, S. F., Jia, Y. F., Wang, S. Y., Wang, X., Wang, H., Zhao, Z. X., & Liu, B. Z. (2009b). Total mercury and monomethylmercury in water, sediments, and hydrophytes from the rivers, estuary, and bay along the Bohai Sea coast, northeastern China. *Applied Geochemistry, 24*, 1702–1711.

136. Wang, C., Wang, Z., Ci, Z., Zhang, X., & Tang, X. (2016). Spatial-temporal distributions of gaseous element mercury and particulate mercury in the Asian marine boundary layer.*Atmospheric Environment, 126*, 107–116.

137. Weber, J. (1993). Review of possible paths for abiotic methylation of mercury(II) in the aquatic environment. *Chemosphere, 26*(11), 2063–2077.

138. Weber, J. H., Evans, R., Jones, S. H., & Hines, M. E. (1998). Conversion of mercury (II) into mercury (0), monomethylmercury

cation, and dimethylmercury in saltmarsh sediment slurries.*Chemosphere, 36,* 1669–1687.

139. Whalin, L., Kim, E.-H., & Mason, R. P. (2007). Factors influencing the oxidation, reduction, methylation and demethylation of mercury species in coastal waters. *Marine Chemistry, 107,* 278–294.

140. Wiener, J., Krabbenhoft, D. P., Heinz, G., & Sheuhammer, A. (2003). Ecotoxicology of mercury. In D. J. Hoffman (Ed.), *Handbook of ecotoxicology* (2nd ed., pp. 407–461). Boca Raton: CRC Press.

141. Wilken, R. D., & Hintelmann, H. (1991). Mercury and methylmercury in sediments and suspended particles from the river Elbe, North Germany. *Water, Air, and Soil Pollution, 56,* 427–437.

142. Winfrey, M. R., & Rudd, J. W. M. (1990). Environmental factors affecting the formation of methylmercury in low pH lakes: a review. *Environmental Toxicology and Chemistry, 9,* 853–869.

143. Wong, A. H. K., Mcqueen, D. J., Williams, D. D., & Demers, E. (1997). Transfer of mercury from benthic invertebrates to fish in lakes with contrasting fish community structures. *Canadian Journal of Fisheries and Aquatic Sciences, 54,* 1320–1330.

144. Zaib, M., Athar, M. M., Saeed, A., & Farooq, U. (2015). Electrochemical determination of inorganic mercury and arsenic—a review. *Biosensors and Bioelectronics, 74,* 895–908.

3

CHAPTER

Mercury in forest mushrooms and topsoil from the Yunnan highlands and the subalpine region of the Minya Konka summit in the Eastern Tibetan Plateau

Jerzy Falandysz, Martyna Saba, Hong-Gao Liu, Tao Li, Ji-Peng Wang, Anna Wiejak, Ji Zhang, Yuan-Zhong Wang, Dan Zhang

1. *Laboratory of Environmental Chemistry and EcotoxicologyGdańsk UniversityGdańskPoland*
2. *College of Agronomy and BiotechnologyYunnan Agricultural UniversityKunmingChina*
3. *College of Resources and EnvironmentYuxi Normal UniversityYuxiChina*
4. *Institute of Mountain Hazards and EnvironmentChinese Academy of SciencesChengduChina*
5. *Institute of Soil and FertilizerSichuan Academy of Agricultural SciencesChengduChina*
6. *Institute of Medicinal PlantsYunnan Academy of Agricultural SciencesKunmingChina*
7. *Yunnan Technical Center for Quality of Chinese Materia MedicaKunmingChina*

ABSTRACT

This study aimed to investigate and discuss the occurrence and accumulation of mercury in the fruiting bodies of wild-growing fungi (Macromycetes) collected from montane forests in two regions of southwestern China with differences in soil geochemistry, climate and geographical conditions. Fungal mycelia in soils of the subalpine region of the Minya Konka (Gongga Mountain) in Sichuan and in the highlands of Yunnan efficiently accumulated mercury in fruiting bodies (mushrooms). The examined sites in Yunnan with highly mineralized red and yellow soils showed Hg contents ranging from 0.066 to 0.28 mg kg^{-1} dry biomass (db) which is roughly similar to the results obtained for samples collected from sites with dark soils relatively rich in organic matter from a remote, the subalpine region of Minya Konka. Due to the remoteness of the subalpine section of Minya Konka, as well as its elevation and climate, airborne mercury from long-range transport could be deposited preferentially on the topsoil and the Hg levels determined in soil samples taken beneath the fruiting bodies were up to 0.48 mg kg^{-1} dry matter. In Yunnan, with polymetallic soils (Circum-Pacific Mercuriferous Belt), *Amanita* mushrooms showed mercury in caps of fruiting bodies of up to 7.3 mg kg^{-1} dry biomass. Geogenic Hg from the mercuriferous belt seems to be the overriding source of mercury accumulated in mushrooms foraged in the regions of Yunnan, while long-range atmospheric transport and subsequent deposition are the mercury sources for specimens foraged in the region of Minya Konka.

Keywords: Asia, Foraging, Heavy metals, Mushrooms, China, Yunnan

Responsible editor: Philippe Garrigues

INTRODUCTION

The anthropogenic emissions of Hg are a process that has been ongoing for decades now and has caused increases in topsoil Hg contents, thereby creating a pool of Hg with the likelihood of the formation of MeHg that accumulates in the aquatic food web (Olivero-Verbel et al. 2002,2015). Geogenic Hg is usually low in soils worldwide due to the low Hg content of the parent bedrock material, which is, on average, <0.05 mg kg^{-1} dry matter (dm) in the top 0–10 cm layer of forest soils collected beneath mushrooms across Poland, and with an average of about 0.08 mg kg^{-1} dm in the Earth's crust (Falandysz and Bielawski 2007; Falandysz and Brzostowski 2007; Rytuba 2003). In the regions of the Circum-Pacific Mercuriferous Belt in China, geogenic Hg from mineral deposits is usually elevated or even high in soils in the Yunnan and Guizhou Provinces in the southwestern parts of China (Fan 1991; Gustin et al. 1999; Shi et al. 2013; Wen and Chi 2007).

In the regions of Yunnan Province, the Hg content of the top (0–10 cm) layer of red and yellow soils in forested areas varied widely: from 0.096 to 0.91 mg kg^{-1} dm (Beicheng, Caoba, Daying, Eshan, Huangcaoba, Jiangchuan, Jiulongchi, Xinping and Yimen); from 0.15 to 3.4 mg kg^{-1} dm in Dayingjie in the Yuxi city; 0.25 mg kg^{-1} dm for Qilin in the Qujing city; 0.22 mg kg^{-1} dm for Dongshan in the Wenshan prefecture; 0.20 to 0.48 mg kg^{-1} dm for Pudacu in the Diqing prefecture; 1.3 mg kg^{-1} dm for Weixi in the Diqing prefecture; 0.065 to 0.91 mg kg^{-1} dm for Nanhua and Yuanmou in the Chuxiong prefecture; 0.19 to 1.1 mg kg^{-1} dm for Anning, Fumin and Shilin in the Kunming city; 0.33 to 0.35 mg kg^{-1} dm for Shiping in the Honghe prefecture; 0.13 to 2.6 mg kg^{-1} dm for Heqing, Midu and Yunlong in the Dali prefecture; 0.43 mg kg^{-1} dm for Yuanmou; from 0.083 to 1.2 mg kg^{-1} dm for Lufeng in the Chuxiong prefecture; 0.22 to 1.2 mg kg^{-1} dm for Dongshan in the Wenshan prefecture; 0.68 mg kg^{-1} dm for Yongren in the Chuxiong prefecture; 0.073 to 0.21 mg kg^{-1} dm for Malong in the Qujing prefecture; 0.31 to 2.1 mg kg^{-1} dm for Longyang region, Baoshan city; 0.13 mg kg^{-1} dm for Zhenyuan; from 0.43 to 0.53 mg kg^{-1} dm for Pu'er city; 2.4 mg kg^{-1} dm for Lanping in the Nujiang prefecture; and 0.42 to 0.98 mg kg^{-1} dm for Gejiu and Shiping in the Honghe prefecture (Falandysz et al. 2015a, b; Kojta et al. 2015).

The mycelial network is a device used by mushrooms in the absorption of chemical elements from substratum wherever the mycelium lives (soil, decaying litter, decaying wood, etc.). The mycelium could form rhizomorphs but no other mechanism (absorbing body) exists. Mushrooms differ in the substrate they grow on as well as in the depth where the mycelia live in soil. Mushrooms have no known device with which to absorb Hg compounds from the atmosphere, and any adsorption of Hg compounds or atmospheric particles containing this element at the surface (cap) of an ephemeral fruiting body would be negligible and without any practical significance for the background forested areas. For some mushrooms with small fruiting bodies, cleaning the litter and soil debris from the fruiting bodies using a plastic knife and brush may not be satisfactory/effective. Hence, effective cleaning can be achieved by washing the fruiting bodies with distilled water, as this is necessary to remove all traces of soil particles and litter debris before further processing and chemical examination.

Little is known of the significance of the mercuriferous belt areas on the occurrence of Hg and other metallic elements and metalloids in mushrooms. Little or almost nothing is known about the impact of the global mercury fallout, which is deposited and retained in the upper organic layer of soils (especially forest soils), which subsequently moves slowly down the soil horizon where it could become available to the mycelia. Suggestions that atmospheric particles could be a significant source of Hg to mushrooms (deposited on and then adsorbed through the cap) lack evidence for the mushrooms typically foraged in the forests and fields, especially considering the short lifetime (1 day to 2 weeks) of the ephemeral fruiting bodies. These issues are addressed shortly below.

At sites polluted by cinnabar mining or mine tailings, Hg could be accumulated by some mushrooms in amounts that far exceed those sequestered by the same species when growing in background areas. For example bay bolete from the genus *Imleria* (*Imleria badia* (Fr.) Vizzini, previous names *Xerocomus badius* (Fr.) E.-J. Gilbert, and *Boletus badius* (Fr.) Fr.) from a medieval gold and copper mine dump area near the town of Złotoryja in southern Poland contained Hg at 0.71 ± 0.49 mg kg^{-1} dry biomass (db) in caps and 0.41 ± 0.29 mg kg^{-1} db in stipes; the values were 1.3 ± 1.1 mg kg^{-1} db in whole fruiting bodies of *I. badia* from a site adjacent to the historical silver-mining area in South Bohemia in the Czech Republic; at 0.4 ± 0.1–0.6 ± 0.1 mg kg^{-1} db in individuals from a site polluted by flotation tailings in the southern part of Poland; and at 15–35 mg kg^{-1} db in a site of a former cinnabar mining area in Germany (Fischer et al. 1995; Kojta et al. 2012; Mleczek et al. 2015; Svoboda et al. 2006). A highly elevated concentration level of Hg occurred in mushrooms that grew in former mercury-mining areas: 110 mg kg^{-1} db in *Melanoleuca grammopodia* (Bull.) Murrill, from the Mountain Amiata in Italy; 52 ± 61 (up to 180) mg kg^{-1} db in *Suillus grevillei* (Klotzsch) Singer; and 48 ± 122 (up to 470) mg kg^{-1} db in *Lactarius quietus* from the Rudñany site in southeast Slovakia (Árvay et al. 2014; Bargagli and Baldi 1984).

Mushrooms collected at sites polluted with Hg from long-term industrial activities (e.g. mercury and copper ore smelting industry) have been observed to contain Hg contents as high as 5.2 ± 3.5 (up to 11) mg kg^{-1} db for *I. badia* (*B. badius*), 32 ± 10 (up to 62) mg kg^{-1} db for *Boletus edulis* Bull. and 85 ± 36 (up to 110) mg kg^{-1} db for *Lepista nuda* (Bull.) Cooke (Kalač et al.1996).

Southwestern China is one of the regions of the world within the mercuriferous belt (Gustin et al. 1999). Soils in this region are highly weathered and mineralized (Fan 1991). Nevertheless, there is currently dearth of data on Hg accumulation in wild-growing mushrooms from the Circum-Pacific Mercuriferous Belt region of China. Yunnan, China has one of the largest bio-diversities of mushrooms in the world. Wild-growing mushrooms are popular and traditionally widely foraged in the regions of Asia. In Yunnan, China, fungi are an important regional organic food that needs to be better characterized regarding the composition and accumulation of trace elements and other compounds in fruiting bodies (Wang et al. 2014).

In southwestern China, some montane regions are subjected to elevated deposition rates of airborne Hg because of long-range atmospheric transportation and deposition due to specific microclimatic and geographical conditions (Falandysz et al. 2014a; Zhang et al. 2013). Airborne Hg on the forest floor is bound by humus (organic) layer constituents, which efficiently retain it for subsequent infiltration down the soil horizons (Suchara and Sucharová 2002).

Both saprobic and mycorrhizal fungi with shallow mycelia depend more or less on decaying litter and organic layer of soils as a source of mineral compounds, while those with mycelia deeper in the soil horizon can depend on organic (humus) and/or mineral fractions of the forest soil. Results from a recent study suggest that for fungi of genus *Boletus*, *Leccinum* and*Xerocomus* which are of the mycorrhizal (symbiotic) type and have mycelia deeper in soil, the geogenic Hg in the mercuriferous belt could be an overriding source of this element accumulated in fruiting bodies (Falandysz et al. 2015a, b; Kojta et al. 2015; Ostos et al. 2015).

The top organic layer of forest soils is often enriched with Hg largely because of aerial deposition from anthropogenic sources over time. For example the Hg content of the top (0–1 cm) organic layer of the red and yellow lateritic soil collected from a forest within the outskirts of Yuxi in Yunnan was 0.16 mg kg^{-1} dm, while the Hg content of the bedrock (at depth of 2 to 3 m) below this soil varied from 0.0078 to 0.020 mg kg^{-1} dm (JF, unpublished). The dark montane forest soils of the subalpine region of Minya Konka in the Eastern Tibetan Plateau is of a different type and bedrock when compared to the red and yellow lateritic soils that dominate the central and southern part of Yunnan. However, both regions differ substantially in their elevation above sea level as well as in their climatic conditions; therefore, their susceptibility to long-range atmospheric pollution could differ.

The aim of this study was to fill the knowledge gap and provide some basic information on the occurrence, accumulation and distribution of Hg in fruiting bodies of several species of fungi collected from the Yunnan highlands which is subjected to direct impact from the Circum-Pacific Mercuriferous Belt and the subalpine region of the Minya Konka summit. The latter locations are within Sichuan Province and are considered pristine and outside the mercuriferous belt.

MATERIALS AND METHODS

Mushrooms and topsoil samples

Twenty three species of edible mushrooms were collected from spatially distributed locations across Yunnan Province, China during the collection seasons (June–September) of a period ranging from 2011 to 2014. Another six species were collected from the Minya Konka summit at an elevation of from 2800 to 3480 m above sea level in 2012 (Fig. 1). Macromycetes develop fruiting bodies seasonally (largely in late summer and early autumn which is the best for reproduction) but not necessarily annually. A pack of dried fruiting bodies of cultivated*Auricularia polytricha* (Mont.) Sacc. (current name *Auricularia nigricans* (Sw.) Birkebak, Looney & Sánchez-García) was also bought from a grocery store in Chengdu in Sichuan in 2012, and this was divided into 15 parts and then examined individually. Alongside the mushrooms from these locations, 21 soil samples (the upper 0–10 cm layer) beneath the fruiting bodies were also collected. No specific permits were required for sampling within these described areas. Similarly, no endangered or protected species were sampled, and these localities where were not protected in any way.

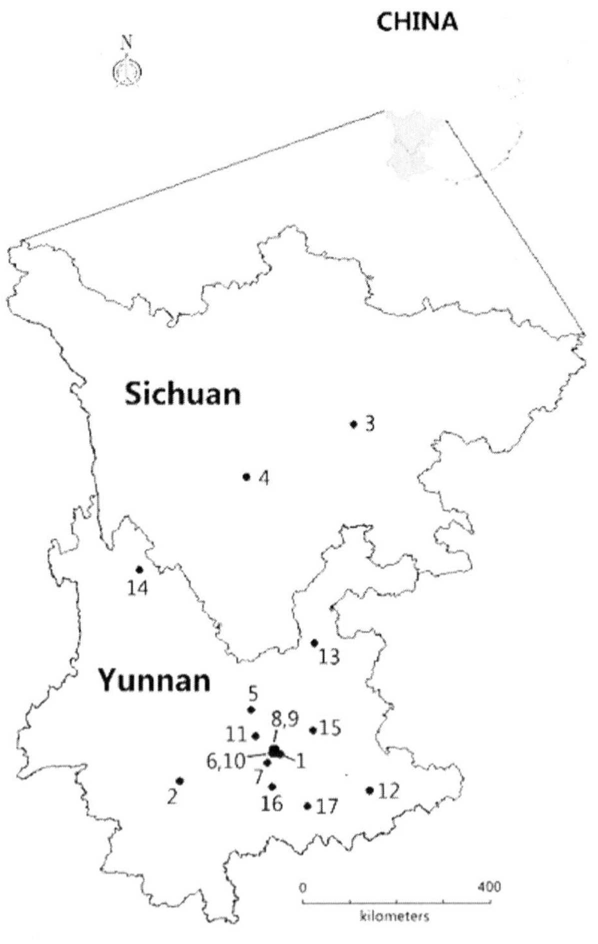

Fig. 1 The sampling sites of mushrooms in the Yunnan and Minya Konka in the Sichuan Province of China (*1* Jiuxi, *2* Zhenyuan, Pu'er, *3* Chengdu, *4* Minya Konka at 3480 m a.s.l., *5* Lufeng, Chuxiong, *6* Dayingjie, *7* Eshan, Yuxi, *8* Jiulongchi, Yuxi, *9* Beicheng, Yuxi, *10* Hongta region, Yuxi, *11* Yimen, Yuxi, *12* Yanshan, WenShan, *13* Huize, Qujing, *14*Pudacuo, Diqing, *15* Shilin, Kunming, *16* Shiping, Honghe, *17* Gejiu, Honghe)—*numbers*inside are explained in Table 1

Previous studies have shown that Hg content of fruiting bodies of fungi such as *B. edulis,Macrolepiota procera* (Scop.) Singer, *P. involutus* collected successively from the same areas over a period of between 2 and 4 years fluctuated but without any consistent pattern. This could therefore be considered as a natural phenomenon related to biological condition and weather (Brzostowski et al. 2011; Gucia et al. 2012; Zhang et al. 2010). Hence, a similar or related pattern could be expected in case of the fungi collected in this study.

The scientific names for species collected from Yunnan are as follows: *Amanita echinocephala*(Vittad.) Quél, *Amanita manginiana* Hariot et Patouillard, *Amanita vaginata* (Bull.:Fr) Vitt,*Clitocybe xanthophylla* Bres, *Heimioporus retisporus* (Pat. & C.F. Baker) E. Horak, *Laccaria amethystina* (Scop.) Cooke, *Laccaria laccata* (Scop.:Fr) Berk.et Br., *Laccaria proxima* (Boud.) Pat., *Laccaria vinaceoavellanea* Hongo, *Lactarius deliciosus* (L.:Fr.) Gray, *Rubinoboletus balloui* (Peck) Heinem. & Rammeloo, *Scleroderma citrinum* Pers., *Scleroderma flavidum* Ellis & Everh., *Sulllus collinitus* Fr., *Suillus pictus* (Peck) A.H. Smith et Thiers, *Tapinella panuoides*(Batsch) E.-J. Gilbert, *Tricholoma matsutake* (S. Ito et Imai) Sing., *Tylopilus felleus* (Bull.) P. Karst., *Tylopinus nigerrimus* (Heim) Hongo & Endo, *Tylopilus plumbeoviolaceoides* T.H. Li, B. Song & Y.H. Shen, *Tylopilus plumbeoviolaceus* (Snell.) Sing, *Tylopilus roseolus* (Chiu) Tai, and from the Minya Konka area: *Cortinarius bovinus* Fr., *Cortinarius collinitus* Fr., *Hygrocybe conica* (Schaeff.) P. Kumm., *Leccinum scabrum* (Bull.) Gray, *Russula grisea* Fr., *Russula puellaris* Fr. and *Tricholoma pessundatum* (S. Ito et Imai) Sing. (Table 1) (Index Fungorum2015).

Table 1 Mercury content in dried mushrooms and soil substratum from Yunnan and Minya Konka in China, quotient of Hg content in cap to stipe ($Q_{C/S}$), and quotient of Hg content in cap/stipe to soil beneath the fruiting bodies (BCF)

Species, place and year of collection	Number	Hg (mg kg^{-1})			$Q_{C/S}$	BCF	
		Whole fruiting bodies		Soils		Whole fruiting bodies	
		Caps	Stipes			Caps	Stipes
Amanita echinocephala (Vittad.) Quél.							
[1][a] Jiuxi, Yuxi; 2014	(3)[b]	5.0	1.8	WD	2.8	WD	WD
Amanita manginiana Hariot & Patouillard							
[2] Zhenyuan, Pu'er; 2014	(5)	7.3	4.2	WD	1.7	WD	WD
Amanita vaginata (Bull.:Fr) Vitt							
[1] Jiuxi, Yuxi; 2014	(3)	2.9	1.8	WD	1.6	WD	WD
Auricularia nigricans							
[3] Chengdu, shop; 2012	15[c]	0.033 ± 0.014 0.028 (0.016–0.068)		NA	NA	NA	NA
Clitocybe xanthophylla Bres							
[2] Zhenyuan, Pu'er; 2014	(20)	2.5	1.5	0.097	1.7	26	16
Cortinarius bovinus Fr.							
[4] Minya Konka at 3480 m a.s.l.; 2012	(> 10)	0.14		0.42	NA		0.33
[4] Minya Konka at 3480 m a.s.l.; 2012	(> 10)	0.11		0.18	NA		0.61
Cortinarius collinus Fr.							
[4] Minya Konka at 3000 m a.s.l.; 2012	(> 5)	0.045	0.53	0.015	0.085	3.0	35
[4] Minya Konka at 3000 m a.s.l.; 2012	(> 5)	0.074	0.26	WD	0.28	WD	WD
Heimioporus retisporus (Pat. & C.F. Baker) E. Horak							
[5] Lufeng, Chuxiong; 2013	(22)	0.73	0.34	0.13	2.1	5.6	2.6
Hygrocybe conica (Schaeff.) P. Kumm							
[4] Minya Konka at 2800 m a.s.l.; 2012	(9)	0.75		WD	NA	WD	
[4] Minya Konka at 2800 m a.s.l.; 2012	(10)	1.7		WD	NA	WD	
[4] Minya Konka at 3000 m a.s.l.; 2012	(10)	1.6		0.28	NA	5.7	
[4] Minya Konka at 3050 m a.s.l.; 2012	(17)	1.6		WD	NA	WD	
Laccaria amethystine (Scop.) Cooke							
[6] Dayingjie, Yuxi; 2014	(55)	0.10	0.056	WD	1.8	WD	WD
[2] Zhenyuan, Pu'er; 2014	(170)	0.034	0.032	WD	1.1	WD	WD
Laccaria laccata (Scop.:Fr) Berk. & Br.							
[7] Eshan, Yuxi; 2014	(176)	0.18	0.082	WD	2.2	WD	WD
[8] Jiulongchi, Yuxi; 2014	(380)	0.12	0.068	WD	1.8	WD	WD
[6] Dayingjie, Yuxi; 2014	(201)	0.15	0.060	WD	2.5	WD	WD
Laccaria proxima (Boud.) Pat.							
[1] Jiuxi, Yuxi; 2014	(80)	0.11	0.086	0.043	1.3	2.6	2.0
Laccaria vinaceoavellanea Hongo							
[9] Beicheng, Yuxi; 2014	(84)	0.13	0.050	WD	2.6	WD	WD
[2] Zhenyuan, Pu'er; 2014	(170)	0.034	0.032	0.12	1.1	0.28	0.27
Lactarius deliciosus (L.:Fr.) Gray							
[2] Zhenyuan, Puer; 2014	(35)	0.83	0.78	WD	1.1	WD	WD
Leccinum scabrum (Bull.) Gray							
[4] Minya Konka at 3000 m a.s.l.; 2012	(2)	0.22	0.12	0.15	1.8	1.5	0.80
[4] Minya Konka at 3000 m a.s.l.; 2012	(6)	0.37	0.29	0.33	1.3	1.1	0.88
[4] Minya Konka at 2900–3000 m a.s.l.; 2012	(4)	0.13	0.37	0.13	0.35	1.0	2.8
Rubinoboletus balloui (Peck) Heinem. & Rammeloo							
[1] Jiuxi, Yuxi; 2014	(17)	3.8	2.0	WD	1.9	WD	WD

Russula grisea Fr.							
[4] Minya Konka at 3000 m a.s.l.; 2012	(>10)	0.15	0.088	0.23	1.7	0.65	0.38
Russula puellaris Fr.							
[4] Minya Konka at 3000 m a.s.l.; 2012	(11)	0.077	0.12	0.37	0.64	0.21	0.32
Scleroderma citrinum Pers.							
[7] Eshan, Yuxi; 2014	(25)	0.090		WD	NA	WD	
Scleroderma flavidum Ell. & Ev.							
[1] Jiuxi, Yuxi; 2014	(9)	0.078		WD	NA	WD	
[10] Hongta region, Yuxi; 2014	(15)	0.19		WD	NA	WD	
[6] Dayingjie, Yuxi; 2014	(18)	0.047		WD	NA	WD	
Suillus collinitus Fr.							
[1] Jiuxi, Yuxi; 2013	(11)	0.089	0.040	0.045	2.2	2.0	0.90
[1] Jiuxi, Yuxi; 2014	(41)	0.32	0.20	0.11	1.6	2.9	1.8
Suillus pictus (Peck) A.H. Smith & Thiers							
[2]Zhenyuan, Pu'er; 2014	(16)	0.42	0.14	WD	3.0	WD	WD
[1] Jiuxi, Yuxi; 2014	(19)	0.22	0.10	WD	2.2	WD	WD
Tapinella panuoides (Batsch) E.-J. Gilbert							
[2] Zhenyuan, Pu'er; 2014	(23)	0.085		WD	NA	WD	
Tricholoma matsutake (S.Ito & Imai) Sing.							
[11] Yimen, Yuxi; 2012	(10)	0.73	0.44	WD	1.7	WD	WD
[12] Yanshan, WenShan; 2012	(10)	1.0	0.62	0.27	1.6	3.7	2.3
Tricholoma pessundatum (S. Ito & Imai) Sing.							
[4] Mt. Gongga at 2800 m a.s.l.; 2012	(24)	1.2	0.61	0.17	2.0	7.0	3.6
Tylopilus felleus (Bull.:Fr.) Karst							
[10] Hongta region, Yuxi; 2012	(10)	0.29	0.26	0.066	1.1	4.4	3.9
[13] Huize, Qujing; 2013	(5)	2.6	2.7	WD	0.96	WD	WD
[13] Huize, Qujing; 2013	(10)	2.2	1.5	WD	1.5	WD	WD
Tylopilus nigerrimus (Heim) Hongo & Endo							
[14] Pudacuo, Diqing; 2012	(10)	2.8	1.4	0.28	2.0	10	5.0
Tylopilus plumbeoviolaceoides T.H. Li, B. Song & Y.H. Shen							
[15] Shilin, Kunming; 2012	(10)	0.68	0.49	0.15	1.4	4.5	3.3
[13] Huize, Qujing; 2013	(10)	1.3	1.0	WD	1.3	WD	WD
Tylopilus plumbeoviolaceus (Snell.) Sing							
[15] Shilin, Kunming; 2012	(10)	0.68	0.49	0.15	1.4	4.5	3.3
Tylopilus roseolus (Chiu) Tai.							
[16] Shiping, Honghe; 2012	(10)	0.61	0.12	WD	1.1	WD	WD
[17] Gejiu, Honghe; 2012	(10)	0.41	0.31	WD	1.3	WD	WD

Mean value ± SD, median value, and range

BCF bioconcentration factor, *WD* without data, *NA* not applicable

[a] In bracket, a number of the sampling site is given (see Fig. 1)

[b] Number of individuals in a composite sample

[c] For *Auricularia nigricans*, 15 pooled samples have been examined (total quantity of mushrooms was 0.5 kg)

Fruiting bodies that were in good condition were collected and then cleaned carefully from any visible plant vegetation and soil debris with a plastic knife. To get insight into the distribution of total mercury (THg) between the two major morphological parts of the fruiting bodies of mushrooms (for some of the species), specimens were separated into cap (with skin) and stipe and further pooled accordingly with the aim of obtaining representative composite samples representing each species, sampling location and time of collection (Table 1). Thereafter, the mushroom samples were placed into electrically heated plastic trays used commercially in drying vegetables and then dried at 65 °C (Yunnan) or into a paper bag and dried in electrically heated laboratory driers at 65 °C (Minya Konka) to constant mass. Drying of mushrooms at 65 °C to constant weight is a standard procedure, e.g. for production of official certified fungal reference materials used in AC/AQ protocols in this study. Some authors dry mushrooms at 40 °C, while drying at 105 °C is not recommended because some organic compounds evaporate alongside water. Dried fungal materials were pulverized in a porcelain mortar and kept in sealed polyethylene bags under dry conditions. The soil samples, free of any visible organisms, small stones, sticks and leaves, were air dried at room temperature for several weeks under clean (dust free) conditions in the laboratory. Next, the soil samples were ground in a porcelain mortar, sieved through a 2-mm plastic sieve and kept in sealed polyethylene bags under dry conditions.

MERCURY DETERMINATIONS

All the reagents used in this study were of analytical reagent grade, unless otherwise stated. Double distilled water was used for the preparation of the solutions. Mercury standard solution of 1.0 mg Hg mL^{-1} was obtained from a 10 mg Hg mL^{-1} standard stock solution. Blank and 100, 150 and 200 µL of 1.0 mg mL^{-1} Hg standard solutions were injected into the analyser for the construction of a calibration curve, which was prepared fresh each week.

The determinations of total Hg content of fungal and soils samples were performed using cold-vapour atomic absorption spectroscopy (CV-AAS) by direct sample thermal decomposition coupled with gold wool trap of Hg and its further desorption and quantitative measurement at wavelength of 296 nm. The analytical instrument used was mercury analyser (MA-2000, Nippon Instruments Corporation, Takatsuki, Japan) equipped with auto sampler and operated at low mode (Brzostowski et al. 2011; Nnorom et al. 2013).

A running quality assurance/quality control (QA/QC) was performed through the analysis of blank samples and certified reference materials such as CS-M-4 (dried fruit bodies of King Bolete *B. edulis* mushroom) and CS-M-1 (dried fruit bodies of Cow Bolete *Suillus bovinus*mushroom) produced by the Institute of Nuclear Chemistry and Technology, Warsaw, Poland. The declared total Hg content of material CS-M-4 is 2.849 ± 0.104 mg kg^{-1} db, while present determinations ($n = 5$) showed 2.833 ± 0.023 mg kg^{-1} db. In the case of the material CS-M-1 the declared content is 0.164 ± 0.004 mg Hg kg^{-1} db, while present determination ($n = 13$) showed 0.185 ± 0.011 mg kg^{-1} db. The limit of detection (LOD) of this study was 0.0015 mg Hg kg^{-1} db, and the quantification limit (LOQ) was 0.005 mg Hg kg^{-1} db. One blank sample and one certified reference material sample were examined with each set of 3–5 samples studied. To estimate the potential of the fruiting bodies of macromycetes to sequester Hg or any other chemical element, it is common to calculate the bio-concentration factor (BCF).

RESULTS AND DISCUSSION

Mercury in soil substrata

The Hg content of the 0–10 cm layer (no litter) of soil samples from stands in Minya Konka for most of the samples ranged from 0.13 to 0.42 mg kg^{-1} dry matter (dm) and from a stand of *C. collinus* was an order of magnitude lower, i.e. at 0.015 mg kg^{-1} dm (Table 1). The examined sites in Yunnan with highly mineralized red and yellow soils showed Hg contents ranging from 0.066 to 0.28 mg kg^{-1} db which is roughly similar to the results obtained for samples collected from sites with dark soils relatively rich in organic matter from a remote, subalpine region of Minya Konka.

Mercury in fruiting bodies of mushrooms

Seven mushrooms species collected from the subalpine region in Minya Konka belonged to 5 genera and the other 22 species from Yunnan were from 11 genera. In all cases, the amounts/quantities of each sample, whether whole fruiting bodies or separated caps and stipes, were relatively large and sufficient for chemical analyses. For several species of mushrooms, it was possible to collect fruiting bodies from more than one site/location.

Mushrooms such as *C. bovinus*, *C. collinus*, *L. scabrum*, *R. grisea* and *R. puellaris* from Minya Konka showed an order of magnitude of lower contamination levels of Hg (<1.0 mg kg^{-1} db) than *H. conica* which showed 1.6 mg kg^{-1} db and *T. pessundatum* with 1.2 mg Hg kg^{-1} db in caps and 0.61 mg Hg kg^{-1} db in stipes (Table 1). The pattern of Hg distribution between cap and stipe in fruiting bodies of *C. collinus* was different from that usually observed in mushrooms. The stipes of *C. collinus* showed Hg at an order of magnitude greater concentration than the caps (value of the $Q_{C/S}$ index between 0.085 and 0.28; $Q_{C/S}$ is a value of cap to stipe Hg concentration quotient).

Mushroom species collected from Yunnan with mycelia living deeper in the soil (including species such as *Amanita* spp., *C. xanthophylla*, *R. balloui*, *Tricholoma* spp. and *Tylopilus* spp.) showed higher contamination with Hg compared to species having shallow mycelia (such as*Laccaria* spp. and *Scleroderma* spp). Mushrooms collected from Yunnan on average contained higher Hg values in the fruiting bodies compared to samples from Minya Konka. Species such as*A. echinocephala*, *A. manginiana* and *A. vaginata* with Hg levels in caps of 5.0, 7.3 and 2.9 mg kg^{-1} db and Hg in stipes of 1.8, 4.2 and 1.8 mg kg^{-1} db respectively, need to be considered as substantially contaminated.

The amounts of Hg observed in two sets of *T. felleus* collected from Huize (in the Qujing region) are orders of magnitude that are higher than the values observed in individual samples from the Hongta region (in Yuxi), which indicates that the elevated geogenic Hg content of the soil substratum could be the source/reason. Similarly, *T. plumbeoviolaceoides* from Huize (in Qujing) showed higher Hg contamination compared to samples from the distant site of Shilin in the region of Kunming (Table 1).

The *Laccaria* spp. are mycorrhizal mushrooms with shallow mycelia and they feed largely on decomposing litter. All four *Laccaria* spp. in this study showed low contamination with Hg. Nevertheless, the Hg content of individual samples of *L. amethystina* and *L. vinaceoavellanea*collected from montane sites in the rural Zhenyuan region in Pu'er were 0.034 and 0.034 mg Hg kg^{-1} db in caps and 0.032 and 0.032 mg Hg kg^{-1} db in stipes which indicate less contamination compared to samples from other sites within the urbanized and industrial areas of Yuxi. *S. citrinum* (with Hg content of 0.090 mg Hg kg^{-1} db) and *S. flavidum* (with Hg range from 0.047 to 0.078 mg kg^{-1} db) can be considered to be less contaminated with Hg compared to other mushrooms from Yunnan Province.

Fruiting bodies of cultivated Wood Ear fungus *A. nigricans* (a very popular mushroom in China readily eaten raw or fried) bought from a shop in Chengdu in Sichuan were less contaminated with mercury (median value of 0.028 mg kg^{-1} db) (Table 1).

Bio-concentration of Hg by mushrooms in fruiting bodies

The BCFs values for Hg were < 1 for mushrooms from Minya Konka (such as *C. bovinus*, *R. grisea* and *R. puellaris*); about ~1 for *L. scabrum*; > 1 for other species and up to 35 for *C. collinus* (stipes). The BCF values for most mushrooms from Yunnan, for which there were available data on Hg in soil, showed values > 1 and up to 26 (for caps) and 16 (for stipes) for *C. xanthophylla.* An exception was observed for the *Laccaria* spp.: the BCF for *L. vinaceoavellanea* was < 1 while the BCF for *L. proxima* was ~2. Most of the mushrooms in this study with available data on Hg in their soil substrate bio-included Hg.

Mercury in forest soils

The forest soils sampled in this study contained Hg in the top 0–10 cm layer at a relatively narrow range with elevated levels of up to 0.42 mg kg^{-1} dm observed for Minya Konka and up to 0.28 mg kg^{-1} dm for Yunnan. The topsoil sampled in the subalpine region of Minya Konka with one exception contained Hg at concentrations above 0.13 mg kg^{-1} dm, which is considered a critical level because of the likely harmful effects on soil organisms (Tipping et al. 2010).

The eastern slope of Minya Konka at the subalpine part is characterized by natural forests starting from an elevation of 2000 to 3600 m above sea level (a.s.l.). In this region, between 2000 and 2400 m a.s.l., the dominant tree species are *Lithocarpus cleistocarpus* (Seemen) Rehder & E.H. Wilson and *Acer flabellatum* J. D. Hooker & Thomson ex Hiern; from 2400 to 2900 m a.s.l. are the broad-leaf and dark coniferous mixed forests and the dominant trees are*Abies fabri* (Mast.) Craib and *A. flabellatum*; and from 2900 to 3650 m a.s.l., in the subalpine dark coniferous forest, the dominant tree species is *A. fabri*. The Minya Konka site differed from the sites sampled in Yunnan because of not only the type of forest but also the type of soil, parent rock material and soil geochemical background, as well as climate which is colder at this location with greater humidity and wet precipitation (Wu et al. 2013). The annual mean precipitation at an altitude of 3000 m a.s.l. is 1949 mm and this increases to between 3500 and 4200 mm as the altitude increases up to 3500 m a.s.l. (Wu et al. 2013). The soil organic matter content (%) in Minya Konka at the altitudes above sea level of 2361, 2777, 3317 and 3775 m are 7.06, 5.54, 5.95 and 5.04 respectively (Sun et al. 2013). As mentioned earlier, the soils in Yunnan and the neighbouring provinces, especially the eastern cities are latosols and lateritic red earths and red and yellow soils, are weathered and mineralized (polymetallic) with elevated contents of geogenic Hg and Sb in the region of the Circum-Pacific Mercuriferous Belt (Fan1991; He et al. 2004; Qiu et al. 2012; Wen and Chi 2007).

In the Minya Konka region in the Eastern Tibetan Plateau and elsewhere in the Tibetan Plateau, depositions of airborne Hg are considered to be the primary mechanism of Hg contamination of the ground, including litter and forest soils as well as some mushrooms (Falandysz et al. 2014a; Fu et al. 2008; Zhang et al. 2008). The anthropogenic, airborne Hg is an important fraction of Hg accumulated in montane soils in southwestern China and atmospheric input of anthropogenic Hg increases with altitude above sea level as evidenced for Mt. Leigong (Zhang et al. 2013). Hence, an elevated content of Hg observed in topsoil in forests on the eastern slope of Minya Konka, which is often dark and rich in organic matter and without known sources of geogenic Hg, could be a result of enhanced atmospheric depositions which are subsequently accumulated by mushrooms into their fruiting bodies. Nevertheless, information of vertical deposition of Hg in soil horizons for forests of Minya Konka is not available. However, it is known that the ability of mushroom mycelia to take up Hg and other mineral constituents and sequester them in the fruiting bodies is, apart from genetic factors, highly related to the mushroom's life style (feeding behaviour) and depth of mycelia in the soil horizon.

The Hg content of a limited number of the samples of red and yellow soils in Yunnan examined in this study (Table 1) was above 0.08 mg kg^{-1} dm, which, as has been cited earlier, is considered typical for the Earth's crust. The red and yellow soils beneath the fruiting bodies of many mushrooms of the genera *Boletus* and *Leccinum* from the forests of Yunnan showed Hg in the top 0–10 cm layer of up to 3.4 mg kg^{-1} dm, which is higher than the results of this study (Table 1) (Falandysz et al. 2015a, b).

Mercury in mushrooms

There are no previous data on Hg contamination of the fruiting bodies of most of the mushroom species/genera studied that were from Yunnan or Minya Konka. Species from the genus*Amanita* are well available in Yunnan and worldwide, and some are edible (Falandysz and Drewnowska 2015a; Mao 2009). Both *A. echinocephala* and *A. manginiana* with 5.0 and 7.3 mg Hg kg^{-1} db in caps respectively, indicated a substantial Hg uptake but there are no other published data available for these two species for comparison. In *A. vaginata*, the median value for caps was 2.9 mg Hg kg^{-1} db, which is an order or two higher than values observed for several sample sets collected from distantly distributed sites in Poland (with medians values ranging from 0.079 to 0.45 mg kg^{-1} db), and 0.75 mg kg^{-1} db for a sample set from the outskirts of Umeå in Sweden (Drewnowska et al. 2014; Falandysz et al. 2001a). *Amanita fulva* Fr., from the background areas in Poland, was much less contaminated with Hg (median mercury in caps ranging from 0.13 to 0.67 mg kg^{-1} db).

Amanita muscaria is a toadstool that has psychedelic effects on humans and was popular among the shamans in some locations in ancient Europe and Euro-Asia and maybe until recently in some parts of Siberia (Michelot and Melendez-Howell 2003). Apart from the organic constituents with psychedelic action, the species is known as a specific accumulator of vanadium (V) and zirconium (Zr) (Cenci et al. 2009). The reported Hg content of *A. muscaria*from the background areas in Central Europe was 0.68 mg kg^{-1} db, while values ranging from 0.19 to 1.2 mg kg^{-1} dm (and up to 1.8 mg kg^{-1} dm) were reported for individual samples collected from within the World War II battle area and from sites close to military (shooting) area and War World II battle grounds in the region of the Niedźwiednik near to the city of Gdańsk (Drewnowska et al. 2013; Falandysz et al. 2003a). A Hg content in caps of 0.39 mg kg^{-1}db, (on the average) was reported for *A. muscaria* collected from the suburbs of the city of Umeå, while other locations within northern Sweden showed substantially smaller concentrations than those observed for *Amanita* spp. collected from the mercuriferous belt in this study (Falandysz et al. 2001a).

Amanita rubescens Pers. showed a similar capacity to accumulate Hg as *A. muscaria* and the reported values of Hg accumulated in caps of this species collected from background areas varied from 0.22 ± 0.6 (0.11–0.34) to 1.0 ± 0.6 (0.67–1.9) and 0.83 ± 1.00 (0.26–3.6) mg kg^{-1}db (Falandysz et al. 2003b). Also, *Amanita phalloides* (Vaill. ex Fr.) Link with 1.2 ± 0.4 (0.62–1.8) mg Hg kg^{-1} db seems to have a similar capacity for sequestration of Hg as mentioned above for *A. muscaria* and *A. rubescens* (Drewnowska et al. 2012b; Falandysz et al. 2003a).

Mushrooms of the genus *Cortinarius* collected from Minya Konka showed weak contamination with Hg and possible explanations could be a weak ability to accumulate Hg and the low Hg contents of the soil substratum of *C. collinus*. Meanwhile, the Hg levels in the soil where the mycelium grows were not reported. In a study by Seeger and Nützel (1976), seven species of mushroom genus *Cortinarius* showed Hg contents ranging from 0.16 to 2.3 mg kg^{-1} db, while the Hg content of one samples of *Cortinarius elatior* Fr., was 16 mg kg^{-1} db (Seeger and Nützel1976). In a recent study, a large collection of *C. caperatus* showed wide differences in Hg content of caps depending on sampling locations and this ranged from 0.91 ± 0.62 to 2.5 ± 0.7 mg kg^{-1} db (median values ranged from 0.81 to 2.4 mg kg^{-1} db; Falandysz 2016). The observed differences in Hg contents of *Cortinarius* mushrooms as reported by other authors compared to this study can be explained by a species-specific ability to accumulate Hg, which seems to differ for various species of this genus on one side and on the abundance of geogenic or anthropogenic Hg from soil to mycelia on the other side.

The mushroom *H. conica* from Minya Konka showed elevated Hg contents and this can be attributed to elevated Hg content of soils beneath the fruiting bodies, which was dark in colour and rich in organic matter.

There is no previous data published on Hg contents of *Laccaria* spp. from Yunnan or elsewhere in China. The species *L. laccata* (whole fruiting bodies) sampled in Hungary, Germany and Turkey in Europe showed Hg contents ranging from 0.059 ± 0.030–0.067 ± 0.01, 0.13 (0.08–0.25) and 0.12 ± 0.01 mg kg^{-1} db respectively, which are within the range of values observed in this study for China and may indicate similar rates of airborne

Hg deposition and contamination of the uppermost organic layer of soils (Seeger and Nützel 1976; Tüzen and Soylak 2005; Vetter and Berta 1997).

For *L. amethystina* Cooke, from regions uncontaminated with Hg in Europe, the values reported ranged from 0.042 ± 0.003 and 0.078 ± 0.008 mg Hg kg^{-1} db for two sites in Hungary, 0.10 (0.08–0.11) mg kg^{-1} db ($n = 6$) for a site in Germany and 0.19 ± 0.08 (0.11–0.27 mg kg^{-1} db) for a site in Poland (Falandysz et al. 2001b; Seeger and Nützel 1976; Vetter and Berta 1997). The Hg content in *L. deliciosus* from Pu'er in Yunnan was similar to the values reported for this mushroom in Europe: Germany (0.35 mg kg^{-1} dm), Hungary (0.45–0.61 mg kg^{-1} db), Poland (0.94 ± 0.13 mg kg^{-1} db) and Spain (0.51–0.77 mg kg^{-1} db) (Melgar et al. 2009; Mleczek et al. 2013; Seeger and Nützel 1976; Vetter and Berta 1997).

As can be observed from data reported by Falandysz et al. (2015a, b) in some regions of the Pu'er Prefecture, the geogenic Hg is lower than in many other regions of Yunnan Province and this can explain the relatively low observed contents of Hg in *L. deliciosus* from Pu'er in this study. Nevertheless, locally, the soils and their parent rock polymetallic background material are diverse both in the mountainous region of Pu'er and other parts of Yunnan and more detailed sampling would be required to gain a better insight of the accumulation of Hg and its relationships with the soil background for mushrooms in Yunnan.

The mushroom *L. scabrum* from the sites in Minya Konka showed moderate contamination with Hg and some potential for accumulation of this element. The observed Hg levels were comparable and often times lower than data for these species reported in literature: at 0.70 ± 0.27 (0.14–1.4) mg kg^{-1} db and from 0.38 ± 0.24 to 1.2 ± 0.4 mg kg^{-1} db in Poland; at 0.40 ± 0.31–0.57 ± 0.23 mg kg^{-1} db in Spain and at 0.18 ± 0.13 (0.0033–0.70) mg kg^{-1} db in Sweden (Melgar et al. 2009; Falandysz and Bielawski 2007; Falandysz et al. 2001a, 2007).

The *Scleroderma* spp. sampled in Yunnan were low in Hg when compared with several other species in this study. In an earlier study, *S. citrinum* sampled in Poland bio-excluded Hg and its carpophores showed very low contamination with Hg, with a mean of 0.0056 ± 0.0021 (0.0026–0.0094) mg kg^{-1} db (Falandysz and Chwir 1997).

Among the mushrooms of the genus *Suillus* in this study, both *S. collinitus* and *S. pictus* from Yunnan have no previous data on the content and bio-concentration potential of Hg in fruiting bodies. The species *S. collinitus* showed some potential to bio-concentrate Hg, but it seems to be less efficient at this when compared to other species of the genus *Suillus* from other studies. For example the species *S. bovinus* (L.) Roussel, *S. granulatus* (L.) Roussel, *S. grevillei*, *S. luteus* (L.) Roussel and *S.*

variegatus (Sw.) Richon & Roze, for which sufficiently data are available, both showed BCF values for Hg at around 10 (caps) (Chudzyński et al. 2009, 2011; Saba et al. 2016a, b, c).

For mushrooms of the genera *Russula*, *Tapionella*, *Tricholoma* and *Tylopilus* respectively, from Yunnan or Minya Konka, there are no available previous data on accumulation and distribution of Hg in the fruiting bodies. Both *T. matsutake* from Yunnan and *T. pessundatum* from Minya Konka showed some ability to bio-concentrate Hg (BCF > 1) and the efficiency was similar to what was observed for *Tricholoma flavovirens* (Pers.) S. Lundel (current name *Tricholoma equestre* (L.) P. Kumm.) from sites in Europe with Hg in the soil substratum of 0.059 ± 0.028 mg kg^{-1} db (Maćkiewicz and Falandysz 2012). The species *T. flavovirens* from the European locations was very efficient in the absorption of Hg from low contaminated soils, i.e. soil Hg of 0.019 ± 0.003 to 0.046 ± 0.007 mg Hg kg^{-1} db and with a BCF for caps as high as 22 ± 9 to 75 ± 13 (overall range 9.0–90) (Falandysz et al. 2003c; Maćkiewicz and Falandysz2012). These data on *Tricholoma* mushrooms showed their good ability for absorption of Hg from soil substrate. An open question for *Tricholoma* mushrooms is the difference in availability between the geogenic Hg in lateritic red earths and red and yellow earths and the anthropogenic Hg deposited in the upper layer of soils.

Possible implications for consumers

Several species of edible wild-growing mushrooms from Yunnan and Minya Konka in Sichuan in China for which data on contamination with Hg were available for the first time showed elevated contents of this element in the fruiting bodies, which are one to two orders of magnitude greater when compared to staple vegetable foods. Important factors to consider are the culinary processing used in Yunnan to prepare a dish which may impact the fate of the Hg contained in mushrooms and also the bio-availability and bio-accessibility of Hg from a meal made of mushrooms high in Hg as well as the role of Se in mushrooms naturally rich in Hg. Consumption of mushrooms foraged by people of Yunnan is one of the highest in the world and in some locations exceeds 20 kg per capita annually (Zhang et al. 2010). Yunnan province is exceptionally rich in species of macrofungi and is called "the kingdom of mushrooms". Mushrooms are commonly foraged everywhere across Yunnan and reach local regional and distant markets in China and abroad (especially in the remote region of the Shangri-la in the north) (Wang et al. 2014). If based on standard measures of Hg intake and risk such as the value of the provisionally tolerable weekly intake (PTWI) or the reference dose (RfD), the many edible mushrooms from Yunnan provide a high dose of Hg when consumed at a rate of 300 g or more per week.

CONCLUSIONS

Mercury content in the red and yellow lateritic soils in the forested regions of Yunnan Province in southwestern China is elevated over the background level due to their location in the Circum-Pacific Global Mercuriferous Belt. Mercury content in forest soil from the subalpine section of the Minya Konka summit in the Eastern Tibetan Plateau is considered to be elevated over the background levels because of the susceptibility of the region to deposition of airborne Hg originating from distant sources. A large majority of the fungal species examined have no previous data on accumulation of Hg in fruiting bodies from either China, Asia or elsewhere.

The BCF values for Hg for most mushrooms from Yunnan, for which there were available data on Hg in soil, showed on bio-inclusion and values up to 26 (for caps) and 16 (for stipes) for *C. xanthophylla.* Most of the mushrooms in this study with available data on Hg in their soil substrate bio-included Hg. Geogenic Hg from the mercuriferous belt seems to be the overriding source of mercury accumulated in mushrooms foraged in the regions of Yunnan, while long-range atmospheric transport and subsequent deposition are the mercury sources for specimens foraged in the region of Minya Konka.

ACKNOWLEDGMENT

This study has been supported in part by the Chinese Academy of Sciences under grant no. 2010T1Z26.

REFERENCES

1. Árvay J, Tomáša J, Hauptvogl M, Kopernická M, Kováčik A, Bajčan D, Massányi P (2014) Contamination of wild-grown edible mushrooms by heavy metals in a former mercury-mining area. J Environ Sci Health Part B 49:815–827

2. Bargagli R, Baldi F (1984) Mercury and methyl mercury in higher fungi and their relation with the substrata in a cinnabar mining area. Chemosphere 13(9):1059–1071

3. Brzostowski A, Jarzyńska G, Kojta AK, Wydmańska D, Falandysz J (2011) Variations in metal levels accumulated in Poison Pax (*Paxillus involutus*) mushroom collected at one site over four years. J Environ Sci Health Part A 46:581–588

4. Cenci RM, Cocchi L, Petrini O, Sena F, Siniscalco C, Vescovi L (2009) Elementi chimici nei funghi superiori I funghi di riferimento come strumento di lavoro per la bioindicazione e la biodiversità. JRC Sci Tech Rep EUR 24415 IT 2009

5. Chudzyński K, Bielawski L, Falandysz J (2009) Mercury bio-concentration potential of larch bolete, *Suillus grevillei*, mushroom. Bull Environ Contam Toxicol 83:275–279

6. Chudzyński K, Jarzyńska G, Stefańska A, Falandysz J (2011) Mercury content and bio-concentration potential of Slippery Jack, *Suillus luteus*, mushroom. Food Chem 125:986–990

7. Drewnowska M, Jarzyńska G, Kojta AK, Falandysz J (2012b) Mercury in European Blusher,*Amanita rubescens*, mushroom and soil. Bioconcentration potential and intake assessment. J Environ Sci Health Part B 47:466–474

8. Drewnowska M, Lipka K, Jarzyńska G, Danisiewicz-Czupryńska D, Falandysz J (2013) Investigation on metallic elements in fungus *Amanita muscaria* (fly agaric) and the forest soils from the Mazurian lakes district of Poland. Fresenius Environ Bull 22:455–460

9. Drewnowska M, Nnorom IC, Falandysz J (2014) Mercury in the Tawny Grisette, *Amanita vaginata* Fr. and soil below the fruiting bodies. J Environ Sci Health Part B 49:521–526

10. Falandysz J (2016) Distribution of mercury in Gypsy *Cortinarius caperatus* mushrooms from several populations: an efficient accumulator species and estimated intake of element. Ecotox Environ Saf 110:68–72

11. Falandysz J, Bielawski L (2007) Mercury and its bioconcentration factors in Brown Birch Scaber Stalk (*Leccinum scabrum*) from various sites in Poland. Food Chem 105:635–640

12. Falandysz J, Brzostowski A (2007) Mercury and its bioconcentration factors in Poison Pax (*Paxillus involutus*) from various sites in Poland. J Environ Sci Health Part A 42:1095–1100

13. Falandysz J, Chwir A (1997) The concentrations and bioconcentration factors of mercury in mushrooms from the Mierzeja Wiślana sand-bar, northern Poland. Sci Total Environ 203:221–228

14. Falandysz J, Drewnowska M (2015) Distribution of mercury in *Amanita fulva* (Schaeff.) Secr. mushrooms: accumulation, loss in cooking and dietary intake. Ecotox Environ Saf 115:49–54

15. Falandysz J, Gucia M, Frankowska A, Kawano M, Skwarzec B (2001a) Total mercury in wild mushrooms and underlying soil substrate from the city of Umeå and its surroundings, Sweden. Bull Environ Contam Toxicol 67:763–770

16. Falandysz J, Szymczyk K, Ichihashi H, Bielawski L, Gucia M, Frankowska A, Yamasaki S (2001b) ICP/MS and ICP/AES elemental analysis (38 elements) of edible wild mushrooms growing in Poland. Food Addit Contam 18:503–513

17. Falandysz J, Lipka K, Kawano M, Brzostowski A, Dadej M, Jędrusiak A, Puzyn T (2003a) Mercury content and its bioconcentration factors at Łukta and Morąg, northeastern Poland. J Agric Food Chem 51:2835–2836

18. Falandysz J, Gucia M, Brzostowski A, Kawano M, Bielawski L, Frankowska A, Wyrzykowska B (2003b) Content and bioconcentration of mercury in mushrooms from northern Poland. Food Addit Contam 20:247–253

19. Falandysz J, Brzostowski A, Kawano M, Kannan K, Puzyn T, Lipka K (2003c) Concentrations of mercury in wild growing higher fungi and underlying substrate near Lake Wdzydze, Poland. Water Air Soil Pollution 148:127–137

20. Falandysz J, Kunito T, Kubota R, Bielawski L, Mazur A, Falandysz JJ, Tanabe S (2007) Selected elements in Brown Birch Scaber Stalk *Leccinum scabrum*. J Environ Sci Health Part A 42:2081–2088

21. Falandysz J, Dryżałowska A, Saba M, Wang J, Zhang D (2014a) Mercury in the fairy-ring of*Gymnopus erythropus* (Pers.) and *Marasmius dryophilus* (Bull.) P. Karst. mushrooms from the Gongga Mountain, Eastern Tibetan Plateau. Ecotoxicol Environm Saf 104:18–22

22. Falandysz J, Zhang J, Wang Y, Krasińska G, Kojta A, Saba M, Shen T, Li T, Liu H (2015a) Evaluation of the mercury contamination in mushrooms of genus *Leccinum* from two different regions of the world: accumulation, distribution and probable dietary intake. Sci Total Environ 537:470–478

23. Falandysz J, Zhang J, Wang Y, Saba M, Krasińska G, Wiejak A, Shen T, Li T (2015b) Evaluation of mercury contamination in fungi boletus species from latosols, lateritic red earths, and red and yellow earths in the Circum-Pacific Mercuriferous Belt of southwestern China. PLoS One 10:e0143608.

24. Fan C (1991) A study on the origin of ore-forming materials in the antimony and mercury multiple-metal ore zone at Weisha, Yunnan Province, China (*in Chinese*). Geochemica 12:399–405

25. Fischer RG, Rapsomanikis S, Andreae MO, Baldi F (1995) Bioaccumulation of methylmercury and transformation of inorganic mercury by macrofungi. Environ Sci Technol 29:993–999

26. Fu X, Feng X, Zhu W, Wang S, Lu J (2008) Total gaseous mercury concentrations in ambient air in the eastern slope of Mt. Gongga, south-eastern fringe of the Tibetan plateau, China. Atmosph Environ 42:970–979

27. Gucia M, Jarzyńska G, Kojta AK, Falandysz J (2012) Temporal variability in twenty chemical elements content of Parasol Mushroom (*Macrolepiota procera*) collected from two sites over a few years. J Environ Sci Health Part B 47:81–88

28. Gustin M, Lindberg S, Marsik F, Casimir A, Ebinghaus R, Edwards G, Hubble-Fitzgerald C, Kemp R, Kock H, Leonard T, London J, Majewski M, Montecinos C, Owens J, Pilote M, Poissant L, Rasmussen P, Schaedlich F, Schneeberger D, Schroeder W, Sommar J, Turner R, Vette A, Wallschlaeger D, Xiao Z, Zhang H (1999) Nevada STORMS project: measurement of mercury emissions from naturally enriched surfaces. J Geophys Res 104:21831–21844

29. He Z, Zhang M, Wilson MJ (2004) Distribution and classification of red soils in China. In: Wilson MJ (ed) The red soils of China. Springer, Netherlands, pp. 29–33

30. Index Fungorum (2015) http://www.indexfungorum.org /Names/Names.asp; retrieved on Feb. 30, 2015

31. Kalač P, Nižnanská M, Bevilaqua D, Stašková I (1996) Concentration of mercury, copper, cadmium and lead in fruiting bodies of edible mushrooms in the vicinity of a mercury smelter and a copper smelter. Sci Total Environ 177:251–258

32. Kojta AK, Jarzyńska G, Falandysz J (2012) Mineral composition and heavy metal accumulation capacity of Bay Bolete (*Xerocomus badius*) fruiting bodies collected near a former gold and copper mining area. J Geochem Explor 121:76–82

33. Kojta AK, Zhang J, Wang Y, Li T, Saba M, Falandysz J (2015) Mercury contamination of fungi genus *Xerocomus* in the Yunnan Province in China and the region of Europe. J Environ Sci Health Part A 50:1342–1350

34. Maćkiewicz D, Falandysz (2012) Total mercury in Yellow Knights (*Tricholoma equestre*) mushrooms and beneath soils. Bull Environ Contam Toxicol 89:755–758

35. Mao X (2009) Macromycetes of China. China Press, ISBN 10:7030244133/ISBN 13:9787030244130

36. Melgar MJ, Alonso J, García MÁ (2009) Mercury in edible mushrooms and soil. Bioconcentration factors and toxicological risk. Sci Total Environ 407:5328–5334

37. Michelot D, Melendez-Howell LM (2003) *Amanita muscaria*, chemistry, biology, toxicology and ethnomycology. Mycol Res Pt 2(107):131–146

38. Mleczek M, Siwulski M, Stuper-Szablewska K, Rissmann I, Sobieralski K, Goliński P (2013) Accumulation of elements by edible mushroom species: part I. Problem of trace element toxicity in mushrooms. J Environ Sci Health Part B 48:69–81

39. Mleczek M, Siwulski M, Mikołajczak P, Gasiecka M, Sobieralski K, Szymańczyk M, Goliński P (2015) Content of selected elements in *Boletus badius* fruiting bodies growing in extremely polluted wastes. J Environ Sci Health Part A 50:767–775

40. Nnorom IC, Jarzyńska G, Drewnowska M, Kojta AK, Pankavec S, Falandysz J (2013) Trace elements in sclerotium of *Pleurotus tuber-regium* (Ósu) mushroom—dietary intake and risk in southeastern Nigeria. J Food Compos Anal 29:73–81

41. Olivero-Verbel J, Johnson B, Arguello E (2002) Human exposure to mercury in san Jorge river basin, Colombia (South America). Sci Total Environ 289:41–47

42. Olivero-Verbel J, Cabarello-Gallardo K, Turizo-Tapia A (2015) Mercury in the gold mining district of San Martin de Loba, South of Bolivar (Colombia). Environ Sci Poll Res 22:589–5907

43. Ostos C, Perez-Rodriguez F, Moreno Arroyo B, Moreno-Rojas R (2015) Study of mercury content in wild edible mushrooms and its contribution to the provisional tolerable weekly intake in Spain. J Food Comp Anal 37:136–142

44. Qiu G, Feng X, Wenig B, Sommar J, Gu C (2012) Environmental geochemistry of an active Hg mine in Xunyang, Shaanxi Province, China. Appl Geochem 27:2280–2288

45. Rytuba JJ (2003) Mercury from mineral deposits and potential environmental impact. Environ Geol 43:326–338

46. Saba M, Falandysz J, Nnorom IC (2016a) Accumulation and distribution of mercury in fruiting bodies by fungus *Suillus luteus* foraged in Poland, Belarus and Sweden. Environ Sci Poll Res 23:2749–2757

47. Saba M, Falandysz J, Nnorom IC (2016b) Evaluation of vulnerability of Suillus variegatus and Suillus granulatus mushrooms' to sequester mercury in fruiting bodies. J Environ Sci Health Part B 51:540–545

48. Saba M, Falandysz J, Nnorom IC (2016c) Mercury determination in *Suillus bovinus* mushroom: accumulation, distribution, probable dietary intake. Environ Sci Poll Res 23:14549–14559

49. Seeger R, Nützel R (1976) Quecksilbergehalt der Pilze. Zeitschr Lebensm Untersuch Forsch 160:303–312

50. Shi J, Meng M, Shao M, Zhang Z, Zhang Q, Jiang G (2013) Spatial distribution of mercury in topsoil from five regions of China. Environ Sci Poll Res 20:1756–1761

51. Suchara I, Sucharová J (2002) Distribution of sulphur and heavy metals in forest floor humus of the Czech Republic. Water Air Soil Pollution 136:289–316

52. Sun H, Wu Y, Yu D, Zhou J (2013) Altitudinal gradient of microbial biomass phosphorus and its relationship with microbial biomass carbon, nitrogen, and rhizosphere soil phosphorus on the eastern slope of Gongga mountain, SW China. PLoS One 8:e72952.

53. Svoboda L, Havličkova B, Kalač P (2006) Contents of cadmium, mercury and lead in edible mushrooms growing in a historical silver-mining area. Food Chem 96:580–585

54. Tipping E, Lofts S, Hooper H, Frey B, Spurgeon D, Svendsen C (2010) Critical limits for Hg (II) in soils, derived from chronic toxicity data. Environ Pollut 158:2465–2471

55. Tüzen M, Soylak M (2005) Mercury contamination in mushroom samples from Tokat, Turkey. Bull Environ Contam Toxicol 5:968–972

56. Vetter J, Berta E (1997) Mercury content of some wild edible mushrooms. Zeitschr-Lebensm Untersuch-Forschung 205:316–320

57. Wang Y, Zhang J, Wu L, Zhao Y, Li T, Li J, Wang Y, Liu H (2014) A mini-review of chemical composition and nutritional value of edible wild-grown mushroom from China. Food Chem 151:279–285

58. Wen X, Chi Q (2007) Geochemical spatial distribution of mercury in China. Geochimica 36:633–637

59. Wu Y, Li W, Zhou J, Cao Y (2013) Temperature and precipitation at two meteorological stations on eastern slope of Gongga Mountain, SW China in the past two decades. J Mt Sci 10:370–377

60. Zhang Q, Huang J, Wang F, Xu J, Mark L, Armstrong D, Li C, Zhang Y, Kang S (2008) Mercury distribution and deposition in glacier snow over western China. Environ Sci Technol 46:5404–5413

61. Zhang D, Frankowska A, Jarzyńska G, Kojta AK, Drewnowska M, Wydmańska D, Bielawski L, Wang J, Falandysz J (2010) Metals of King Bolete (*Boletus edulis*) collected at the same site over two years. African J Agricult Res 5:3050–3055

62. Zhang H, Yin R, Feng X, Sommar J, Anderson CWN, Sapkota A, Fu X, Larssen T (2013) Atmospheric mercury inputs in montane soils increase with elevation: evidence from mercury isotope signatures. Scientific Reports 3:3322

4

CHAPTER

Influence of soil contamination with PAH on microbial community dynamics and expression level of genes responsible for biodegradation of PAH and production of rhamnolipids

Zuzanna Szczepaniak, Jakub Czarny, Justyna Staninska-Pięta, Piotr Lisiecki, Agnieszka Zgoła-Grześkowiak, Paweł Cyplik, Łukasz Chrzanowski, Łukasz Wolko, Roman Marecik, Wojciech Juzwa, Katarzyna Glazar, Agnieszka Piotrowska-Cyplik

1. Institute of Food Technology of Plant OriginPoznan University of Life SciencesPoznańPoland

2.Institute of Forensic GeneticsBydgoszczPoland

3.Department of Biotechnology and Food MicrobiologyPoznan University of Life SciencesPoznańPoland

4.Faculty of Chemical TechnologyPoznan University of TechnologyPoznańPoland

5.Department of Biochemistry and BiotechnologyPoznan University of Life SciencesPoznańPoland

6.Department of Forest TechnologyPoznan University of Life SciencesPoznanPoland

ABSTRACT

The aim of this study was to evaluate the effect of bioaugmentation and addition of rhamnolipids on the biodegradation of PAHs in artificially contaminated soil, expression of genes crucial for the biodegradation process (PAHRHDαGN, PAHRHDαGP), and the synthesis of rhamnolipids as well as population changes in the soil bacterial metabiome. The positive effect of bioaugmentation and addition of rhamnolipids on the bioremediation of the majority of PAHs was confirmed during the early stages of treatment, especially in case of the most structurally complicated compounds. The results of metagenomic analysis indicated that the initial changes in the soil metabiome caused by bioaugmentation diminished after 3 months and that the community structure in treated soil was similar to control. The survival period of bacteria introduced into the soil via bioaugmentation reached a maximum of 3 months. The increased expression of genes observed after addition of PAH into the soil also returned to the initial conditions after 3 months.

Keywords: Bioaugmentation Dioxygenase Mi Seq technology PAHs Rhamnolipids Rhamnosy lotransferase

INTRODUCTION

The high interest in the bioremediation of polycyclic aromatic hydrocarbons is directly associated with their increasing occurrence in the environment related to civilization progress. Many PAHs exhibit toxic, carcinogenic, mutagenic and/or teratogenic properties; therefore, it is justified to search for effective methods to eliminate such compounds from the environment (Gupta et al. 2015).

Removal of organic pollutants, such as polycyclic aromatic hydrocarbons, from the environment is carried out using both physicochemical and biological methods. However, the latter are described in the literature as time-, energy-, and resource-saving as well as cost-effective (Haritash and Kaushik 2009). Bioremediation is one of the biological methods based on the use of microorganisms to decompose organic compounds. However, it has several limitations, especially in the case of water-insoluble hydrocarbons. The low bioavailability of hydrophobic compounds to microorganisms reduces their biodegradability. Generally, it is well known that in the case of biodegradation of mixtures, such as diesel oil, the aliphatic fraction is most susceptible to biological decomposition, while the fractions containing complex compounds (e.g., PAHs) are biotransformed at a relatively slow rate (Kostka et al. 2011; Pasumarthi et al.2013).

For several years, scientists have been searching for methods which may increase the efficiency of bioremediation processes. Recently, much attention has been paid to strategies such as bioaugmentation or the addition of surfactants (Lladó et al. 2013; Szulc et al. 2014; Teng et al.2010).

There are numerous literature reports which confirm the positive effect of surfactants on the biodegradation process. Researchers use both synthetic and natural surfactants, including biosurfactants, which are considered as a more eco-friendly alternative to chemical surface active agents (Sachdev and Cameotra 2013). According to Ławniczak et al. (2013), the stimulation of biodegradation by surfactants is associated with increased bioavailability of the substrates by mechanisms such as emulsification and solubilization. Moreover, the authors also mention the possible changes in cell surface properties of microorganisms, such as hydrophobicity or permeability. Additionally, the presence of surfactants may affect the oxygen content in contaminated soil, which in turn improves the conditions for growth and reproduction of microorganisms (Choi et al. 2009; Park et al. 2009). On the other hand, the inhibitory effect of surfactants on biodegradation, mainly associated with their toxicity (especially when used at high concentrations), is also observed. This results from the interaction of surfactants with the lipid components of the membrane, or the reaction with proteins necessary for proper functioning of cells (Ławniczak et al. 2013; Volkering et al. 1998).

However, the low abundance of microorganisms with appropriate metabolic capabilities to biotransform the pollutant at the contaminated site may result in low bioremediation efficiency, regardless of the use of surfactants in order to increase the bioavailability of the substrate. Many researchers suggest using bioaugmentation (defined as the addition of individual strains or microbial consortia with desired catabolic capabilities) in order to increase the biodegradation potential in the polluted area (Fantroussi and Agathos 2005; Silva et al. 2009). Owsianiak et al. (2009) emphasize that consortia isolated from contaminated sites are particularly valuable in terms of degradation of xenobiotics. On the other hand, Silva et al. (2009) reported no positive impact of bioaugmentation using individual fungi and bacteria as well as fungal consortia on the biodegradation of a mixture of PAHs. Such diverse results are associated with a multitude of biotic and abiotic factors which may affect the success of bioaugmentation.

Among the biotic factors, the competition between indigenous microorganisms and those introduced via bioaugmentation is mentioned most frequently (Mrozik and Piotrowska-Seget2010). The adaption of the introduced microorganisms to the environmental conditions and maintenance of high metabolic activity seem to be key factors in the success of bioremediation. Groups of microorganisms present in the environment are characterized by a complexity of interrelationships which determine the biodegradation potential of the system (Zhang et al.2006). Metagenomic analyses allow to determine the structure of the indigenous population and track its changes after bioaugmentation, due to the fact that they are culture-independent. Moreover, they also allow to conduct a specific diagnosis through qualitative evaluation of polymorphisms of genes which are crucial in terms of enzymatic degradation of contaminants.

The aim of this study was to define the effect of selected microorganisms, rhamnolipids and a combined approach (bioaugmentation + rhamnolipids) on the biodegradation of PAHs and changes of bacterial metabiome in PAH-contaminated soil. Metapopulation studies using NGS technologies (MiSeq Technology, Illumina) allowed to characterize the roles of different species in the microbial consortium and their functioning in the environment with autochthonous microorganisms. The expression of genes crucial for the biodegradation process (PAHRHDαGN, PAHRHDαGP) and the synthesis of rhamnolipids (RhlA, RhlC) as well as the metabolic activity of microorganisms, the efficiency of PAHs biodegradation and the rhamnolipids residues were determined. Analyses were carried out in soil systems for 365 days, after 0, 1, 3, 6, and 12 months of the experiment.

MATERIALS AND METHODS

Soil, microbial consortium, and rhamnolipids

The soil used in the experiments was collected from the center of Poznań, Poland (N 52.428674, E 16.900955). According to the soil texture analysis, it could be classified as silt loam. The soil had the following characteristics: sand 40 %, silt 52 %, clay 8 %, moisture 55 %, organic matter 16 %, pH = 6.4, and included the following main elements ($mg*kg^{-1}$): Ca 18.6, Mg 8.3, N 598, P 1.2, K 24.0.

Microorganisms were added into the soil systems in the form of consortia derived from an area in the Polish Carpathian Mountains which is permanently contaminated by petroleum hydrocarbons. Prior to their introduction, the microorganisms had been cultivated for 5 days in 100-ml Duran-Schott bottles containing 250 ml of mineral medium (Cyplik et al. 2011). Diesel oil (2 %) was used as the sole carbon source.

A commercial solution of high-purity rhamnolipids from AGAE Technologies (USA) was used in the research.

Chemicals

The following amounts of respective polycyclic aromatic hydrocarbons (Sigma Aldrich) were used for each sample during the experiments: fluorene 348.4 mg; phenanthrene 348.4 mg; anthracene 348.4 mg; pyrene 300 mg; fluoranthene 98.8 mg; acenaphthylene 98.8 mg; acenaphthen 98.8 mg; chrysene 4 mg; benzo[a]anthracene 4 mg; benzo[b]fluoranthene4 mg; benzo[k]fluoranthene4 mg; benzo[a]pyrene 4 mg.

Experimental variants

The study concerning the biodegradation of PAHs was carried out under natural conditions in containers (volume of 4 L) filled with 2 kg of soil (with a geofiber-based bottom in order to allow the runoff water to freely percolate into the deeper soil layers). The geofiber surface at the bottom of each container was covered with a layer of activated carbon and an additional layer of geofiber. This ensured that there was no direct contact between the soil and activated carbon layer and enabled the evaluation of analyte loss. Appropriate amount of each PAH was dissolved in ethyl acetate and the solution was added to 200 g of the soil sample. Upon evaporation of ethyl acetate, the soil was used as a PAH carrier, which was introduced into natural soil (Szczepaniak et al. 2015).

The following experimental systems were studied: (i) 2 kg of soil (reference sample); (ii) 2 kg of soil + polycyclic aromatic hydrocarbons; (iii) 2 kg of soil + polycyclic aromatic hydrocarbons + additional microbial consortium; (iv) 2 kg of soil + polycyclic aromatic hydrocarbons + rhamnolipids; (v) 2 kg soil + polycyclic aromatic hydrocarbons + additional microbial consortium + rhamnolipids.

The experiment was conducted in containers, which were placed in soil in order to ensure maximum imitation of natural conditions. The experiment was started in March 2014 and it was conducted during a 1-year period. The temperature was not regulated. Six subsamples per plot were collected with a hand auger and mixed to yield one mean sample per plot. The soil samples were directly sieved to 1 mm and parts of the samples were used for chemical analyses.

Analyses were carried out (i) at the beginning and after: (ii) 1 month; (iii) 3 months; (iv) 6 months; (v) 12 months of treatment.

Analyses of PAH biodegradation

Preparation of samples for GC-MS analyses

In order to determine the dissipation of PAHs after 0, 1, 3, 6, and 12 months, one replicate of each setup was sacrificed for extraction and analysis. The samples were prepared in the following manner: First, an aliquot of deuter-labeled internal standards (acenaphthene-d10, anthracene-d10 and chrysene-d12) was added; next, 80 ml of acetone were added to the samples, which were shaken on a vortex mixer and placed into ultrasonic bath. After 10 min, the samples were shaken vigorously one more time to mix the sample matrix on the bottom of the flask, sonicated for another 10 min and was shaken for 20 min at 250 rpm on a horizontal shaker. Then, 20 g of anhydrous $MgSO_4$ were added into the samples and the procedure of sonication and mechanical shaking was repeated as described earlier. Finally, 40 ml of hexane were added and the procedure of sonication and mechanical shaking was repeated once again.

After the extraction, 20 ml of the obtained extract were collected into a vial and subjected further to clean-up procedure. Firstly, acetone was removed. For this purpose, a 2.5 ml aliquot of the extract was shaken with 10 ml of 10 mM NaOH and 0.1 M of NaCl. After phase separation, 100 μl of the hexane phase were applied on the Florisil column. The analytes were eluted with a 400 μl portion of hexane/MTBE mixture (1:3 *v*/*v*). Finally, an aliquot of 2-ethylnaphthalene solution (serving for recovery control) was added to the collected eluate and the samples were subjected to gas chromatographic analysis.

GC-MS parameters

The GC-MS analyses were performed on a Shimadzu 17A chromatograph coupled with a 9200 mass spectrometer. The samples were injected at the temperature of 250 °C and a ratio of 1:20. The analytes were separated on a RTX-5 column (30 m length, 0.25 mm internal diameter and 0.25 μm film thickness), using helium as the carrier gas (flow rate of 1 ml/min). The temperature program was as follows: initial temperature of 50 °C held for 1 min, ramped to 270 °C at 15 °C/min, and then ramped to 300 °C at 5 °C/min. The MS operated in SIM mode with ion source temperature of 250 °C and ion source voltage of 70 eV, solvent delay 290 s and acquisition rate 10 spectra.

Analyses of rhamnolipid residuals

Preparation of samples for HPLC analyses

For determination of rhamnolipids, 20 g of dried soil was subjected to a four step ultrasound-assisted extraction with ethanol. The obtained extract (50 ml) was filtered through 0.45 μm pore-size PTFE syringe filter, diluted with a water to acetonitrile mixture (1:4 *v*/*v*) and subjected to HPLC-MS/MS analysis. The extraction and analysis were performed in three replicates for each microcosm.

HPLC parameters

The HPLC-MS analysis was carried out using the UltiMate 3000 RSLC chromatographic system from Dionex (Sunnyvale, CA, USA). Samples (5 μL) were injected into a Gemini-NX C18 column (100 mm 2.0 mm I.D.; 3 μm) from Phenomenex (Torrance, CA, USA). The mobile phase used for the analysis consisted of 5 mmol l^{-1} ammonium acetate in water and acetonitrile at a flow rate of 0.3 ml min^{-1}. The following gradient was used: 0 min 55 %; 4 min 55 %; 4.5 min 90 %; and 6 min 90 % of acetonitrile. The LC column effluent was directed to the API 4000 QTRAP triple quadrupole mass spectrometer from AB Sciex (Foster City, CA, USA) through the electrospray ionization source (Turbo Ion Spray). The ionization source operated in negative ion mode. The dwell time for each mass transition detected in the MS/MS multiple reaction monitoring mode was set to 20 ms. All rhamnolipids were detected using the following settings for the ion source and mass spectrometer: curtain gas 10 psi, nebulizer gas 40 psi, auxiliary gas 45 psi, temperature 450 °C, ion spray voltage −3500 V, and collision gas set to medium. The declustering potential was set to −80 V for monorhamnolipids and −85 V for dirhamnolipids. The collision energy for quantitative transitions was set to −28 V for monorhamnolipids and −35 V for dirhamnolipids. The collision energy for confirmatory transitions was set to −21 V for monorhamnolipids and −29 V for dirhamnolipids.

Determination of the half-lives of PAHs and rhamnolipids

The half-lives of PAHs and rhamnolipids were calculated using a single, three-parameter equation (exponential decay).

Flow cytometry analyses

In order to determine the metabolic activity of microorganisms during the experiment, flow cytometry analyses were conducted. Samples from soil microcosms (10 g) were washed with distilled water (100 ml) and shaken on the MS3 vortex (Ika, Germany) for 5 min at 2500 rpm. To avoid binding of fluorescent dyes to non-cellular particle, 2 ml of the suspension was filtered through a filter with a 20-m nylon net syringe filter (Merck Millipore, Germany). Then, all the samples were centrifuged at 4000 rpm for 6 min (Heraeusbiofuge primo R), the supernatant was decanted, and the precipitate was washed with 0.5 ml PBS. Green reagent (1.5 l) and propidium iodide (0.8 l) were added and the samples were incubated at 38 °C for 10 min. Green reagent is a fluorescent dye for measuring redox, while propidium iodide is used to measure cell viability on the basis of intensive incorporation into cells with damaged cell wall and lack of penetration (and hence the absence of dying) in intact cells. The analyses on flow cytometer were performed during the last step.

The redox potential of microbial cells and viability assessment were evaluated using the BacLight™ Redox Sensor™ Green Vitality Kit by Invitrogen. Samples were analyzed using a BD FACS Aria™III (Becton Dickinson, USA) flow cytometer (cell sorter). The populations were defined by gating in the dot plots of green fluorescence (FITC-A) versus red fluorescence (PE-Texas Red-A). Each sample was analyzed in triplicate. The estimation of cellular redox potential was performed using medians of green fluorescence (FITC-A) signals of gated populations defined on a bivariate dot plot (FITC-A vs. PE-Texas Red-A) (Szczepaniak et al. 2015).

IDENTIFICATION OF MICROORGANISMS

DNA extraction

Total DNA was extracted from 500 mg of each soil sample using Genomic Mini AX Soil kit (A&A Biotechnology) according to manufacturer's instruction. The extracted DNA was quantified using Quant-iT HS ds.-DNA assay kit (Invitrogen) on Qubit2 fluorometer; 2 μl of extracts were examined on a 0.8 % agarose gel.

PCR amplification

Region IV of bacterial 16S RNA gene was amplified using universal primers 515F and 806R: containing reverse complement of 3′ Illumina adapter, golay barcode, reverse primer pad, reverse primer linker, and reverse primer (Table 1). Genomic DNA (100 ng) was used for PCR amplification in 50 μl volume in reaction containing the following: 1× PCR reaction mix, 0.25 μM each primer, and 5 U Taq DNA polymerase (A&A Biotechnology). Following conditions were used for PCR amplification: initial denaturation 95 °C for 3 min; 35 cycles of denaturation 30 s at

94 °C, annealing 30 s at 52 °C and extension 2 min at 72 °C and a final extension at 72 °C for 10 min. Products were purified in Clean-Up columns (A&A Biotechnology) according to manufacturer's protocol. The libraries were constructed from amplicons using NEBNext® DNA Library Prep Master Mix Set for Illumina (New England Biolabs UK). Then, the libraries were polled at equimolar concentration. Pooled library was quantified using a Qubit Fluorometer and dsDNA HS assay kit (Life Technologies, USA). The library was denatured with 0.2 N NaOH and diluted with HT1 buffer (Illumina, USA) to a final concentration of 8 pM. To balance the overall lack of sequence diversity, a spike-in of denatured Phix was added to the concentration of 40 %. Sequencing was conducted on an Illumina MiSeq (Illumina, USA) using paired-end (2 × 250) MiSeq Reagent Kits v2 (Illumina, USA). Three sequencing primers were based on Caporaso et al. 2012 (Table 1). The sequencing reaction was performed with MiSeq Illumina instrument and MiSeq Reagent Kit v2 (2 × 250 bp).

Bioinformatic analysis

The sequencing data was processed using CLC Genomic Workbench 8.5 and CLC Microbial Genomics Module 1.2. (Qiagen, USA). Total number of reads ranged from 438,216 to 461,123. After sequencing, the reads were demultiplexed to the probes and the overlapping paired-end reads were merged (70 % of total reads) and trimmed to yield fragments of 291 nt. Just fragments which passed the merging were retained for downstream processing. Chimeric reads (from 27,121 to 29,785) were filtered and remaining sequences were assigned to operational taxonomic units (OTUs). Number of reads which passed merging and trimming ranged from 136,576 to 154,294. Reads were clustered against the SILVA v119 99 % 16S rRNA gene database (July 24, 2014, Quast et al. 2013). Rarefaction analysis with a depth of 80,000 sequences per sample was used to calculate alpha diversity measured using OTU abundance, Shannon's index, Chao 1 bias-corrected, and phylogenetic diversity.

RT-PCR

Total RNA was extracted from 1 g of soil samples by RNA PowerSoil Total RNA Isolation Kit (MO BIO Laboratories, Inc.). Genomic DNA was digested using TURBO DNA-free Kit (Ambion by Life Technologies). RNA concentration was measured using Qubit RNA HS Assay Kit on Qubit Fluorometer (Invitrogen/Life Technologies). Random-primed cDNA synthesis was conducted using High-Capacity cDNA Reverse

Table 1: Characteristics of primer for PCR amplification, sequencing (MiSeq Technology) and real-time PCR

Primers	Sequence (5′ to 3′)	Reference
PCR amplification		
Forward 515F	AATGATACGGCGACCACCGAGATCTACACTATGG TAATTGTGTGCCAGCMGCCGCGGTAA[1]	Caporaso et al. (2012)
Reverse 806R	CAAGCAGAAGACGGCATACGAGATXXXXXXAGT CAGTCAGCCGGACTACHVGGGTWTCTAAT[2]	
Sequencing		
Read 1	(TATGGTAATT)[6a] (GT)[7a] (GTGCCAGCMGCCGCGGTAA)[8a]	
Read 2	(AGTCAGTCAG)[6b] (CC)[7b] (GGACTACHVGGGTWTCTAAT)[8b]	
Index	(ATTAGAWACCCBDGTAGTCC)[6c] (GG)[7c] (CTGACTGACT)[8c]	
RT-PCR RhlA		Schmidberger et al. 2013
Forward	F 5′-GATCGAGCTGGACGACAAGTC-3′	
Reverse	R 5-GCTGATGGTTGCTGGCTTTC-3′	
RhlC		
Forward	F 5′-ATCCATCTCGACGGACTGAC-3′	
Reverse	R 5′-GTCCACGTGGTCGATGAAC-3′	
PAH-RHDα GN		Cébron et al. 2008
Forward	GAG ATG CAT ACC ACG TKG GTT GGA	
Reverse	AGC TGT TGT TCG GGA AGA YWG TGC MGT T	
PAH-RHDα GP		
Forward	CGG CGC CGA CAA YTT YGT NGG	
Reverse	GGG GAA CAC GGT GCC RTG DAT RAA	
16 S rDNA		
968 Forward	AAC GCG AAG AAC CTT AC	
1401 Reverse	CGG TGT GTA CAA GAC CC	

[1] Containing 5′ Illumina adapter, forward primer pad, forward primer linker, and forward primer sequence

[2] Containing reverse complement of 3′ Illumina adapter, golay barcode, reverse primer pad, reverse primer linker, and reverse primer

[6a] forward primer pad

[6b] reverse primer pad

[6c] reverse complement of reverse primer

[7a] forward primer linker

[7b] reverse primer linker

[7c] reverse complement of reverse primer linker

[8a] forward primer

[8b] reverse primer

[8c] reverse complement of reverse primer pad

Transcription Kit and MultiScribeMuLV Reverse Transcriptase. Gene expression was analyzed using a Power SYBR Green PCR Master Mix (Life Technologies) on ABI 7500 SDS (Applied Biosystems). Primers used for real-time PCR are listed in Table 1. Total bacterial RNA was quantitated by real-time PCR amplification of fragment of bacterial 16S ribosomal RNA with universal bacterial primers and TaqMan MGB probe using TaqMan Universal Master Mix II (Life Technologies) on ABI 7500 SDS (Applied Biosystems). Sequences of primers and probe used are listed in Table 1. All analysis was done in triplicates. In order to compare the gene expression in each sample, the mean expression index was calculated according to formula: C_T target/C_T 16S using data from three analyses. This parameter reflects the expression level of a specific gene compared to the expression level of the universal gene (16S RNA) in the whole metabiome.

RESULTS AND DISCUSSION

PAH biodegradation

GC-MS analyses were performed in order to determine the biodegradability of the individual PAHs. The results indicated a close correlation between the structure of the compounds and the rate of biodegradation. Calculated half-lives of PAHs showed that the increase in the number of rings resulted in the increase of half-lives of the compounds from 0.2 months for the simplest compounds to 14 months in the case of pentacyclic hydrocarbons. In all cases (except acenaphthene and acenaphthylene), the half-lives were shorter in variants containing additional microorganisms or rhamnolipids and the combined method (bioaugmentation + rhamnolipids). However, no differences between the mentioned variants were observed in general. In the case of the combined method, it was not possible to calculate the half-lives for three PAHs (benz[b]fluoranthene, benz[k]fluoranthene, benzo[a]pyrene), since the residues accounted for more than half of the initial mass (Table 2). The results which indicate the increase of half-lives as a result of increased structural complexity of the compounds are consistent with those obtained by Silva et al. (2009). The researchers noted that the LMW-PAHs (naphthalene, phenanthrene, and anthracene) and the HMW-PAH (pyrene) were removed more rapidly compared to the HMW-PAHs (benz[a]anthracene and benz[a]pyrene).

Table 2: Half-lives in months of studied PAHs and rhamnolipids

		Soil + PAH	Soil + PAH + M	Soil + PAH + R	Soil + PAH + M + R
monorhamnolipids		Not detected	Not detected	0.20 ± 0.4	0.24 ± 0.3
dirhamnolipids		Not detected	Not detected	0.48 ± 0.3	0.04 ± 0.2
	Acenaphthylene	0.21 ± 0.13	0.19 ± 0.10	0.42 ± 0.12	0.32 ± 0.06
	Acenaphthene	0.18 ± 0.30	0.13 ± 0.08	0.50 ± 0.19	0.37 ± 0.09
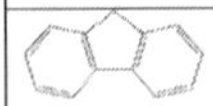	Fluorene	0.42 ± 0.05	0.19 ± 0.06	0.34 ± 0.06	0.30 ± 0.04
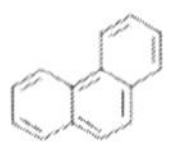	Phenanthrene	0.54 ± 0.07	0.15 ± 0.07	0.34 ± 0.07	0.29 ± 0.05
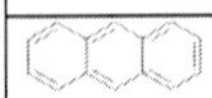	Anthracene	1.77 ± 1.24	0.69 ± 0.07	0.86 ± 0.04	0.74 ± 0.01
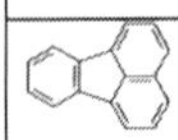	Fluoranthene	1.27 ± 0.18	0.18 ± 0.12	0.68 ± 0.16	0.57 ± 0.07
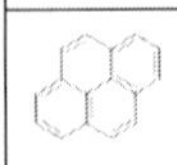	Pyrene	1.39 ± 0.40	0.29 ± 0.11	0.74 ± 0.11	0.61 ± 0.05
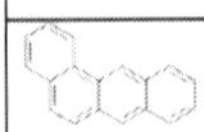	Benz[a]anthracene	7.15 ± 1.53	1.72 ± 0.34	3.78 ± 1.50	3.29 ± 1.02
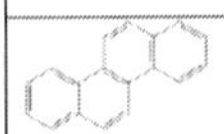	Chrysene	8.05 ± 11.56	2.81 ± 0.94	3.84 ± 0.92	3.71 ± 0.75
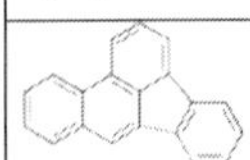	Benzo[b]fluoranthene	14.17 ± 1.21	14.82 ± 03.60	11.88 ± 3.92	Not determined
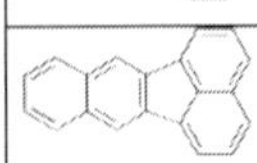	Benzo[k]fluoranthene	12.14 ± 1.24	7.77 ± 6.56	6.39 ± 3.06	Not determined
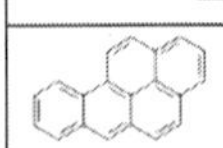	Benzo[a]pyrene	12.14 ± 1.36	14.94 ± 7.65	8.62 ± 5.78	Not determined

The obtained results indicated that bioaugmentation, addition of rhamnolipids and the use of the combined approach (bioaugmentation + rhamnolipids) had a positive effect on the biodegradation efficiency compared to natural attenuation during the early stages of the process (with the exception of acenaphthylene and acenaphthene), especially for the most structurally complex compounds. However, no differences in PAHs content in soil were observed between the natural attenuation and other variants after 12 months of treatment. Moreover, the time needed to equate natural attenuation efficiency with the other treatments methods varied depending on the molecular structure of PAHs (from 1 to 3 months for tricyclic hydrocarbons up to 12 months for pentacyclic compounds). The obtained results were presented in Fig. 1.

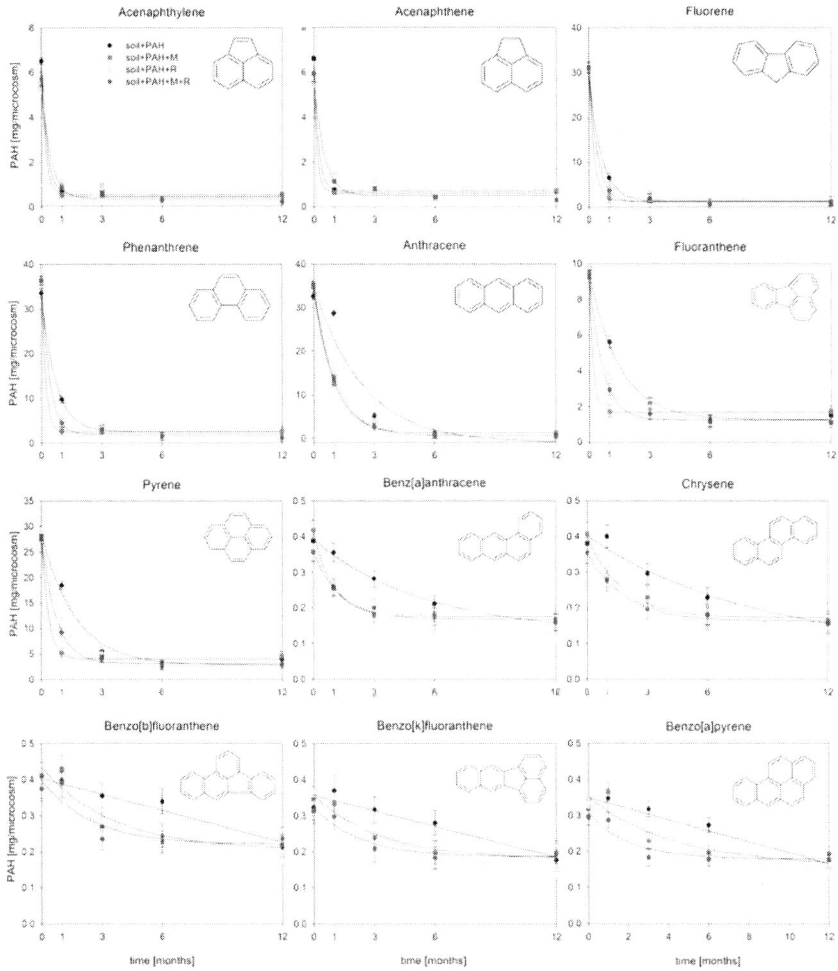

Fig. 1 Biodegradation of polycyclic aromatic hydrocarbons during 12 months of process

In order to identify the bacterial consortium (M), a metagenomic analysis of the gene encoding 16S rRNA was conducted on the basis of V4 hypervariable region of the 16S rRNA gene (Fig. 2a). The use of next generation sequencing allowed to identify the bacterial consortium (M) as members of seven classes, with the dominance of *Gammaproteobacteria* (77.9 %),*Betaproteobacteria* (7.7 %),

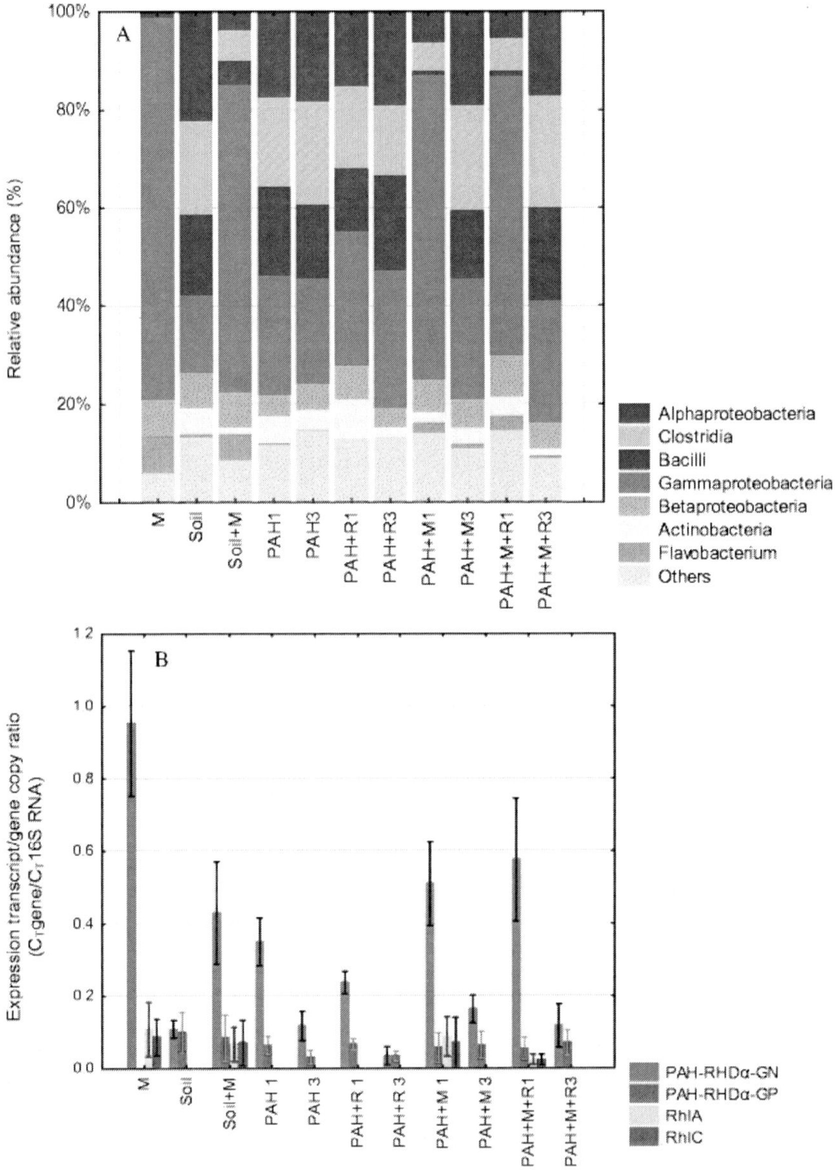

Fig. 2 Relative abundance of bacterial classes (**a**) and quantitative changes in RhlA, RhlC, PAHRHDαGN, and PAHRHDαGP expression level genes (**b**) present in the soil after 1 and 3 months of biodegradation

and *Flavobacteria* (7.3 %). Three genera dominated among*Gammaproteobacteria*: *Pseudomonas* (58 %), *Achromobacter* (14 %), and *Stenotrophomonas*(12 %). On the other hand, the results of the 16S metagenomic analysis of the soil metabiome indicated that the soil was more diverse. A total of 57 classes were identified in the bacterial microbiome, with the dominance of*Alphaproteobacteria* (22.5 %), *Clostridia* (19.1 %), *Bacilli*(16.4), *Gamma proteobacteria* (15.8 %), *Betaproteobacteria* (7.2 %), and *Actinobacteria*(5.4 %).

The contamination of soil with PAHs caused significant changes in the soil bacterial community. The dominance of species belonging to *Gammaproteobacteria* could be observed after the first month of biodegradation. Their percentage share in the population amounted to 24.4 % after the first month and did not change significantly ($p > 0.05$) until the third month of the process. The remaining dominant classes included *Clostridia* (21.36 %),*Alphaproteobacteria* (18.3 %), and *Bacilli* (15 %). Similar changes in the community structure were observed after the addition of rhamnolipids. In this case, the *Gammaproteobacteria* also dominated the system (28.2 %) and the share of remaining classes (*Alphaproteobacteria*,*Clostridia*, and *Bacilli*) ranged from 14.3 to 19.3 %. Notable differences in the community structure were observed in case of PAH-contaminated soil with the addition of the microbial consortium (PAH + M). A significant increase of the abundance of bacteria added in the form of a preparation (PAH + M) could be observed in the soil metabiome. *Gammaproteobacteria* was the dominant class, which accounted for 57 to 62 % of the total bacterial population. The abundance of the remaining identified classes (*Alphaproteobacteria*, *Clostridia*, *Bacilli*,*Betaproteobacteria*, and *Actinobacteria*) ranged from 0.8 to 8.4 %. However, the community structure returned to the initial state (the structure determined for control soil) after 3 months of the biodegradation process.

Attachment of oxygen to the aromatic ring catalyzed by dioxygenases (PAHRHDα) is the key step for biodegradation of PAH, which results in the formation of *cis*-dihydrodiol. The enzyme catalyzing this reaction (dioxygenase) is composed of a large α subunit and a smaller β subunit. The α subunit contains a region named [Fe2-S2] Rieske center. This center was found in numerous enzymes with a similar catabolic activity to PAHRHDα but with different substrate specificity. The presence of various genes encoding similar enzymes among Gram positive as well as Gram negative bacteria marks the significance of the presence of both these groups for a proper assessment of biodegradation efficiency among the whole microbial consortium (Cébron et al. 2008). An indirect method was employed in order to determine the expression of all the studied genes, which was based on the determination of the number of cycles for the analyzed gene (C_{Tgene}) and its comparison with an internal standard (the number of cycles for the gene encoding the 16S rDNA subunit—C_{T16S}). The employed method allowed to establish the relative changes in the expression of genes with respect to the abundance of bacteria in soil. The determined C_{T16S} value for the 16S RNA gene (internal standard) ranged from 5.7 to 7.4 between the samples. During the first month of biodegradation, the expression of gene encoding dioxygenases in Gram negative bacteria (PAH-RHDα GN) was highest in soil with the addition of both rhamnolipids and the bacterial consortium (PAH + M + R) as well as in bioaugmented soil (PAH + M). The relative expression of these genes reached 0.723 ± 0.245 and 0.693 ± 0.194, accordingly, and these values were not statistically different ($p > 0.05$). On the other hand, the expression in untreated soil as well as soil with the addition of rhamnolipids (PAH + R) was lower and ranged from 0.365 ± 0.178 to 0.395 ± 0.087. These values were statistically different compared to bioaugmented soil ($p < 0.05$). After 3 months, the expression returned to the level observed in control soil. In contrast, the expression of gene encoding dioxygenases in Gram positive bacteria (PAH-RHDα GP) did not change significantly and ranged from 0.092 ± 0.037 to 0.120 ± 0.072.

The expression of genes encoding rhamnosyltransferase I (RhlA) and III (RhlC) was observed only in case of soils subjected to bioaugmentation (PAH + M and PAH + M + R); however, it was inhibited after the third month. No expression of genes associated with the biosynthesis of rhamnolipids was observed in case of control soil and soil with the addition of rhamnolipids (PAH + R) (Fig. 2b). Therefore, the presence of rhamnosyltransferases in soil was associated with their presence in the bacterial consortium, which was introduced to the soil. It is worth noticing that the addition of rhamnolipids did not inhibit the expression of RhlA and RhlC in the environment, and its decrease was associated with the decreased abundance of bacteria introduced into the soil via bioaugmentation (M).

Principal component analysis (PCA) was employed in order to evaluate the changes in the soil bacterial metabiome. The results which reflect the grouping of the analyzed bacterial metabiomes are shown in Fig. 3a. The two first principal components are crucial and describe the variability of the initial data in 89.9 %. The first primary component carries 67.13 % of the data regarding microbial populations contained in the initial variables. It includes the positively correlated *Bacilli*, *Clostridia*, and *Alphaproteobacteria* as well as negatively correlated*Betaproteobacteria*, *Flavobacterium*, and *Gammaproteobacteria*. The second component is mainly associated with the presence of *Actinobacteria* and others and describes the variability of the analyzed data in 22.67 % (Fig. 3b). The high diversity of the soil bacterial microbiome and the introduced bacterial consortium (M) is also visible in Fig. 3a. The contamination of soil with PAHs as well as the addition of the selected microbial consortium and rhamnolipids caused significant changes in the soil metabiome, which are reflected as a visible cluster in the upper-right corner of the figure. However, after 3 months, the metabiomes were similar in terms of structure to control soil.

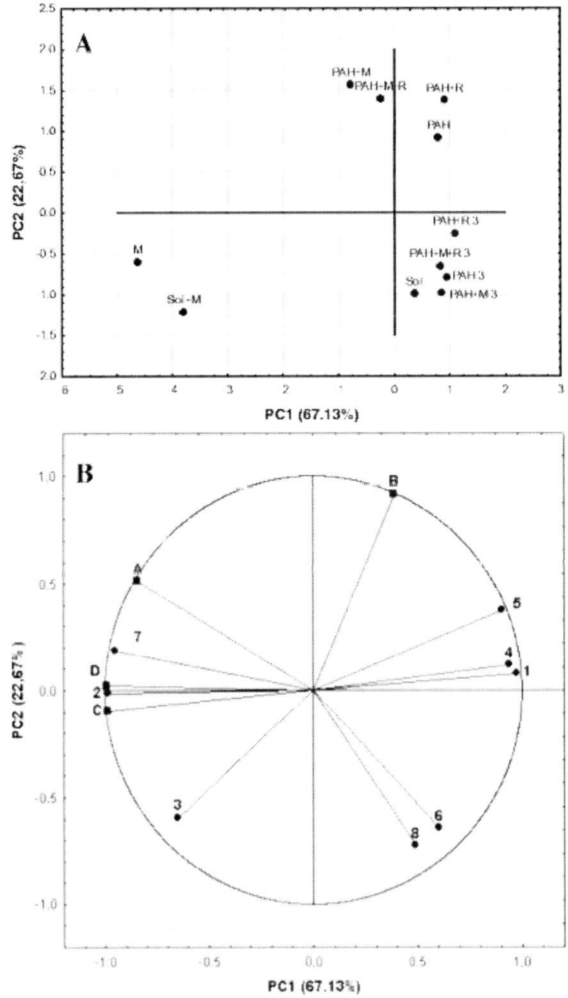

Fig. 3 Principal component analysis (PCA) of taxonomy profile (**a**) and correlations between relative abundance of bacterial classes and genes: RhlA, RhlC, PAHRHDαGN, PAHRHDαGP expression (**b**) during PAHs biodegradation

Analysis of mutual relationships between the specific classes of microorganisms and the expression of genes suggests a strong correlation between the expression of genes associated with the synthesis of rhamnolipids (RhlA and RhlC) and bacteria belonging to the*Gammaproteobacteria* class (which includes the *Pseudomonas* genus). A weak correlation between these genes and the expression of dioxygenases in Gram negative bacteria was established, whereas no correlation was found in Gram positive bacteria. The PCA analysis confirmed the positive correlation of dioxygenase G+ with *Bacilli* and *Clostridium* classes as well as dioxygenase G− with Gram negative bacteria belonging to *Betaproteobacteria*,*Flavobacterium*, and *Gammaproteobacteria*.

The alpha diversity indicators determined after 3 months of biodegradation did not differ significantly, which confirms the high potential of autochthonic microorganisms to remove the contaminants from soil via natural attenuation and to recover the initial community structure (Table 3).

Table 3: Alpha diversity measured using number of OTUs, Shannon's index, Chao 1 bias-corrected, and phylogenetic diversity

	Soil	Soil + PAH	Soil + PAH + R	Soil + PAH + M	Soil + PAH + M + F
OTU observed	1984 ± 41	1930 ± 48	1956 ± 54	1990 ± 43	2014 ± 61
Shannon's index	5.7 ± 0.2	5.4 ± 0.3	5.5 ± 0.2	5.5 ± 0.2	5.6 ± 0.3
Chao 1 bias-corrected	1464 ± 251	1432 ± 279	1451 ± 197	1487 ± 211	1496 ± 173
Phylogenetic diversity	4.89 ± 0.34	4.76 ± 0.28	4.95 ± 0.39	4.68 ± 0.19	4.79 ± 0.29

The use of bioaugmentation for biodegradation of PAHs has been described in detail by many authors (Fantroussi and Agathos 2005; Fernández-Luqueño et al. 2011; Mohan et al. 2006). Positive results were obtained using single strains (Colombo et al. 2011; Teng et al. 2010), as well as microbial consortia (Owsianiak et al. 2009). The bioaugmentation strategy turned out to be useful not only in the case of PAH-contaminated soil but also for diesel oil pollution (Bento et al. 2005; Szulc et al. 2014).

On the other hand, Silva et al. (2009) did not observe a positive impact of bioaugmentation in case of low molecular weight (LMW) and high molecular weight (HMW) PAHs, using both individual fungi as well as bacterial and fungal consortia, with the exception of *Aspergillus sp*. Similar results regarding no positive effect of the introduction of additional microorganisms on biodegradation were also observed by other researchers (Launen et al. 2002; Saponaro et al.2002).

The obtained results indicated the positive effect of the use of biosurfactants in the early stages of the experiment (with the exception of acenaphthene and acenaphthylene). The ability to stimulate the biodegradation of PAHs by the use of surfactants has been confirmed by numerous researchers and is associated, e.g., with the mechanisms of increasing the bioavailability of hydrophobic compounds by their emulsification and/or solubilization, or altering the properties of the microorganisms cell surfaces (Ławniczak et al. 2013). Rodrigues et al. (2013) have reported that low concentrations of surfactants not only stimulated the decomposition of anthracene and fluoranthene but also indicated that the effect varied depending on the type and concentration of surfactant and the type of PAHs. An exceedingly high concentration or the wrong choice of surfactant can be toxic toward to microorganisms cells, resulting in a decrease of their number and, consequently, lower efficiency of the biodegradation process. Moreover, despite the natural origin of rhamnolipids, it is necessary to consider the need to select the optimal concentration for terrestrial environment. There are literature reports which indicate the possibility of decreasing of plant biomass (Millioli et al.2009) and even increasing the toxicity of contaminants such as diesel oil by increasing their bioavailability in the presence of rhamnolipids (Marecik et al. 2012).

We hypothesized that the combined treatment method (which involved bioaugmentation and addition of rhamnolipids) would improve the biodegradation due to the advantages of both techniques and also solve the problem of inoculation/introduction of microorganisms into the soil According to Mrozik and Piotrowska-Seget (2010), the use of surfactants and foams can facilitate the dispersion of inoculants into subsurface environment. However, the obtained results did not confirm this theory since no differences between the studied treatment methods (bioaugmentation, addition of rhamnolipids, and combined method) were observed. Similarly, Szulc et al. (2014) did not observe a significant positive impact of the combined approach compared to other strategies during research regarding biodegradation of diesel oil using bioaugmentation, biosurfactants, and the combined approach.

The obtained results indicated that it is reasonable to use bioremediation strategies such as bioaugmentation or rhamnolipid addition, when rapid purification of the area is necessary. According to Forsyth et al. (1995), the use of bioaugmentation is justified in the case of low or non-detectable amount of microorganisms that are able to degrade contaminants, in fields contaminated with compounds requiring multi-process remediation (e.g., a toxic to microbes), and for small-scale areas. No difference between the treatment methods after 12 months of experiment suggested that in the absence of time pressure, the introduction of additional microorganisms or surfactants may not be justified due to a sufficient number of contaminant-degrading microorganisms and possible natural attenuation. This is in agreement with the results obtained by Mesbaiah et al. (2014). The strain used in the studies was able to degrade the LMW PAHs without additional intervention. Moreover, Silva et al. (2009) clearly demonstrated the possibility of immediate reaction of native microorganisms to the addition of PAH and their ability to degrade both LMW and HMW added to soil.

Rhamnolipid residues

The mixture of rhamnolipids used in the experiment contained monorhamnolipids and dirhamnolipids. Analyses were performed for variants with addition of rhamnolipids (soil + PAH + R; soil + PAH + M + R) as well as for the variants without rhamnolipids (soil + PAH; soil + PAH + M) in order to determine whether microorganisms in the studied soil were able to produce biosurfactants.

The obtained results demonstrated that monorhamnolipids dominated over dirhamnolipids in the solution of biosurfactants used in the experiment, which corresponded to the initial content of rhamnolipids in variants with their addition. In case of variants without biosurfactants addition, their contents for all treatment time remained at the background level. This suggested that microorganisms in studied systems did not produce rhamnolipids.

In addition, the analyses carried out after the first month of treatment proved that during the early stage of the experiment (up to the first month) rhamnolipids were almost entirely biodegraded. Similar results of rapid biodegradation of rhamnolipids were obtained by Szulc et al. (2014). During research regarding the effects of addition of rhamnolipids on the bioremediation of diesel-oil contaminated soil, the authors established that rhamnolipids should be reintroduced after 30 days (Fig. 4).

The increasing use of biosurfactants in many industries as well as in environmental protection is associated with, e.g., their natural origin resulting in their relatively low toxicity and biodegradability and therefore greater environmental benefits (Sachdev and Cameotra 2013). Natural surfactants can be successfully used as alternatives to synthetic surfactants. Singh and Cameotra (2014) suggested that the use of rhamnolipids contributed to higher effectiveness compared to Triton X-100 during biodegradation of atrazine herbicide by A6, belonging to the genus *Acinetobacter*.

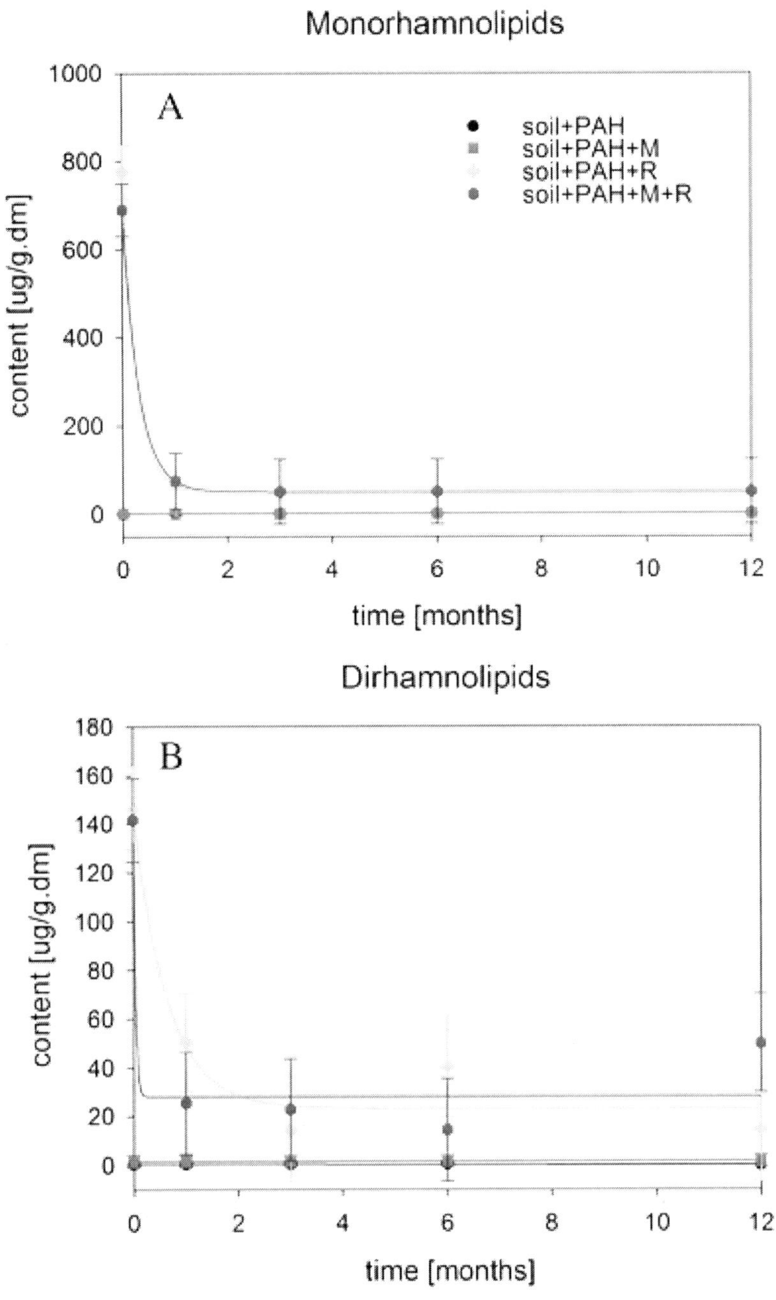

Fig. 4 Biodegradation of monorhamnolipids (**a**) and dirhamnolipids (**b**) during 12 months of process

The half-life of rhamnolipids was shorter compared to most of the PAHs in variants with the addition of biosurfactants. This means that rhamnolipids may not affect the bioavailability of PAHs in the later stages of bioremediation.

Metabolic activity of microorganisms

The changes of metabolic activity in the individual variants of method treatments are presented in Fig. 5. Analyzes of the results were based on the parameter %Q2, which specifies the percentage of the most metabolically active population. In terms of the differences in %Q2 between the variants after first month of the experiment, it was noted that bioaugmentation did not increase the number of the most metabolically active microorganisms—%Q2 in variants with additional microorganisms was statistically similar to control. Moreover, samples containing PAHs but no additional microorganisms were characterized by the highest number of the most metabolically active cells (6.5 % for soil + PAH and 7 % for soil + PAH + rhamnolipids). The obtained results suggest that the addition of the consortium decreased %Q2 in systems containing PAHs. The differences in the Q2 value between soil inoculated with microorganisms and soil without inoculation resulted from the removal rate of PAHs in the period between the beginning of the experiment and the first month of biodegradation. The addition of microorganisms with a high biodegradation potential and high metabolic activity caused a more rapid removal of PAHs; therefore, upon depletion of the carbon source a decrease of metabolic activity was observed after 1 month of biodegradation in case of inoculated soil, in contrast to non-inoculated soil. In non-inoculated soil, the microorganisms had to undergo an adaptation phase caused by the addition of PAHs, which resulted in the extension of the biodegradation time; hence, the metabolic activity of autochthonic microorganisms during the first month was higher. During the third month, the metabolic activity of microorganisms in all tested soils was similar, which was caused by the complete removal of PAHs. According to Mrozik and Piotrowska-Seget (2010), the competition between indigenous and exogenous microorganisms for limited carbon sources is one of the most important biotic factors affecting the success of bioaugmentation. It is plausible that such competition occurred in the studied case. On the other hand, the higher value of %Q2 observed in systems containing PAHs compared to the variant without PAHs may be associated with the use of hydrocarbons as a carbon source, as confirmed by Silva et al. (2009). Analyzes carried out after the first month of treatment excluded the toxic effects of PAHs on microorganisms in the studied systems—the activity of microorganisms in the samples containing PAHs was not reduced compared to the activity of microorganisms in a systems without PAHs.

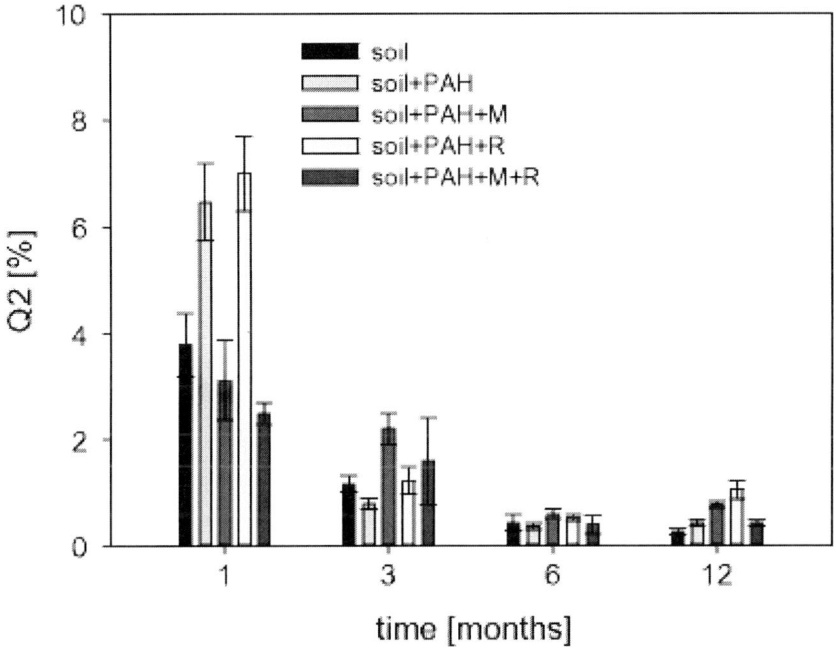

Fig. 5 The percentage of the most metabolically active microorganisms (%Q2) during PAHs biodegradation

A decrease of the %Q2 parameter was observed during the treatment, which means that the number of metabolically active microorganisms decreased with the passage of time. Bento et al. (2005) suggest that this effect may be associated with possible formation of the toxic intermediates during the degradation of higher molecular weight hydrocarbons. However, no differences between control (without hydrocarbons) and soil with PAHs suggest a lack of inhibitory effect of intermediates on microbes. Similar results were observed by Silva et al. (2009). The total PAH content of 600 mg kg^{-1} soil did not contribute to an inhibitory effect toward endogenous and exogenous microorganisms. Thus, the results suggest that the decrease in the %Q2 value was most likely related to the depletion of carbon source and biogenic elements needed for the development of microorganisms. This also resulted in no statistically significant differences between the individual variants after 3, 6, and 12 months of treatment.

CONCLUSIONS

The addition of rhamnolipids and bioaugmentation increased the bioremediation efficiency during the early stage of treatment. The differences between the natural attenuation and other approaches increased with the number of aromatic rings. In addition, the biodegradation rate depended on the structure of PAHs. However, no differences between natural attenuation and other strategies were noticed after 12 months of treatment. This suggests the existence of a sufficient number of contaminant-degrading autochthonous microorganisms and the capacity for natural attenuation in the studied soil. Therefore, in the case of the lack of time pressure, the use of bioaugmentation or rhamnolipids may not be justified.

ACKNOWLEDGMENTS

This research work is funded by the National Science Centre in Poland in the years 2012–2015 as a research project Opus 2 no. 2011/03/B/NZ9/00274 and partially by funds granted by the National Science Centre in Poland conferred on the basis of the decision DEC-2012/07/D/NZ9/00981.

REFERENCES

1. Bento FM, Camargo FAO, Okeke BC, Frankenberger WT (2005) Comparative bioremediation of soils contaminated with diesel oil by natural attenuation, biostimulation and bioaugmentation. Bioresour Technol 96:1049–1055

2. Caporaso JG, Lauber CL, Walters WA, Berg-Lyons D, Huntley J, Fierer N, Owens SM, Betley J, Fraser L, Bauer M, Gormley N, Gilbert JA, Smith G, Knight R (2012) Ultra-high-throughput microbial community analysis on the Illumina HiSeq and MiSeq platforms. ISME J 6:1621–1624

3. Cébron A, Norini M-P, Beguiristain T, Leyval C (2008) Real-time PCR quantification of PAH-ring hydroxylating dioxygenase (PAH-RHDα) genes from gram positive and gram negative. J Microbiol Meth 73:148–159

4. Choi YJ, Kim Y-J, Nam K (2009) Enhancement of aerobic biodegradation in an oxygen-limiting environment using a saponin-based microbubble suspension. Environ Pollut 157:2197–2202

5. Colombo M, Cavalca L, Bernasconi S, Andreoni V (2011) Bioremediation of polyaromatic hydrocarbon contaminated soils by native microflora and bioaugmentation with *Sphingobium chlorophenolicum* strain C3R: a feasibility study in solid- and slurry-phase microcosms. Int Biodeter Biodegr 65:191–197

6. Cyplik P, Marecik R, Piotrowska-Cyplik A, Olejnik A, Drożdżyńska A, Chrzanowski Ł (2011) Biological denitrification of high nitrate processing wastewaters from explosives production plant. Water, Air and Soil Poll 223:1791–1800

7. Fantroussi SE, Agathos SN (2005) Is bioaugmentation a feasible strategy for pollutant removal and site remediation? Curr Opin Microbiol 8:268–275

8. Fernández-Luqueño F, Valenzuela-Encinas C, Marsch R, Martínez-Suárez C, Vázquez-Núñez E, Dendooven L (2011) Microbial communities to mitigate contamination of PAHs in soil—possibilities and challenges: a review. Environ Sci Pollut Res 18:12–30

9. Forsyth JV, Tsao YM, Bleam RD (1995) Bioremediation: when is augmentation needed? In: Hinchee RE, Fredrickso J, Alleman BC (eds) Bioaugmentation for site remediation. Battelle Press, Columbus OH, pp. 1–14

10. Gupta S, Pathak B, Fulekar MH (2015) Molecular approaches for biodegradation of polycyclic aromatic hydrocarbon compounds: a review. Rev Environ Sci Biotechnol 14:241–269

11. Haritash AK, Kaushik CP (2009) Biodegradation aspects of polycyclic aromatic hydrocarbons (PAHs): a review. J Hazard Mater 169:1–15

12. Kostka JE, Prakash O, Overholt WA, Green SJ, Freyer G, Canion A, Delgardio J, Norton N, Hazen TC, Huettel M (2011) Hydrocarbon-degrading bacteria and the bacterial community response in gulf of Mexico beach sands impacted by the Deepwater horizon oil spill. Appl Environ Microbiol 77:7962–7974

13. Launen LA, Buggs VH, Eastep ME, Enriquez RC, Leonard JW, Blaylock MJ (2002) Bioremediation of polycyclic aromatic hydrocarbon - contaminated sediments in aerated bioslurry reactors. Bioremediat J 6:125–141

14. Ławniczak Ł, Marecik R, Chrzanowski Ł (2013) Contributions of biosurfactants to natural or induced bioremediation. Appl Microbiol Biotechnol 97:2327–2339

15. Lladó S, Covino S, Solanas AM, Viñas M, Petruccioli M, D'annibale A (2013) Comparative assessment of bioremediation approaches to highly recalcitrant PAH degradation in a real industrial polluted soil. J Hazard Mater 248-249:407–414

16. Marecik R, Wojtera-Kwiczor J, Ławniczak Ł, Cyplik P, Szulc A, Piotrowska-Cyplik A, Chrzanowski Ł (2012) Rhamnolipids increase the phytotoxicity of diesel oil towards four common plant species in a terrestrial environment. Water Air Soil Pollut 223:4275–4282

17. Mesbaiahab FZ, Mansoura F, Eddouaoudaa K, Badisa A (2014) Surfactant effects on biodegradation of polycyclic aromatic hydrocarbons. Desalin Water Treat 57:5995–6000

18. Millioli VS, Servulo E-LC, Sobral LGS, De Carvalho DD (2009) Bioremediation of crude oil-bearing soil: evaluating the effect of rhamnolipid addition to soil toxicity and to crude oil biodegradation efficiency. Global NEST J 11:181–188

19. Mohan SV, Kisa T, Ohkuma T, Kanaly RA, Shimizu Y (2006) Bioremediation technologies for treatment of PAH-contaminated soil and strategies to enhance process efficiency. Rev Environ Sci Biotechnol 5:347–374

20. Mrozik A, Piotrowska-Seget Z (2010) Bioaugmentation as a strategy for cleaning up of soils contaminated with aromatic compounds. Microbiol Res 165:363–375

21. Owsianiak M, Szulc A, Chrzanowski Ł, Cyplik P, Bogacki M, Olejnik-Schmidt AK, Heipieper HJ (2009) Biodegradation and surfactant-mediated biodegradation of diesel fuel by 218 microbial consortia are not correlated to cell surface hydrophobicity. Appl Microbiol Biotechnol 84:545–553

22. Park JY, Choi YJ, Moon S, Shin DY, Nam K (2009) Microbubble suspension as a carrier of oxygen and acclimated bacteria for phenanthrene biodegradation. J Hazard Mater 163:761–767

23. Pasumarthi R, Chandrasekaran S, Mutnuri S (2013) Biodegradation of crude oil by*Pseudomonas aeruginosa* and *Escherichia fergusonii* isolated from the Goan coast. Marine Poll Bull 76:276–282

24. Quast C, Pruesse E, Yilmaz P, Gerken J, Schweer T, Yarza P, Peplies J, Glöckner FO (2013) The SILVA ribosomal RNA gene database

project: improved data processing and web-based tools. Nucl Acids Res 41:590–596

25. Rodrigues A, Nogueira R, Melo LF, Brito AG (2013) Effect of low concentrations of synthetic surfactants on polycyclic aromatic hydrocarbons (PAH) biodegradation. Int Biodeter Biodegr 83:48–55

26. Sachdev DP, Cameotra SS (2013) Biosurfactants in agriculture. Appl Microbiol Biotcchnol 97:1005–1016

27. Saponaro S, Bonomo L, Petruzzelli G, Romele L, Barbafieri M (2002) Polycyclic aromatic hydrocarbons (PAHs) slurry phase bioremediation of a manufacturing gas plant (MGP) site aged soil. Water Air Soil Pollut 135:219–236

28. Schmidberger A, Henkel M, Hausmann R, Schwartz T (2013) Expression of genes involved in rhamnolipid synthesis in *Pseudomonas aeruginosa* PAO1 in a bioreactor cultivation. Appl Microbiol Biotechnol 97:5779–5791

29. Silva ÍS, Santos EC, Menezes CR, Faria AF, Franciscon E, Grossman M, Durrant LR (2009) Bioremediation of a polyaromatic hydrocarbon contamined soil by a native soil microbiota and bioaugmentation with isolated microbial consortia. Bioresour Technol 100:4669–4675

30. Singh AK, Cameotra SS (2014) Influence of microbial and synthetic surfactant on the biodegradation of atrazine. Environ Sci Pollut Res 21:2088–2097

31. Szczepaniak Z, Cyplik P, Juzwa W, Czarny J, Staninska J, Piotrowska-Cyplika A (2015) Antibacterial effect of the *Trichoderma viride* Fungi on soil microbiome during PAH's biodegradation. Int Biodeter Biodegr 104:170–177

32. Szulc A, Ambrożewicz D, Sydow M, Ławniczak Ł, Piotrowska-Cyplik A, Marecik R, Chrzanowski Ł (2014) The influence of bioaugmentation and biosurfactant addition on bioremediation efficiency of diesel-oil contaminated soil: feasibility during field studies. J Environ Manag 132:121–128

33. Teng Y, Luo Y, Sun M, Liu Z, Li Z, Christie P (2010) Effect of bioaugmentation by *Paracoccus*sp. strain HPD-2 on the soil microbial community and removal of polycyclic aromatic hydrocarbons from an aged contaminated soil. Bioresour Technol 101:3437–3443

34. Volkering F, Breure AM, Rulkens WH (1998) Microbiological aspects of surfactant use for biological soil remediation. Biodegradation 8:401–417

35. Zhang X-X, Cheng S-P, Zhu C-J, Sun S-L (2006) Microbial PAH-degradation in soil: degradation pathways and contributing factors. Pedosphere 16:555–565

5

CHAPTER

Sustainability Aspects of In-Situ Bioremediation of Polluted Soil in Developing Countries and Remote Regions

Anders Jonsson[1] and Henrik Haller[1]

[1] Department of Ecotechnology and Sustainable Building Engineering, Mid Sweden University, Sweden

1. INTRODUCTION

Xenobiotic and other toxic compounds from industrial activities are presently being accumulated in the ecosystems at an unsustainable rate. From an anthropocentric point of view such contamination is unsustainable because it threatens basic ecosystem functions and services that are necessary to maintain food security and provision of potable ground water. Furthermore natural habitats for plants and animals are threatened. Thousands of chemical compounds in varying concentrations and compositions have contaminated a wide range of habitats [1]. These contaminated sites demand diverse strategies and development of new technologies that are efficient and cheap enough to make remediation feasible even in remote areas and where financial support is limited.

Soil remediation has been a developing field for several decades and a common assumption is that it is a sustainable industry but very often clean-ups demand vast amounts of energy and lead to severe physical damage to the landscape. Traditional remediation technologies include soil excavation and landfill disposal. Sustainable remediation technology has been defined as a *"remedy or combination of remedies whose net benefit on human health and the environment is maximized through the careful use of limited resources"* [1]. Such technologies are only modestly implemented at present but new landfill legislations against "dig and dump" technologies can favour green solutions over chemical and energy-intensive techniques in the future [2].

Microorganisms have colonized some of the most extreme environments on the earth and some of them are efficient degraders of the pollutants that our industrial society produces. Bioremediation is a technology that uses living organisms and their enzymes to degrade, remove or immobilize toxic compounds [3]. Bioremediation have been gaining terrain recently [1] and some authors predict a further increase as technological advances surmount the limitations that exist today.

In this chapter we briefly discuss how principles of ecological engineering in concurrence with the application of basic thermodynamic principles and kinetic modelling can provide useful tools for the development of energy conserving and economically feasible bioremediation projects. We further discuss the potential of organic waste materials and by-products in locally adapted soil bioremediation. Finally we present some illustrative cases of novel research on sustainable bioremediation for tropical developing countries and remote locations and discuss some promising fields of future research and possible future applications.

2. BIOREMEDIATION AND ECOLOGICAL ENGINEERING

Ecological engineering has been proposed as a theoretical framework to design *"sustainable ecosystems that integrate human society with its natural environment for the benefit of both"* [4]. Bioremediation is generally considered an ecological engineering practice but even if it addresses one of the core goals of ecological engineering, i.e. *restoration of damaged ecosystems* [5], bioremediation can be energy-intensive and have a serious impact on the remediated ecosystems particularly if excavation and ex situ methods are employed.

Mitsch and Jørgensen [5] identify 5 basic concepts that have been developed to distinguish ecological engineering from other approaches such as industrial ecology, biotechnology or environmental engineering. Ecological engineering:

1. is based on the self-designing capacity of ecosystems
2. can be the acid test of ecological theories

3. relies on systems approaches
4. conserves non-renewable energy sources
5. supports ecosystem conservation

If used collectively, these 5 concepts can provide guidelines for design of sustainable bioremediation projects. In a sustainable society the ultimate goal is to limit the rate of pollution to what ecosystems can assimilate and break down without long term negative effects on ecosystem functions or human health. Meanwhile that goal is attained we will have to give the ecosystems a hand to boost their intrinsic biodegradation capacity.

2.1. Self Design

Self-design is the property of a system to reorganise in an unstable non-homogeneous environment [5]. The inclusion of self design in a project can facilitate the implementation of flexible strategies that adapt to new conditions. The powerful capacity of ecosystems to reorganise after change can be used in bioremediation projects as a means to reduce costs and energy use. Consortiums of soil microorganisms naturally restructure to adapt to new conditions including contamination and are therefore good candidates for self-design strategies. A typical microbial community in soil consist of less than 1 % hydrocarbon degraders but that fraction can increase to 10% after exposure to oil pollutants [6]. An example of how engineers can accelerate the ecosystems self-designing capacity is the introduction of a microbial strain with capacity to initiate degradation of high molecular-weight pollutants that later provide food (energy) for other microorganisms that take care of the metabolic by-products as the introduced strain dies off. In conventional engineering, items are often designed to behave in an as predictable way as possible since reliability and robustness are desired criteria of safety and quality. The outcomes of a technology that depend on self-design, on the contrary, are often a lot more difficult to predict. Bioremediation projects that include self-designing strategies thus require thorough monitoring to make sure that the degradation of the pollutants is satisfactory.

2.2. The Acid Test

Just as strong acids were traditionally used to distinguish gold from base metals, a full scale bioremediation project can be used to confirm or reject ecological theories. Ecological theories provide a useful framework that establishes restrictions and opportunities for bioremediation strategies.

The complex nature of full scale in-situ bioremediation however might lead to outcomes somewhat different from what was expected. Since intense use of monitoring is required for full scale implementations of unproven technologies, the result from these can be used to refine the theories on which the project was based.

2.3. A Systems Approach

Reductionist analysis of the parts of a system in isolation from each other can provide important data and basic understanding of significant mechanisms but should be combined with a holistic view of the entire system in order to achieve a sustainable bioremediation project. The site specific conditions of the ecosystem that is subject to remediation must be quantified and analysed with appropriate models. That particular ecosystem however is part of a greater context whose characteristics must also be born in mind. When bioremediation strategies are chosen, socioeconomic and even cultural aspects must be taken into account. Questions like "what is the economic impact of the project?", "can polluted land be reclaimed for food production or construction?", "what resources are available in the human environment around the site?" might supply useful information. The use of locally available agricultural waste products, that are potential polluters by themselves, as amendments (see section about waste recycling) is an example of advantageous design that requires a systems approach. The source of the pollution must also be addressed to make sure that the remediation project isn't an encouragement to further pollution.

2.4. Non-Renewable Resources Conservation

Energy use is one of the major sustainability issues for conventional remediation projects. Ex-situ remediation is typically too energy-intensive to be considered ecological engineering. Furthermore it resembles what Jørgensen & Mitsch [5] call a *shell game*, in which pollutants are moved from one location to another. In sustainable bioremediation external energy input is ideally used only in the initiation phase to start a process that is subsequently powered by solar energy and the embodied chemical energy of the pollutant itself. The engineer's role is to facilitate the proper conditions in which such a process can take place.

2.5. Ecosystem Conservation

The ability to cycle nutrients is a fundamental function of all ecosystems. Within the matter-cycling function lies the catabolic capacity to degrade and mineralise molecules of varying complexity, including pollutants. This function alone gives a nonmarket value to ecosystems, not always recognised by society. The role of ecological engineers is to identify ecosystem functions that are adoptable to human needs and apply them in ways that doesn't create any further degradation of nature. Since bioremediation projects aims at eliminating pollution that exerts stress on the ecosystem, it is by definition an ecosystem conservation approach. However, if large amounts of soil are physically removed from the site in ex-situ operations the remediation itself might be a threat to the ecosystem. Bioremediation projects are typically involved in the restoration of disturbed ecosystems but measures to prevent toxic compounds from entering sound ecosystems are likewise part of the bioremediation agenda.

3. THERMODYNAMIC AND KINETIC CONSIDERATIONS ON BIOREMEDIATION

The fact that most organic pollutants are composed of molecules with high embodied energies is an interesting starting point for any consideration about the best practice for sustainable soil remediation, especially in remote locations and developing countries where energy-intensive project are cost-prohibitive. Economic incentives are generally small for any kind of remediation practice in such regions and, furthermore, time constraints may not be as limiting as in highly developed urban environments. Technologies that manage to optimize the use of ecosystem services provided by indigenous organisms, in combination with a minimum input of extra energy and material resources, are potentially more sustainable in such conditions.

Figure 1 illustrates some basic principles including thermodynamic driving forces (from molecules with high to low embodied energy) and available transformation pathways in air and soil. In the atmosphere, abiotic photochemical reactions rapidly oxidize organic vapours. The rate determining step in these complex photochemical reactions is usually the initial attack on the primary pollutant by hydroxyl radical. Typical turnover times of i.e. reactive hydrocarbons are only a few hours in the sunlit atmosphere while lack of photochemistry in the subsurface of soils leads to significantly lower reaction rates and hence to much longer turnover times of pollutants.

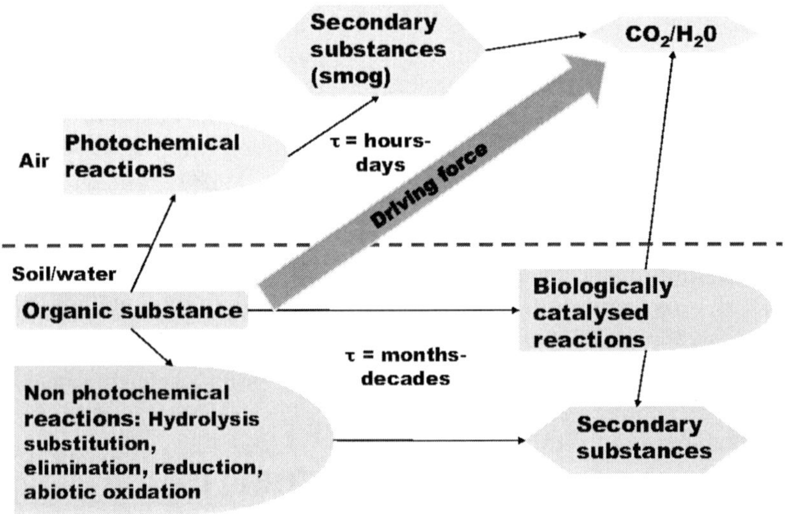

FIGURE 1. Chemical and biochemical transformations and mineralisation of organic substances in polluted soil and in air. The broad arrow indicates the thermodynamic driving force for the complete degradation of the pollutant in an oxidising environment. Typical average residence times of organic pollutants (τ) in air and soil are indicated.

In soils where conditions for microbial activity are unfavourable, degradation relies on abiotic chemical reactions. In such cases organic pollutants may persist for years to decades or longer and secondary, incompletely degraded products may accumulate. On the other hand if conditions are favourable (temperature, electron acceptors, nutrients, moisture, available carbon sources, low toxicity to microorganisms, pH, to mention some of the most determinant variables), microorganisms such as bacteria and fungi may use the organic pollutant as a source of energy and carbon. Enzymatic degradation is generally much faster than abiotic oxidation reactions and the organic pollutant that enters the metabolic pathways of soil microorganisms may be partly or completely mineralized. The driving force for such an enzymatic process is again the free energy potential that can be used by the organism. For typical hydrocarbon and oxygenate pollutants the free energy that is available by complete oxidation with oxygen as electron acceptor, is in the range of 3-12 x 10^3 KJ mol^{-1} or 37-46 MJ kg^{-1}, as can be seen in table 1. This energy potential may be used by a wide range of heterotrophic microorganisms.

In areas of hydrocarbon contamination the most commonly found heterotrophic bacteria are of the genera Pseudomonas, Achromobacter,

Arthrobacter, Micrococcus, Vibrio, Acinetobacter, Brevibacterium, Corynebacterium, Flavobacterium, Mycobacterium and Nocardia [7]. Some soil microorganisms that are incapable of utilizing the full embodied energy potential of the pollutant may still be able to initiate catabolic processes that ultimately lead to complete mineralisation if oxygen is present at sufficient concentration. Filamentous fungi, such as e.g., white-rot fungi, can oxidize recalcitrant high molecular weight polycyclic aromatic hydrocarbons (PAH:s) with their extracellular enzymes to more water soluble substances. Both lignolytic and non lignolytic fungi has been shown to initiate degradation of high molecular-weight PAH and the oxidation products formed are thereby made available for further degradation by bacteria [8]. Such interactions between soil microorganisms is an important factor to consider in sustainable soil bioremediation and remediation practices that favour a high soil microbial biodiversity are likely to be more sustainable.

In oxygen depleted soils microorganisms are unable to mineralise organic compounds to inorganic molecules only. Complete degradation leaves carbon dioxide and methane as end products. Intermediate fermentation products, that may be toxic to microorganisms and plants, may accumulate if they are not converted to methane [7]. Since methane and fermentation products has retained a substantial part of the embodied energy, the energy gain by microorganisms by anaerobic degradation is less than under aerobic conditions and the thermodynamic driving force for biodegradation is thus lower. Methane is also a highly potent greenhouse gas [9] and its unintended production should, from a sustainability point of view, be avoided. It is therefore in many cases preferable to maintain aerobic conditions in the soil during in-situ bioremediation projects. There are however some important exceptions. Xenobiotic and toxic chlorinated organic compounds such as e.g. highly chlorinated PCB:s, pentachlorophenol, tri- and perchloroethylene, gamma hexachlorocyclohexane (lindane) and many others are highly persistent to biodegradation under aerobic conditions. The reason for this may be lack of metabolic pathways available to common soil microorganisms that can generate energy gain for the organism. Persistence is also caused by high lipophilicity of many of these substances which reduce their bioavailability.

Chlorinated substances may be efficiently degraded by microorganisms capable of halorespiration under anaerobic conditions [10]. The degradation mechanism is characterized as reductive dechlorination. In contrast to aerobic co-metabolism which does not provide any energetic benefit for the degrading organism, reductive dechlorination may provide a substantial free energy yield. The thermodynamic driving force of such reactions is demonstrated by the fact that hexachloroethane for example, may be a stronger electron acceptor than oxygen [11].

TABLE 1. Thermodynamic properties and free energy of mineralisation of some representative "high energy" organic substances commonly found in polluted soil. Standard entropy of formation ($\Delta S^o{}_f$) was calculated from standard entropy values (S^o) of the substances and their elements (phase indicated in table). Standard free energy of formation ($\Delta G^o{}_f$) was calculated from standard enthalpy of formation ($\Delta H^o{}_f$) and standard entropy of formation. Standard free energy of mineralisation (ΔG^o) was calculated for the complete oxidative degradation of the organic compounds to the thermodynamically most stable end products according to the reaction: 1 mol of substance + a O_2 = b CO_2 + c H_2O.

Category	Substance	Formula	Phase	M	S^o	$\Delta H^o{}_f$	$\Delta S^o{}_f$	$\Delta G^o{}_f$	b	c	ΔG^o(mineralisation)	
				g mol^{-1}	J $mol^{-1}K^{-1}$	KJ mol^{-1}	KJ mol^{-1}	KJ mol^{-1}			KJ mol^{-1}	MJ kg^{-1}
Gasoline	Benzene	C_6H_6	Liquid	78.11	173.26*	49.0*	-253.78	124.66	6	3	-3202	-41.00
Gasoline	Toluene	C_7H_8	Liquid	92.14	220.96*	12.0*	-342.59	114.14	7	4	-3823	-41.50
Gasoline	m-xylene	C_8H_{10}	Liquid	106.16	253.80*	-25.4*	-446.26	107.65	8	5	-4448	-41.90
Gasoline additive	MTBE	$C_5H_{12}O$	Liquid	88.15	265.3*	-315.4*	-650.52	-121.45	5	6	-3273	-37.14
Diesel	Hexadecane	$C_{16}H_{34}$	Liquid	226.44	586.18*	-456.3*	-1728.71	59.11	16	17	-10401	-45.93
Diesel	Octadecane	$C_{18}H_{38}$	Solid	254.49	480.2*	-457.4*	-2107.71	171.01	18	19	-11776	-46.27
Diesel	Phenanthrene	$C_{14}H_{10}$	Solid	178.23	215.06*	110.1*	-520.00	265.14	14	5	-6972	-39.12
Diesel	Perylene	$C_{20}H_{12}$	Solid	252.31	264.6*	182.4*	-636.14	372.07	20	6	-9682	-38.37
Mineralisation product	Carbon dioxide	CO_2	Gas	44.01	213.79*	-393.5*	2.81	-394.36				
Mineralisation product	Water	H_2O	Liquid	18.02	69.95*	-285.8*	-163.30	-237.19				
Element	Carbon	C	Solid (graphite)	12.01	5.83*	0	0	0				
Element	Hydrogen	H_2	Gas	2.02	130.68*	0	0	0				
Element	Oxygen	O_2	Gas	32.00	205.15*	0	0	0				

**NIST, National Institute of Standards and Technology website. Available at: http://webbook.nist.gov/chemistry/name-ser.html*

3.1. Kinetic Considerations

Biodegradation kinetics tells us not only about the rate of degradation of target compounds and of rate dependence on ambient factors, but by studying the shape of the degradation curves we may also get valuable information about degradation mechanisms. It is preferable to use models that have a theoretical basis rather than just empirical mathematical modelling that have limited predictive ability.

Natural attenuation of organic pollutants are typically mediated by slow growing organisms and if the concentration of pollutants is much lower than the concentration required for maximum growth of degrading organism (which is often the case in many polluted soils) then first order kinetics may apply, i.e. degradation rate is directly dependent on substrate concentration (Equation 1).

$$-dC/dt = kC \tag{1}$$

C is the substrate concentration and k is the first order rate constant. First order kinetics is often applied in degradation studies because of its simplicity but have little predictive power when conditions for growth change.

In situations with rapid growth of degrading organisms or when other factors than the substrate concentration controls the rate of degradation, first order kinetics will not describe the degradation process accurately. According to the earlier mentioned criteria for sustainable soil bioremediation, efforts should be made to stimulate growth of indigenous soil microorganisms that have or can achieve the ability to degrade target pollutants. The addition of organic waste products from the dairy industry that promotes growth of degrading organisms has been proved to be an efficient means to increase both rate and total degradation of some representative diesel range hydrocarbons in soil [12-14]. Although the addition of readily available carbon sources, such as lactose, lactate and amino acids (components of whey) to contaminated soil would be expected to trigger exponential growth of organisms, a common phenomenon in liquid culture media, it seems that the soil matrix itself does not readily promote exponential growth of degrading organisms. Growth of soil microorganisms is limited by a number of factors caused by the properties of the soil matrix. Among the limiting factors is diffusion of substrate and nutrients but also the specific area available for bacterial sorption on to soil particles. When a large fraction of this area is occupied by bacteria, further growth might be increasingly restricted by diffusion of substrate and nutrients into multilayers of bacteria. Kinetic models that take into consideration slow or restricted (linear growth) may more accurately describe degradation patterns in real life situations. Figure 2 illustrates the application of a three-half-order kinetic model (II) originally proposed by Brunner and Focht [15]. The model assumes linear biomass growth rate and simulates the accumulated $^{14}CO_2$-production from phenanthrene-9-^{14}C mineralisation in soil microcosms.

$$P(t) = S_0\left(1 - \exp\left(-k_1 t - k_2 t^2/2\right)\right) + k_0 t \tag{2}$$

P[t] is the cumulative $^{14}CO_2$-production (%) where t is the time (days), S_0 is the amount of substrate (%) that is "highly bioavailable", k_1 is the first-order rate constant (day^{-1}), k_2 is a linear growth term (day^{-2}), and k_0 is the zero order rate constant describing the mineralisation of "less bioavailable" substrate and indigenous mineralisation (% day^{-1}).The lag time was defined as the initial time until 1% of the added ^{14}C was detected as $^{14}CO_2$ [16]. From the end of the lag phase, iterative non-linear least squares regression was performed to fit data to the model by using the SOLVER function in Microsoft Excel 2000 [17].

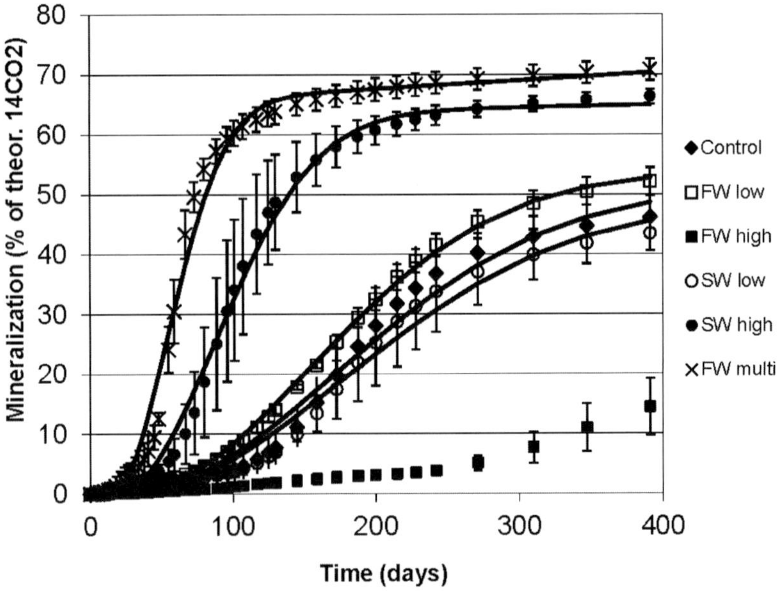

FIGURE 2. Effect of added fermented whey and sweet whey on ^{14}C-phenanthrene mineralisation to $^{14}CO_2$ in soil at 25 °C. FW low = Fermented whey 210 mg dw kg^{-1} soil dw; FW high = Fermented whey 2100 mg dw kg^{-1} soil dw; SW low = Sweet whey 210 mg dw kg^{-1} soil dw; SW high = Sweet whey 2100 mg dw kg^{-1} soil dw, all added at the beginning of the experiment. FW multi = Fermented whey 210 mg dw kg^{-1} soil dw added at the beginning of the experiment and then every 14th day. The curves are fitted to the data by use of the three-half-order kinetic model. Kinetic parameters are given in table 2 (From [18]. Reprinted with permission from Elsevier).

The effects of applications of different amendments on phenanthrene biodegradation are clearly seen infigure 2. Kinetic parameters such as the linear growth term (k_2) and S_0 was significantly affected by the addition of the different amendments (Table 2). It should be noted that the stimulating effect of complex organic amendments such as whey and fermented whey may not only be an effect of the addition of a readily available carbon source (that promotes growth) but also of available micronutrients, such as B-vitamins, that is required to maintain efficient catabolic pathways for degradation of i.e. polycyclic aromatic hydrocarbons.

TABLE 2. Tree-half-order kinetic constants for phenanthrene mineralisation in soil, at 25 °C. Lag time is expressed in units of days, S_0 and mineralisation in %, k_1 in day^{-1}, k_2 in day^{-2}, k_0 in % day^{-1} and R is the correlation coefficient. Values having the same superscript letters are not significantly different at a 95 % level. (From [18]. Reprinted with permission from Elsevier).

	Lag time (days)	S_0 (%)	k_1 (day^{-1})	k_2 (day^{-2})	k_0 (% day^{-1})	R	Mineralisation after 391 days (%)
Control	21	51.3^{a}	0	$4.35^{a} \times 10^{-5}$	0	0.973	46.3^{ab}
FW low	24	53.8^{a}	0	$5.83^{b} \times 10^{-5}$	0	0.997	52.1^{a}
FW high	87	-	-	-	-	-	14.4^{c}
SW low	21	40.6^{a}	0	$4.09^{a} \times 10^{-5}$	0	0.974	43.4^{b}
SW high	18	63.2^{b}	0	$20.8^{c} \times 10^{-5}$	0.0046^{a}	0.983	60.3^{d}
FW multi	15	65.0^{b}	0	$63.6^{d} \times 10^{-5}$	0.014^{b}	0.992	70.8^{e}

4. WASTE RECYCLING

Organic by-products are produced globally on a massive scale. Economic operations such as agriculture, livestock, fishing and forestry [19-21] generate a number of by-products that can be seen inFigure 3. Frequently unexploited organic waste products create serious environmental problems when disposed of [20]. Agriculture is the economic base of most tropical developing countries, accounting for 50 % of the employment [21]. Organic by-products from agriculture, forestry, fishery and livestock operations are thus available in abundance and are typically of limited economical value [19]. Animal feed has been proposed as an outlet for some of these by-products but due to low levels of protein, high levels of moisture and some anti-nutritional factors i.e. presence of tannins and polyphenols, this practice is limited [22]. However, being inexpensive and easily available sources of carbon, nutrients and bioactive compounds, many of these by-products can potentially be used as sources of readily available carbon for bioremediation purposes [23].

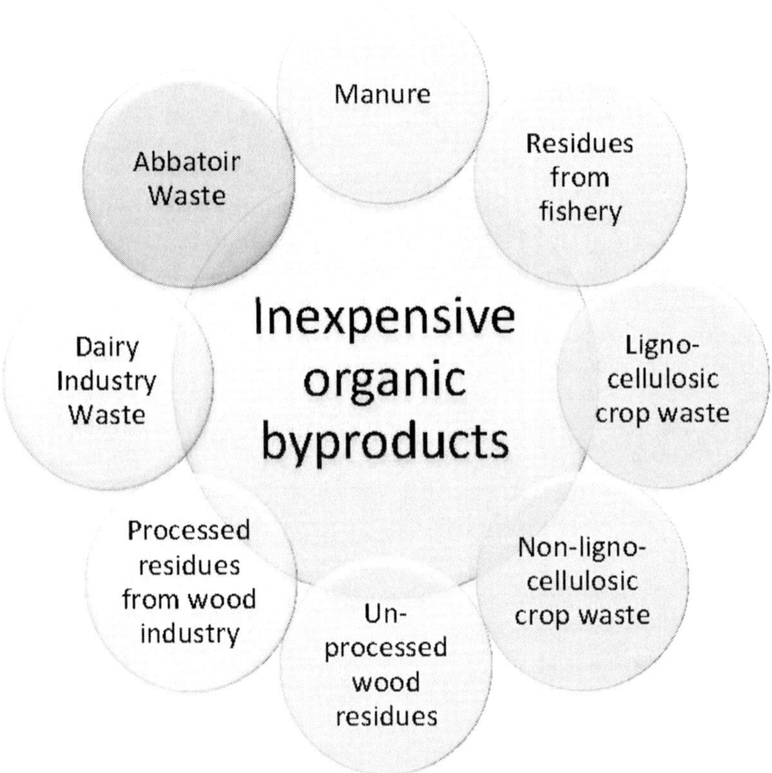

FIGURE 3. Some organic by-products that potentially can be used as amendments for bioremediation.

A number of organic by-products have the ability to enhance the degradation of different toxic compounds including lignocellulosic wastes such as sugarcane, bagasse and sawdust [26], crop residues such as coffee pulp [27] and molasses [28] and whey, a by-product of the dairy industry [12-14]. Three examples of inexpensive by-products with potential or confirmed bioremediation capacity are described in more detail below: pyroligneous acid, compost teas and whey.

4.1. Pyroligneous Acid

Pyroligneous acid (PA) is a by-product of charcoal production [29] with a limited economic value that is also referred to as wood vinegar, pyrolysis oil, smoke extract, bio-oil etc. [30, 31]. PA is typically disposed of as waste [31] but in Japan it has been used for centuries to increase crop production and to combat agricultural pests [29, 32]. At present, the majority of the charcoal consumed worldwide is produced in traditional kilns in developing countries [33]. Such charcoal production lacks appropriate control of the gas and vapours produced in the process and causes land and air pollution [34]. Production systems exist however that recover these condensate co-products or use them as fuel in the pyrolysis process, reducing the emission to a minimum [30]. Several low tech kilns that recover the PA from the charcoal production have been developed including *the advanced Brazilian beehive kiln* and the *Casamance kiln* [35]. In recent years the research and technology concerning PA have advanced significantly [30], and the current trend in charcoal production is to improve the environmental performance while maintaining or improving charcoal yield and quality. The chemical composition of PA varies depending on factors such as type of feedstock, pyrolysis temperature and duration etc. [25,30]. The most frequently identified compounds in PA include methanol, acetic acid, methyl acetone, acetaldehyde, furan, furfural and volatile organic acids. PA contains phenolic compounds that are known to have antimicrobial properties [36] and PA has successfully been used to control fungal growth on wood [30] and bacterial decay of foods [36, 37]. At low concentrations however, PA has been shown to stimulate germination, growth and yield in a wide range of plants [37]. Although the growth promoting effect of PA is not yet clearly understood it has been attributed to nutritious components such as nitrates of Ca, Mg, K, magnesium sulphate etc. but also to its stimulatory effect on symbiotic fungi and bacteria [31]. The formation of fruit bodies of hydrocarbon-degrading fungi such as *Lentinus edodes*and *Pleurotus ostreatus* was significantly increased after applications of PA to the growth medium [38]. Furthermore PA has been shown to significantly increase basal respiration and microbial biomass in highly weathered tropical soils suggesting that microbes used PA for their metabolism [29].

4.2. Compost Teas

Compost teas (CT) are liquid compost extracts obtained when compost is soaked in water for a determined number of days. Optionally additives (derived from by-products) such as molasses, fish hydrolysate, rock dust, yeast extract, humic acids etc. are used to add nutrients and stimulate microbial activity. To date virtually all research on CT has focused on its potential to control plant diseases in agriculture. However, the fact that CT have proved to significantly increase soil microbial respiration and dehydrogenase activity [39] makes it an interesting candidate for bioremediation. A number of production parameters (e.g. aeration, compost source, nutrient additives etc.) have been manipulated in order to optimize plant disease suppression [40]. To date, consensus on optimum production parameters for disease suppression hasn't been reached and some studies report inconsistency in the performance [41]. However a number of reports have shown that aeration and additions of nutrients lead to significant increases of the cell mass of active bacteria [39, 40, 42] and the addition of humic acids and yeast extracts significantly increased the fungal populations [43]. Plate counts and microscopic examination studies have revealed that aerated CT has bacterial population dominated by *Bacillus sp.,Lactobacillus sp., Micrococcus lutues, Staphylococcus sciuri, Pseudomonas putida, Burkholderia glumae* and *Clavibacter agropyri* while species of *Aspergillus, Penicillium* and *Trichoderma*dominated the fungal communities [43]. Many of these organisms have the capacity to degrade a number of toxic substances including PAH:s and organochlorine compounds [44, 45].

In a recent study, applications of compost teas made from vermicompost caused significantly greater dehydrogenase activity and microbial respiration [39] which makes it an interesting candidate for bioremediation. Vermicompost generally has higher concentrations of available nutrients (NO_3^-, exchangeable Ca, P and soluble K) and significantly larger and more diverse microbial populations than common compost. Microorganisms found in worm castings of *Eisenina fetida* have caused increased catabolic activity and enhanced degradation of PAHs in a number of experiments [46, 47].

4.3. Whey

Whey is produced when milk casein is removed from the milk in dairy operations to make cheese and other products. This liquid by-product constitutes between 85-95% of the milk volume and 55% of milk nutrients remain in the whey. About half of the yearly global production of 145 million tonnes is utilized for animal feed etc. The remaining large volumes end up as potential contaminants in the environment [48]. The chemical content of whey is characterised by lactose, a number of essential and non-essential amino acids in different proportions, vitamin B 1,2,6,7,12, folic acid and lactic acid [12,13]. Relatively few studies have been performed on the effect of whey on biodegradation of organic pollutants [12]. However, research at Mid Sweden University has showed that the degradation of aliphatic and aromatic hydrocarbons can be significantly enhanced by the addition of whey and fermented whey [12, 13]. The biodegradation-enhancing effect of whey was primarily attributed to an increased microbial biomass stimulated by the readily available carbon source [12]. The degradation studies also showed a more complex dependence of carbon sources and growth factors, such as B-vitamins, on the degradation of an aromatic compound (phenanthrene) compared to an aliphatic compound (hexadecane) [18]. These results indicate that the presence of co-factors such as vitamins and micronutrients may be important to consider when evaluating the suitability of organic amendments for bioremediation of soil contaminated by various types of organic pollutants.

5. CASE STUDIES

Different technologies and strategies are necessary to remediate the present and future soil pollution. As our understanding of the degradation mechanisms of different pollutants evolves, so do our chances to develop alternative and complementary strategies that can take the concentrations of toxic compounds to a sustainable level. Since 2006, a research group at Mid Sweden University has undertaken research dedicated to the development of bioremediation methods for low priority remediation sites including, remote locations and tropical developing countries. Ecological engineering provide an appealing theoretical framework in such locations where energy and material costs are limiting factors. The research has implied laboratory experiments as well as pilot scale and in-situ projects in tropical and temperate climate. It includes pilot scale experiments in an experimental station that was developed in Nicaragua (se figure 4) and field trials in remote parts of northern Sweden where transport distances are long and the cold climate constitutes an additional challenge to in-situ bioremediation projects.

5.1. Field Station In Nicaragua, Designed For Bioremediation Experiments In A Tropical Climate

Waste management programs in developing countries are often deficient and the inappropriate disposal of hazardous waste creates pollution problems. Furthermore the use of pesticides and other toxic products is minimally restricted in many developing countries which leads to extensive

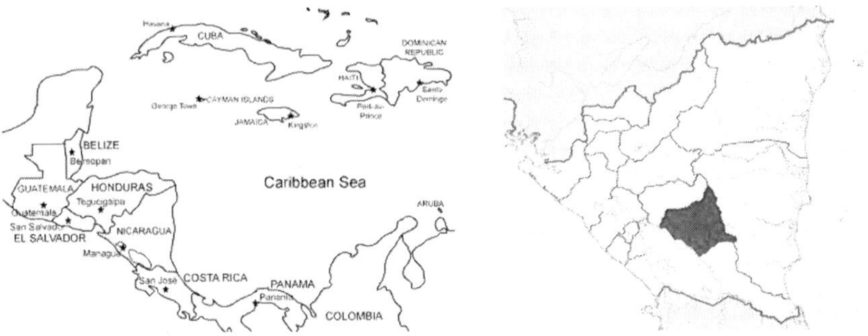

FIGURE 4. *Left picture: Map of Central America and Nicaragua. Right picture: The red marked area is Chontales, the region where Mid Sweden University's pilot scale experimental station is located. (Map of Central America by Викmop B licensed under Creative Commons Attribution-Share Alike 2.0 Generic, Mapa de Nicaragua mostrando al departamento de Chontales by Vrysxy licensed under Creative Commons Attribution-Share Alike 3.0 Unported.)*

FIGURE 5. Mid Sweden University´s experimental station at Casa Montesano in Nicaragua.

contamination with persistent substances such as polycyclic aromatic hydrocarbons and organochlorine pesticides [49,50]. The experimental station in Nicaragua (figure 5) is located on the agro ecological farm *Casa Montesano,* (11°59'70″ N 84°53'06″ E), where CIPP, a local NGO, dedicated to sustainable rural development projects has its headquarters. The region has a humid tropical climate with an average precipitation of 2000 mm per year and an annual average temperature between 25 and 28 °C. The soil on the experimental site is classified as ultisol, a common soil in the humid tropics occupying about 8 % of the global ice-free land area and supporting 18 % of the world's population.

Figure 6 shows the 24 compartments that were built in concrete with a smooth cement plaster in two sets of twelve compartments each. The compartments were built to fit 150 L each of agriculture soil. The purpose of the closed compartments is to test sustainable remediation methods in conditions as similar to in-situ remediation as possible without risks for pollutants to escape into nature. All technologies assessed in the experimental station are designed to be low cost/low technology methods, appropriate to economically marginalised people in developing countries. Notable is the use of locally available organic waste or by-products such as whey, pyroligneous acid and compost teas that can be achieved at a very low cost. The pilot scale research aim to find practical solutions to the obstacles that emerge when positive results from the laboratory are to be reproduced in the field. This implies dealing with issues like chemotaxis limitations, i.e. the importance of biodegrading microorganisms to be in close proximity to the pollutant. That can be achieved by the use of surfactants and solvents and by avoiding clogging of the soil pores.

The first experiment conducted in the station was on soil spiked with diesel at a concentration of 5.00 g diesel dw^{-1} kg^{-1} soil dw. Whey, pyroligneous acid and compost teas were applied at a high and a low dose on day 3, 18, 32, 46, 60, 74 and 88 after diesel contamination. The doses were adjusted to reflect the expected ideal dose (high) and an economically more viable dose (low). Soil samples were taken on day 2, 7, 12, 17, 31, 45, 59, 73, 87 and 101 after diesel contamination. Leachate water samples were collected by the end of the experiment in order to detect presence of metabolic by-products.

When this chapter was written the soil samples were not yet analyzed. The soil samples will be analysed by GC/FID to determine the effect on degradation from the different amendments. The amendments' effect on the microbial composition will subsequently be analysed by molecular methods. Whey has engendered significant enhancement of the degradation of aliphatic and aromatic hydrocarbons in previous laboratory experiments within our research group but compost tea and pyroligneous acid have not been previously tested.

FIGURE 6. The compartments filled with soil at the experimental station in Nicaragua.

In an upcoming project the research group envisages to use the experimental station for novel research on relevant pesticides. Many of the least biodegradable pesticides are relatively persistent to aerobic degradation [27] which poses additional challenges for in-situ remediation. These challenges will be addressed constructively in trials with redox barriers and other strategies.

5.2. Experiences From In-Situ Remediation Trials In Remote Areas Of Northern Sweden

5.2.1. Bioremediation field trial using whey treatment of diesel contaminated soil at a former gas station in northern sweden

A field experiment was carried out at a diesel contaminated site in Gäddede, in the county of Jämtland, north-western Sweden (65°30’N, 14°09’E) [51], see figure 7. The contaminated site is an old fuel station, established in the 1960:s. The contaminated site is about 10 m^2 in area and about 0.5 m in depth. Previous investigations had pointed out an area with relatively high loads of aliphatic and aromatic hydrocarbons C_5-C_{35} (approximately 2200 mg kg^{-1} dw soil) dominated by diesel range aliphatics.

This site is representative of the many small contaminated areas that are spread out in rural areas, and are typically considered low priority sites. According to a study by the Swedish EPA in 2006 excavation and transport was the dominating remediation technology in Sweden [52] which is also the case in the county of Jämtland. In this case the distance from Gäddede to Gräfsåsen, the nearest approved treatment plant for contaminated soil, is 230 km. The alternative of excavation and transport will therefore be expensive and contribute to air pollution and greenhouse gas emissions. For such areas, a low cost, easy to apply in-situ remediation technique would be appropriate. In-situ whey treatment might be such a method. Addition of whey to increase biological activity in soil for remediation purposes has been previously demonstrated in laboratory studies [12, 13, 18].

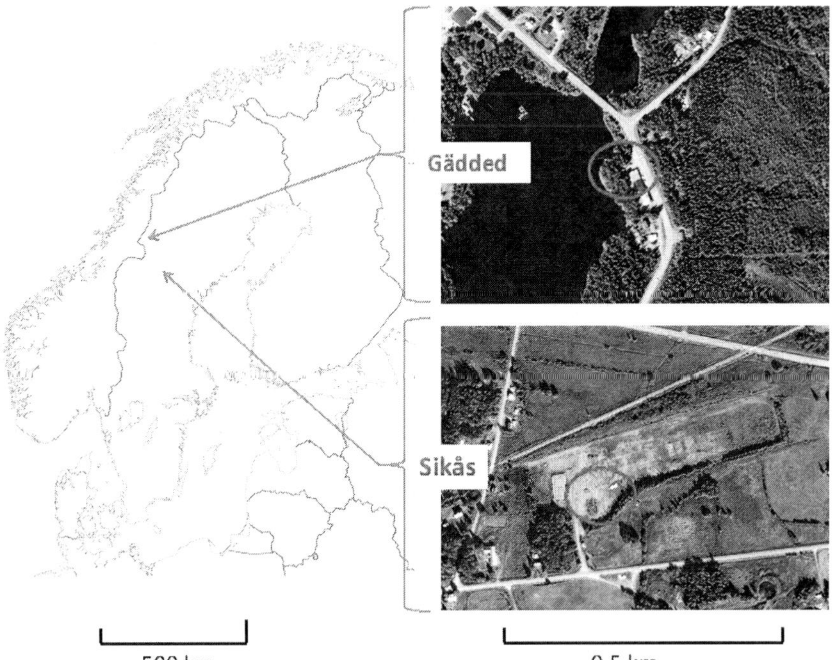

FIGURE 7. The test spots for in-situ bioremediation trials with whey in northern Sweden. (Location map of Scandinavia by NordNordWest licensed under the Creative Commons Attribution-Share Alike 3.0 Unported) Top picture: Diesel contaminated site at former gasoline station in Gäddede (Village). Bottom picture: Oil spot at lumberyard in Sikås (Village). © Lantmäteriet 2012/1067.

5.2.1.1. Life cycle assessment of in-situ bioremediation with whey treatment in gäddede in comparison to traditional excavation and composting

In order to compare the two alternative treatment strategies, a screening life cycle assessment model was set up. The purpose was to investigate the environmental performance of the whey method, to benchmark the whey method towards the excavation and composting practice and to identify environmental hot spots in the whey treatment life cycle [53]. The functional unit of this study was the treatment of 5 m^3 of contaminated soil reaching a diesel concentration of 500 mg kg^{-1} dw (upper limit in generic guidelines for "non sensitive areas"). For impact assessment the results are characterized as global warming potential (GWP), acidification potential (AP), eutrophication potential (EP) and photo oxidant creation potential (POCP). It was assumed that whey is added together with commercial fertilizer to avoid macronutrient deficit. Since the biological activity only takes place when it is warm enough (this mountainous region has a cold climate with at least 6 months of winter and a yearly average temperature of around 0°C) the total treatment time was estimated to two summers. The alternative treatment assumed in this study is to excavate the 5 m^3 of contaminated soil at the site in Gäddede, transport it to the waste treatment site in Gräfsåsen, Östersund, Sweden, and compost the soil in windrows after mixing the soil with manure. The windrows are turned using an excavator and when the concentrations of contaminants are low enough the compost is used as top cover of a landfill.Figure 8 illustrates a simplified model of the life-cycle process flow diagram for the whey treatment scenario.

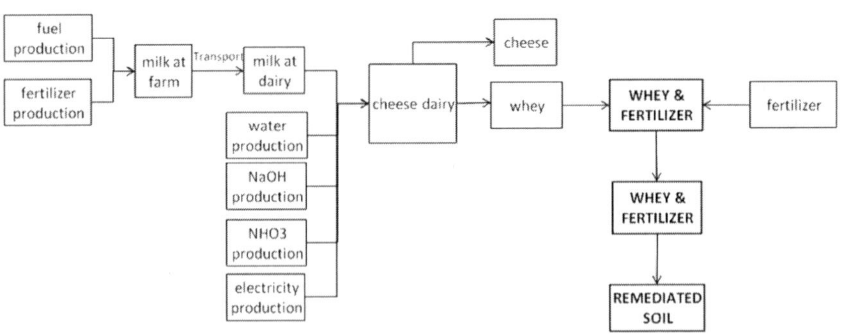

FIGURE 8. The life cycle inventory model of the whey scenario (From [53]).

The hydrocarbon pollution in the 5 m^3 soil to be treated (over an area of 10m^2) was assumed to be the same in the whole mass of soil i.e. 2200 mg kg^{-1} dw. Comparisons with laboratory results [12] indicate that an addition of 210 mg dw whey kg^{-1} diesel polluted soil might give a sufficient reduction in diesel oil contamination to reach the higher target value of 500 mg kg^{-1} dw [54]. This gives a calculated need of about 2 kg dw whey, equivalent to 367 L of whey (5.5 g dw L^{-1}) for the area to be treated. The laboratory tests also indicate that a number of successive additions of whey will give a better degradation compared to one large addition, and eight additions (46 L of wet whey per addition) spaced out with 14 days in between was assumed to be a good practice. The total time needed for the degradation is from the laboratory tests estimated to 167 days. Since on-site biodegradation will only take place when it is warm enough [13] the degradation was assumed to take place during two summer months each year for two years. To ensure that biodegradation is not limited by macronutrient deficiencies addition of commercial fertilizer was assumed. It was estimated that a one-time addition of 1.5 kg YaraMila 18-4-14 would provide the necessary nutrients. During the remediation process whey and fertilizer was assumed to be manually added to the contaminated area. Whey additions of 46 L every 14 day over a period of two subsequent summers were assumed.

For the intended bioremediation a total of 2 kg dw whey is needed. Data on cheese and whey production was obtained from Milko dairy in Östersund. To produce 0.9 ton of Edam cheese and 1 ton of dry whey the cheese dairy on average requires 11000 L of milk. Inventory data was gathered to describe the use of average Swedish electricity, water and auxiliary chemicals [55]. Inventory data for milk production, including farming activities, transports and dairy operations was based on a LCI by the Swedish Diary Association.

Since whey is a by-product of cheese production, economic allocation was used giving an allocation of 80 % on cheese and 20 % on whey. In a sensitivity analysis the results were investigated, assuming that all production environmental impact is allocated to the whey. The fertilizer production is described using emissions data on the production of 1 kg of fertilizer. Whey and fertilizer will have to be transported to the contaminated site. Whey is produced in Sweden by Milko AB is transported over a distance of 230 km from Östersund to Gäddede. Fertilizer is obtained from commercial fertilizer store and transported the same distance. Environmental impacts from the transports were calculated using emissions factors assuming a light distribution truck, Euro 2, vehicle manufactured during 1996-2000.

Figure 9 illustrates a simplified model of the life-cycle process flow diagram for the excavation and composting treatment scenario. Assuming a soil density of 1922 kg m^{-3} for the 5 m^3 soil to be treated (over an area of 10 m^2) the calculated mass of contaminated soil was 9.6 ton.

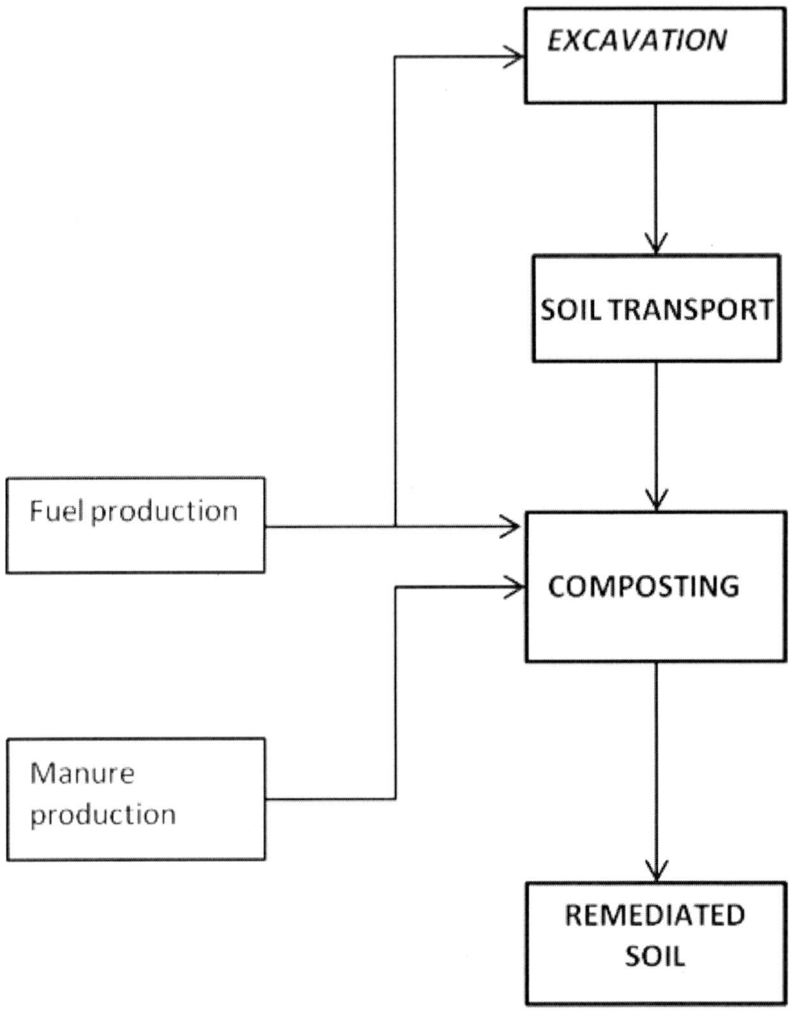

FIGURE 9. The life cycle inventory model of the excavation and composting scenario (From [53]).

It was assumed that the Gäddede case would need a 14 ton excavator for 15 minutes and use 2 L of diesel. Production of the diesel oil was described by inventory data from [56]. The excavated soil will then have to be transported 230 km from Gäddede to Östersund to a composting plant. Impacts from the transport is described using emission factors for a medium sized distribution truck with a payload of 8.5-14 ton, Euro class 2, produced between 1996 –2000.

At Gräfsåsen, the waste treatment and landfill in Östersund, the contaminated soil is mixed with horse manure and placed in windrows in a string-like arrangement. Over the period 2005 to 2009 some 2320 tons of contaminated soil and 1560 tons of manure have been received at the composting plant in Gräfsåsen, giving an average manure use of 0.67 ton/ton. For the manure transport we assume the same emission factors as for the soil transport and an average transport distance of 50 km. The manure generations are totally allocated to the horse keeping. Every one month during the summer the windrows are mixed using a mixer mounted on a wheel loader. For this process half the environmental impact from the excavation in Gäddede was assumed. Use of the treated compost as cover of the land fill is not assumed to be connected to any environmental impacts. The results from the whey treatment scenario and the composting scenario for the four impact categories GWP, AP, EP and POCP is shown in Figure 10, together with the results from the sensitivity analysis where all environmental impacts from milk production and the cheese dairy are allocated to the whey.

For the GWP, AP and POCP impact categories the whey scenarios give lower results than the excavation and off-site composting scenario, whereas the EP category shows higher results. The EP result for the whey treatment is totally dependent on the whey production chain and originates at the farm (such as nutrient runoff, ruminating cattle, vaporization of ammonia from manure) or by the production of materials used at the farm (such as fertilizers). The milk farm activity is the most contributing in the whey scenario for all four impact categories shown.

When comparing the allocated and the unallocated whey treatment scenarios the milk-cheese chain has a rather large impact, especially for eutrophication (EP in diagram). It should be noted that the transport of whey and fertilizer for the whey scenario is described with emission factors of a truck transport, that means it is assumed to have the necessary whey and fertilizer as part of the load of the truck and the off-loading will be part of a delivery route where other parts of the load will be off-loaded in other places. Arranging good logistics is thus important both for the whey treatment and for the excavation and transport to off-site composting.

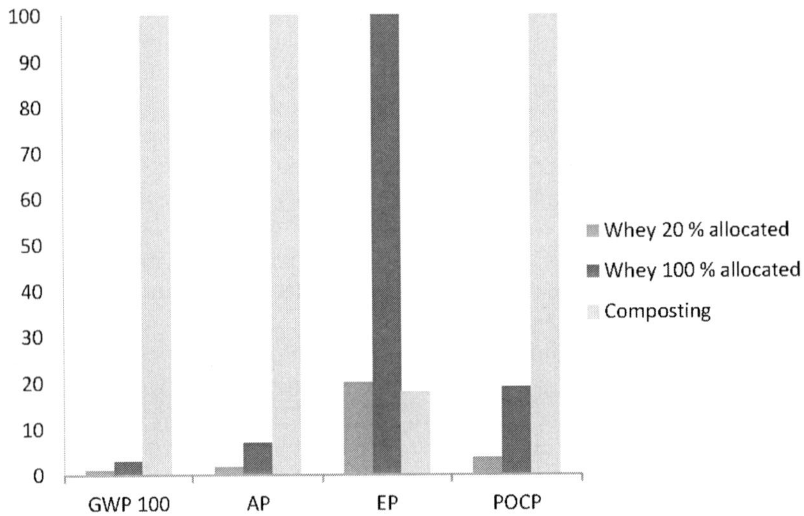

FIGURE 10. Results from the LCA models for the in-situ whey treatment scenario – with two levels of allocation of the whey production – and the excavation and off-site composting scenario. GWP 100, Global Warming Potential, 100 year time horizon; AP, Acidification Potential; EP, Eutrophication Potential; POCP, Photochemical Ozone Creation Potential. The allocation scenarios mean that respectively 20 and 100 % of the environmental effect of the cheese production was allocated to the whey. Normalized data, the highest result in each category is set to 100% (From [53]).

5.2.1.2. Results of bioremediation field trials and microcosm experiments using whey treatment of the contaminated soil in gäddede

The result from whey treatment of the polluted site during a 3 ½ months field experiment was not promising. The experiment was initiated 10 May when the snow had melted and the soil temperature at 18 cm depth was 5 °C. The polluted area was given an initial dose of 545 g m^{-2} NPK fertilizer (18 % N; 4 % P; 10 % K) followed by bi-weekly doses of 3.2 L m^{-2} milk whey [51]. During the course of the summer the soil temperature rose to a maximum of 15 °C in the beginning of August and dropped thereafter. When the experiment was ended in mid August no detectable reduction of soil hydrocarbons was observed.

In parallel to the field experiment, microcosm studies were carried out on soil samples, from the vicinity of the contaminated site that had been artificially contaminated with diesel fuel. In the microcosms the soil was subjected to the same treatment as in the field study with the only difference that the soil was kept at 22 °C. During these conditions the diesel range aliphatics was almost completely degraded within 60 days. This is in accordance with previous studies on hexadecane that showed significant difference in degradation rates at 7 and 22 °C respectively [13]. Soil temperature is probably the most limiting factor for in-situ bioremediation in cold regions and the short period of warmth during 3 summer months is rarely enough to bring the soil temperature to levels that give sufficient rates of biodegradation. A possible way to circumvent this problem would be to heat up the soil with heat injection. Preliminary calculations carried out at our department reveal that the electric power needed to heat up a 100 x 100 x 2 m contaminated volume of polluted soil from 8 to 25 °C by e.g. a sufficient number of heat rods, would be 166 kW during a one month period corresponding to an input heating energy of 98 x 10^3 kWh. The calculation assumed that the soil surface was insulated and that there were no vertical and downwards heat losses during the heat-up period. The energy needed to maintain the temperature at 25 °C for another 11 month period assuming horizontal and vertical (downwards) heat losses would amount to 235 x 10^3 kWh. Total energy need for a one year period is then 333 x 10^3 kWh. The following input data was used: A linear temperature drop distance of 9.1 m to ambient temperatures, heat conductivity 1.5 W m^{-1} K^{-1}, specific heat capacity 800 J kg^{-1} K^{-1} and soil density 1300 kg m^{-3}. These are rough calculations based on some simplified assumptions about heat transfer in soil but may still give an estimate of the magnitude of the input energy needed for a temperature enhanced in-situ soil bioremediation project in cold climate. Downscaled to the 10 m^2 by 0.5 m of the Gäddede diesel oil plot the corresponding figures would be 27 and 387 kWh respectively for the heating and 11 months maintenance input energy need respectively giving a total energy need for a 12 month period of roughly 414 kWh. Part of this energy need could easily be provided by solar panels. Furthermore, taking into consideration the very rapid increase in degradation rate even at a moderately elevated temperature such as 22 °C it is possible that a successful temperature enhanced remediation of the Gäddede plot could even be carried out within 6 months during a period when solar panels can provide the total need of input energy. Solar panels can easily be moved from one project to another so the technology described here should have a potential for application in in-situ soil bioremediation in remote places and in cold regions.

5.2.2. One-time application of whey to oil spot at a lumberyard in sikås

An experiment was carried out to test a low-cost and low-tech solution for remediation of soil contaminated by waste motor oil in a cold region. A 10 m^2 spot that had been polluted by leakage from a 200 l barrel with used motor oil was chosen for the experiment. This spot would certainly come up low in the priority list for remediation projects due to its location in a remote region in the northern parts of the county of Jämtland and also due to the scale of the pollution in proportions to the costs of traditional remediation practices. Such spots are however far from uncommon and the total impact from the highly toxic substances found in waste oil, such as carcinogenic PAH, may be significant. The polluted area was within a timber terminal and the soil was covered with highly compacted sand and gravel. The amount of waste oil that had leaked from the corroding barrel was estimated to 100 L. The permeability of the soil was very low and the oil had penetrated the soil to a depth of approximately10 mm or less. Samples from the plot showed that the oil mainly consisted of high molecular weight aliphatics and aromatics C_{16}-C_{35}.

The oil spot was divided in two parts separated by a shallow cut made by the shovel that can be seen in the top picture in Figure 11. 200 L of fresh milk whey was applied to the left side of the spot. The plot had a slight slope to the left and to the front in the picture. Approximately 50 % of the whey was lost by runoff in these directions due to the low permeability of the soil but some whey crossed the notch made by the shovel and penetrated the right part of the plot. The dashed line in the figure separates the treated left part of the plot from the untreated, right part of the plot. To make sure that lack of macronutrients would not limit the biodegradation a total of 6 kg commercial granulated fertilizer (N/P/K 18/4/10) was evenly applied to the whole spot (treated and control).

The photo from April 2007 was taken just before treatment. The snow had just melted and the average air temperature during April is around 2 °C at this latitude and rose to 15 °C during July, which is also normal for these latitudes. It is plausible that the soil surface could reach higher temperatures during a few hot summer days and that the dark colour of the oil would enhance heat absorption in such cases. Nevertheless, one year later, in May 2008, no difference whatsoever could be observed between treated and untreated spot, see middle picture in Figure 11 (note snow remaining in May).

The bottom picture in Figure 11 shows the plot two years and two months after application of whey to the left part. There is a remarkable difference between the darkness of the sectors and clearly, degradation of the oil in the treated part had commenced. The person in the picture indicating the divine between the light and the dark side of the plot had no a priori knowledge of the treatment strategy. Soil samples taken from the sectors indicate that the TPH concentration in the left part is half the concentration in the untreated part.

FIGURE 11. One time treatment of left part of oil spot with 200 l fresh milk whey.

The long term effect of a one-dose of the liquid amendment (whey) is remarkable. It is worth to note that the summer 2008 was hotter than normal in the region and the monthly average temperature anomaly in July was + 1.5 °C according to the Swedish Meteorological and Hydrological Institute, www.smhi.se/klimatdata/meteorologi. Clearly the low soil temperature in this, just as in the above presented case, severely limits the rate of biodegradation, but when conditions becomes thermodynamically more favourable, the microorganisms are able to benefit from the extra carbon and micronutrients provided by the amendment, even after a one year incubation time.

6. OPPORTUNITIES AND FUTURE PREDICTIONS OF SOIL REMEDIATION

Remediation of contaminated soils is presently a costly venture but as new technologies emerge to treat soils in-situ, expenses can diminish considerably. Technology optimists like Ray Kurzweil forecast that pollution will discontinue to be produced in the future and predicts a set of technologies (including biotechnologies) that will be used to clean up the previously generated contamination. Kurzweil expresses a strong faith in nanotechnology that he believes will provide useful features for remediation of contaminants including; improved catalysis, chemical and atomic bonding, sensing, and mechanical manipulation. Currently, extensive research is undertaken to develop nanoproduced crystalline materials for catalysts that have the potential to improve chemical yields, reduce toxic by-products, and remove contaminants [59].

To overcome some of the difficulties to degrade recalcitrant compounds research on combined biological/abiological degradation technologies might provide useful technologies. Pre-treatments by strong oxidizing agents such as ozone, Fenton´s reagent, potassium permanganate, ferrate etc. have shown promising results and Singh and Ward predict that combined chemical/biological treatment of pollutants will play an important role in the future [45]. Strong chemical oxidants may however severely affect soil microbial populations and furthermore requires input of energy and material resources.

The use of biocatalysts is another interesting method whose practical applications have been predicted to increase in the future. This kind of enzymatic remediation can potentially overcome difficulties such as low survival rate of whole cell organisms and generation of toxic by-products that are associated with bioaugmentation. Research on enzymatic bioremediation is at its infancy and few field studies exist. The major limitation for large-scale applications is currently the prohibitive cost for the production of the enzymes. Molecular tools are increasingly explored to overcome that obstacle which could decrease the costs significantly in the future [60]. As we have demonstrated in this chapter the use of readily available organic waste products as amendments and the appropriate application of principles of ecological engineering is potentially a way to make bioremediation more sustainable.

7. CONCLUSION

It is a common belief that soil remediation is a sustainable industry but many remediation projects use vast amounts of energy and often physically damage the remediation site by removing its topsoil. Ecological Engineering can potentially make remediation projects more sustainable. The large embodied energy in many organic pollutants may be used as a driver of a self-designing in-situ remediation process by indigenous microorganisms in the soil.

The intrinsic capacity of soil microbial communities to initiate and accelerate degradation of soil pollutants rely on the fundamental thermodynamic driving force towards complete degradation of target organic pollutants to low molecular-weight inorganic compounds such as carbon dioxide and water. The degradation process releases quantities of useful "free" energy that can be used by soil microorganisms in their metabolism. Figure 1 illustrates this together with important abiotic and biotic degradation pathways and intermediate storages. Table 1 gives some numbers to illustrate the theoretical "energy gain" by degrading organisms as they catabolise representative soil pollutants. Turnover times for organic pollutants are typically short in the atmosphere, e.g. hours-days [61] but considerably longer in soil, e.g. months-decades [7, 44] mainly because of lack of photochemistry in the subsurface of the soil. The slow biodegradation kinetics in soil can however be considerably improved by the use of locally available waste materials, such as whey and fermented whey that stimulate microbial activity, figure 2 and table 2. There are a number of other waste materials that have been successfully tested by us [12-14, 18] or others [26-28] or remain to be tested (figure 3). Remediation strategies based on amendments from locally available by-products can be tailored to suit site-specific conditions and be adapted to a wide range of climates and contexts as illustrated in the case studies presented in this chapter.

Concepts of Ecological engineering, i.e. measures to stimulate the self-organizing capacity of ecosystems by the use of organic amendments, conservation of non-renewable energy resources and ecosystem conservation by in-situ treatment with locally available waste resources have the potential of making soil remediation projects in low priority sites in developing countries and in remote regions more sustainable than conventional methods that rely on costly and less environmentally benign technology. Life cycle assessment (LCA) was used to evaluate the different remediation technologies from a systems perspective. Results of the screening LCA study of the whey treatment method versus a conventional dig-transport-compotation scenario at a remote diesel oil polluted site in Northern Sweden (See map, Gäddede, figure 7) are shown in figure 10. Clearly, the in-situ whey treatment has an overall lower environmental impact compared to the transport-composting method. The most contributing activity for the composting scenario in all of the environmental impact categories studied was the transport of the excavated soil to the composting site. Thus the transport distance is of large importance. For sites where the transport distance is much shorter, the composting scenario would come out much better. This outcome is in accordance with what has been earlier reported by Suèr et al [57] and Toffoletto et al [58] that transportation and transport distance is a key element when choosing between in-situ and ex-situ remediation technology.

In this chapter we have demonstrated that Ecological engineering in concurrence with the application of basic thermodynamic principles and kinetic modelling of data from laboratory experiments can provide useful guidelines for the development of energy conserving and economically feasible bioremediation technologies. Such technologies have the potential of making in-situ bioremediation of polluted soil in low priority sites in developing countries and in remote regions more sustainable.

8. ACKNOWLEDGEMENTS

We gratefully acknowledge the financial support of the European Union Regional Development Fund. We are indebted to Morgan Fröling and Joseph Akambih Tajam for conducting the LCA screening analysis of the whey method and to Anja Wedermeyer and Sylvia Vespermann for calculating energy need for temperature-enhanced bioremediation in cold regions.

REFERENCES

1. Mudhoo A, Mohee R. Elements of Sustainability and Bioremediation. In: Bioremediation and Sustainability. John Wiley & Sons, Inc.; 2012..s. 1-41.
2. Lynch JM, Moffat AJ, Lynch JM, Moffat AJ. Bioremediation - Prospects for the Future Application of Innovative Applied Biological Research. Ann. Appl. Biol.;146(2:217-21.
3. Vidali M. Bioremediation, An Overview. Pure and Applied Chemistry 2001;73(7):1163-72.
4. Mitsch WJ. What is Ecological Engineering? Ecological Engineering 2012;45(0):5-12.
5. Mitsch WJ, Jørgensen SE. Ecological Engineering and Ecosystem Restoration. Wiley; 2004.
6. Green AE, Voordouw G. Microbial Community Dynamics During Bioremediation of Hydrocarbons. In: Singh A, Ward OP. (eds) Biodegradation and Bioremediation. Berlin: Springer-Verlag; 2004.
7. Riser-Roberts E. Remediation of Petroleum Contaminated Soils. Boca Raton: Lewis Publishers; 1998.
8. Peng RH, Xiong AS, Xue Y, Fu XY, Gao F, Zhao W, et al. Microbial Biodegradation of Polyaromatic Hydrocarbons. FEMS Microbiology Reviews 2008;32(6):927-55.
9. Solomons SD, Qin D, Manning M, Chen Z, Marquis M, Averyt KB, et al. Technical Summary. In: Climate Change 2007: The Physical Science Basis. Contribution of Working Group I to the Fourth Assessment Report of the Intergovernmental Panel on Climate Change. Cambridge, United Kingdom and New York, USA: Cambridge University Press; 2007.
10. Mohn WW. Biodegradation and Bioremediation of Halogenated Organic Compounds. In: Sing A, Ward AP. (eds) Biodegradation and Bioremediation. Berlin, Heidelberg, New York: Springer 2004..p. 125-48.
11. Adriano DC, Bollag JM, Frankenberger WT, Sims RCJ. (eds) Bioremediation of Contaminated Soil. Madison: American Society of Agronomy; 1999.
12. Östberg TL, Jonsson AP, Bylund D, Lundström US. The Effects of Carbon Sources and Micronutrients in Fermented Whey on the Biodegradation of n-Hexadecane in Diesel Fuel Contaminated Soil.

International Biodeterioration & Biodegradation 2007;60(4):334-41.

13. Östberg TL, Jonsson AP, Lundström US. Enhanced Degradation of n-Hexadecane in Diesel Fuel Contaminated Soil by the Addition of Fermented Whey. Soil and Sediment Contamination: An International Journal 2007;16(2):221-32.

14. Östberg TL, Jonsson AP, Lundstrom US. Accelerated Biodegradation of n-Alkanes in Aqueous Solution by the Addition of Fermented Whey. International Biodeterioration & Biodegradation 2006;57(3):190-4.

15. Brunner W, Focht DD. Deterministic Three-Half-Order Kinetic Model for Microbial Degradation of Added Carbon Substrates in Soil. Applied and Environmental Microbiology 1984;47(1):167-72.

16. Mohn WW, Stewart GR. Limiting Factors for Hydrocarbon Biodegradation at Low Temperature in Arctic Soils. Soil Biology & Biochemistry 2000;32:1161-72.

17. Brown AM. A Step-by-step Guide to Non-linear Regression Analysis of Experimental Data Using a Microsoft Excel Spreadsheet. Computer Methods and Programs in Biomedicine 2001;65(3):191-200.

18. Jonsson AP, Östberg TL. The Effects of Carbon Sources and Micronutrients in Whey and Fermented Whey on the Kinetics of Phenanthrene Biodegradation in Diesel Contaminated Soil. Journal of Hazardous Materials 2011;192(3):1171-7.

19. Reddy N, Yang Y. Biofibers From Agricultural Byproducts for Industrial Applications. Trends in Biotechnology 2005;23(1):22-7.

20. Ulloa JB, van Weerd JH, Huisman EA, Verreth JAJ. Tropical Agricultural Residues and their Potential Uses in Fish Feeds: the Costa Rican Situation. Waste Management 2004;24(1):87-97.

21. Ruane J, Sonnino A. Agricultural Biotechnologies in Developing Countries and their Possible Contribution to Food Security. Journal of Biotechnology 2011;156(4):356-63.

22. Aregheore EM. Chemical Composition and Nutritive Value of Some Tropical By-product Feedstuffs for Small Ruminants — In Vivo and In Vitro Digestibility. Animal Feed Science and Technology 2000;85(1–2):99-109.

23. Singh H. Mycoremediation: Fungal Bioremediation. Wiley; 2006.

24. Eggen T. Application of Fungal Substrate from Commercial Mushroom Production - Pleuorotus Ostreatus - for Bioremediation of Creosote Contaminated Soil. International Biodeterioration & Biodegradation 1999;44(2-3):117-26.

25. Jayasinghe P, Hawboldt K. A Review of Bio-Oils from Waste Biomass: Focus on Fish Processing Waste. Renewable and Sustainable Energy Reviews 2012;16(1):798-821.

26. Dzul-Puc JD, Esparza-García F, Barajas-Aceves M, Rodríguez-Vázquez R. Benzo(a)pyrene Removal from Soil by Phanerochaete Chrysosporium Grown on Sugarcane Bagasse and Pine Sawdust. Chemosphere 2005;58(1):1-7.

27. Velázquez-Fernández JB, Ramírez-Sandoval M, Domínguez-Ojeda D, Martínez-Rizo AB. Biodegradation and Bioremediation of Organic Pesticides. Intech: 2012..http://www.intechopen.com/books/pesticides-recent-trends-in-pesticide-residue-assay/biodegradation-and-bioremediation-of-organic-pesticides (acessed 22 October 2013)

28. Al-Hadhrami MN, Lappin-Scott HM, Fisher PJ. Studies on the Biodegradation of Three Groups of Pure n-Alkanes in the Presence of Molasses and Mineral Fertilizer by Pseudomonas Aeruginosa. Marine Pollution Bulletin 1997;34(11):969-74.

29. Steiner C, Das KC, Garcia M, Förster B, Zech W. Charcoal and Smoke Extract Stimulate the Soil Microbial Community in a Highly Weathered Xanthic Ferralsol. Pedobiologia 2008;51(5–6):359-66.

30. Mohan D, Pittman CU, Steele PH. Pyrolysis of Wood/Biomass for Bio-Oil: A Critical Review. Energy & Fuels 2006;20(3):848-89.

31. Zulkarami B, Ashrafuzzaman M, Husni MO, Ismail MR, Zulkarami B, Ashrafuzzaman M, et al. Effect of Pyroligneous Acid on Growth, Yield and Quality Improvement of Rockmelon in Soilless culture.(Report). Australian Journal of Crop Science;5(12):1508.

32. Spokas KA, Novak JM, Stewart CE, Cantrell KB, Uchimiya M, DuSaire MG, et al. Qualitative Analysis of Volatile Organic Compounds on Biochar. Chemosphere;85(5):869-82.

33. Chidumayo EN, Gumbo DJ, Chidumayo EN, Gumbo DJ. The Environmental Impacts of Charcoal Production in Tropical Ecosystems of the World: A synthesis. Energy for Sustainable Development.

34. Zandersons J, Gravitis J, Kokorevics A, Zhurinsh A, Bikovens O, Tardenaka A, et al. Studies of the Brazilian Sugarcane Bagasse

Carbonisation Process and Products Properties. Biomass and Bioenergy;17(3):209-19.

35. FAO. Simple Technologies for Charcoal Making. Food and Agriculture Organization of the United Nations (Mechanical Wood Products Branch); 1983.

36. Loo AY, Jain K, Darah I, Loo AY, Jain K, Darah I. Antioxidant Activity of Compounds Isolated from the Pyroligneous Acid, Rhizophora Apiculata. Food Chemistry;107(3):1151-60.

37. Kadota M, Niimi Y, Kadota M, Niimi Y. Effects of Charcoal with Pyroligneous Acid and Barnyard Manure on Bedding Plants. Scientia Horticulturae;101(3):327-32.

38. Yoshimura H, Washio H, Yoshida S, Seino T, Otaka M, Matsubara K, et al. Promoting Effect of Wood Vinegar Compounds on Fruit-Body Formation of Pleurotus Ostreatus. Mycoscience;36(2):173-7.

39. Pant A, Radovich TJK, Hue NV, Arancon NQ. Effects of Vermicompost Tea (Aqueous Extract) on Pak Choi Yield, Quality, and on Soil Biological Properties. Compost Science & Utilization 2011;19(4):279-92.

40. St. Martin CCG, Brathwaite RAI. Compost and Compost Tea: Principles and Prospects as Substrates and Soil-Borne Disease Management Strategies in Soil-less Vegetable Production. Biological Agriculture & Horticulture 2012;28(1):1-33.

41. Ghorbani R, Koocheki A, Jahan M, Asadi GA. Impact of Organic Amendments and Compost Extracts on Tomato Production and Storability in Agroecological Systems. Agron. Sustain. Dev. 2008;28(2):307-11.

42. Scheuerell SJ, Mahaffee WF, Scheuerell SJ. Variability Associated with Suppression of Gray Mold (Botrytis cinerea) on Geranium by Foliar Applications of Nonaerated and Aerated Compost Teas. Plant Disease;90(9):1201-8.

43. Naidu YM, Sariah Kadir, Jugah; Siddiqui, Yasmeen. Microbial Starter for the Enhancement of Biological Activity of Compost Tea. International Journal of Agriculture and Biology 2010;12(1):51-6.

44. Alexander M. Biodegradation and Bioremediation. 2 ed. San Diego: Academic Press; 1999.

45. Singh A, Ward OP. Biodegradation and Bioremediation. Berlin: Springer; 2004.

46. Tognetti C, Laos F, Mazzarino MJ, Hernandez MT, Tognetti C, Laos F, et al. Composting vs. Vermicomposting: A Comparison of End Product Quality. Compost Sci. Util.;13(1):6-13.

47. Sinha RK, Agarwal S, Chauhan K, Chandran V, Soni BK, Sinha RK, et al. Vermiculture Technology: Reviving the Dreams of Sir Charles Darwin for Scientific Use of Earthworms in Sustainable Development Programs. Technology and Investment;1(3):155.

48. González-Siso MI. The Biotechnological Utilization of Cheese Whey: A Review. Bioresource Technology 1996;57(1):1-11.

49. Crawford R. Bioremediation. In: Dworkin M, Falkow S, Rosenberg E, Schleifer K-H, Stackebrandt E. (eds.) The Prokaryotes. Springer New York; 2006..p. 850-63.

50. Eddleston M, Karalliedde L, Buckley N, Fernando R, Hutchinson G, Isbister G, et al. Pesticide Poisoning in the Developing World—A Minimum Pesticides List. The Lancet 2002;360(9340):1163-7.

51. Vilches AP, Bylund D, Jonsson AP. Enhanced Natural Biodegradation of Diesel Fuel Contaminants in Soil by Addition of Whey and Nutrients. In: Proceedings of Linnaeus ECO-TECH ´10; November 22-24..2010; Kalmar, Sweden; 2010.

52. Naturvårdsverket. Åtgärdslösningar -Erfarenheter och Tillgängliga Metoder. Sweden: Naturårdsverket; 2006.

53. Tajam JA, Fröling M, Jonsson AP. Small Scale In-situ Bioremediation of Diesel Contaminated Soil - Screening Life Cycle Assessment of Environmental Performance. In: Proceedings of Linnaeus ECO-TECH ´10; November 22-24..2010; Kalmar, Sweden; 2010.

54. Naturvårdsverket. Remediation of Contaminated Areas. Sweden: Naturvårsverket; 2008.

55. Berlin J. Environmental Life Cycle Assessment (LCA) of Swedish Semi-Hard Cheese. International Dairy Journal 2002;12(11):939-53.

56. CPM. SPINE LCI dataset: Diesel Productions. http://cpmdatabase.cpm.chalmers.se (accessed June 2013)

57. Suèr P, Nilsson-Påledal S, Norrman J. LCA for Site Remediation: A Literature Review. Soil and Sediment Contamination: An International Journal 2004;13(4):415-25.

58. Toffoletto L, Deschênes L, Samson R. LCA of Ex-Situ Bioremediation of Diesel-Contaminated Soil (11 pp). Int J Life Cycle Assess 2005;10(6):406-16.

59. Kurzweil R. The Singularity Is Near: When Humans Transcend Biology. Penguin (Non-Classics); 2006.

60. Peixoto RS, Vermelho AB, Rosado AS, Peixoto RS, Vermelho AB, Rosado AS. Petroleum-Degrading Enzymes: Bioremediation and New Prospects. Enzyme Research 2011, Article ID:475193..7 pp. doi 10.4061/2011/475193.

61. Altshuller AP. Lifetimes of Organic Molecules in the Troposphere and Lower Stratosphere. In: Pitts JN Jr., Metcalf RL., Grosjean D. (ed.) Advances in Environmental Science and Technology. New York: John Wiley & Sons; 1980..p181-219.

6

CHAPTER

Heavy Metal Contamination in Brazilian Agricultural Soils due to Application of Fertilizers

Affonso Celso Gonçalves Jr.[1], Herbert Nacke[2], Daniel Schwantes[3] and Gustavo Ferreira Coelho[4]

[1] *Western Paraná State University, Center of Agricultural Sciences, Marechal Cândido Rondon, Brazil*

[2] *University Center Dynamic of Cataracts, Technological Center, Foz do Iguaçu, Brazil*

[3] *Pontifical Catholic University of Paraná, Polytechnic School, Toledo, Brazil*

[4] *Federal University of Paraná, Palotina, Brazil*

1. INTRODUCTION

The Brazilian grain production reaches record productivities each year and among the major crops, stand out the maize (*Zea mays* L.) and soybean (*Glycine max* L.). For these two cultures, the obtaining of the maximum yield has a fundamental importance in the current global scenario [1]. In this relentless pursuit of higher productivities, farmers use technologies related to various areas, such as new forms of fertilizer and pesticide application, different fertilizers, crop breeding, equipment and techniques of planting and harvesting. In order to increase the efficiency in this activity, due to the increased requirement for competitiveness of economic globalization, one of the tools found by farmers is the use of fertilizers containing micronutrients in their crops [2]. Among the various factors of production, the need of balanced and sustainable fertilizers occupies a prominent place, which should include not only the primary and secondary macronutrients, but also micronutrients, which are not considered in routine fertilization by most farmers. Currently, with the advancement of the industries of fertilizers, micronutrients are added directly in the formulation of basic fertilizers ($N:P_2O_5:K_2O$), which can be found in the market of domestic and imported agricultural inputs [3]. Among the several micronutrients used in Brazilian and Global agriculture, zinc (Zn) is a major one, having its need scientifically established over 70 years being its main functions related to photosynthesis, respiration, protein synthesis and plant membrane permeability [2].

Nutrient uptake by crops is influenced by various factors, including climatic conditions such as rain and temperature, genetic differences between varieties, type of soil, nutrient content of soil and various cultural practices.

In the strategy of reducing costs, both in manufacture of fertilizers, as well in croplands, the use of industrial waste for the obtaining of micronutrients has been widely used. These wastes, when used appropriately and rationally present themselves as a good solution for agriculture, but their improper use can cause serious damage to soil, plants and human beings. Several researches are touting in scientific works, facts that generate severe criticism to the fertilizer industry, mainly because the involvement of these industries in operations that endanger the environment, where the main factor is the use of raw materials containing toxic metals such as cadmium (Cd), lead (Pb) and chromium (Cr) in the manufacture of fertilizers [4]. In that way, the fertilizers used to supply micronutrients have often, in their composition, in addition to the desirable elements, toxic metals such as Cd, Pb and Cr [5,6].

The purpose of this chapter is to assist in developing a more adequate environmental legislation, that ensures a sustainable environment and a production of quality food. In that way, this chapter highlight the lack of criteria, in the current Brazilian environmental legislation, regarding the allowed limits of toxic heavy metals in agricultural soils and in the use of fertilizers contaminated with heavy metals in agriculture, as well, demonstrate that the use of these contaminated inputs can cause degradation of agricultural soils, making food production unviable at these locations. Furthermore, it was made the study of three cases about the existence of fertilizers contaminated with toxic metals in Brazilian agriculture and the impact caused by its use. The chapter also presents some strategies for recuperation and remediation of soils contaminated with toxic heavy metals.

The presence of heavy metals in fertilizers is a extremely important subject, since investigations in Brazil, coordinated by prosecutors in São Paulo, show evidence of entry loads containing toxic waste through irregular importation and smuggling, several times in Brazilian ports. These investigations shown that companies and manufacturers of agricultural inputs, are importing these toxic chemicals from United States, Canada, Mexico, Spain, Holland and England to be used as raw material in the manufacture of fertilizers to supply micronutrients.

Brazil presents itself as one of most promising nations of contemporary and future world, however, for the effective consolidation of its presence among the great powers is extremely necessary maintain the sustainability of its agroecosystem. Therefore, the Brazilian agricultural activity should be developed without damaging the environment and thus continue to grow in a rational and appropriate way, avoiding harm to future generations. In the world of agribusiness, which needs to be highly competitive, concepts such as traceability, environment and risk assessment are fundamental starting points for those who want to establish and grow.

2. POTENTIAL CONTAMINATION OF SOIL BY AGRICULTURAL ACTIVITIES

Due to its continental dimensions, Brazil has a great diversity of soil types in different regions of the country. Most of these soils have limitations for agriculture, characteristic of tropical regions such as acid, deep and highly weathered soils, with low nutrient availability and high exchangeable acidity (Al^{3+}), resulting in a low cation exchange capacity and low availability of P [7].

These factors highlight the need for adopting management strategies that aim to increase the chances of good crop of the culture when grown in soils of low fertility, giving them the ability to meet the nutritional requirements of these plants. The practices of fertilization and soil amendment aim to make soil suitable for the development of major crops, providing the necessary nutrients for the plants and decreasing the acidity of the soil. These practices are adopted throughout the world, and are based on the prerogative of the need to produce food to sustain the world population that grows rapidly each year.

According to the Food and Agriculture Organization of the United Nations (FAO) [8], in 1980 there were 4.4 billion people on the planet, 5.7 billion in 1995 and 6.9 billion in 2010. Studies estimate that by 2050 there will be 9.4 billion people and in 2200 about 11 billion [9]. Based on this scenario, agricultural activities have expanded greatly since 1945, currently holding approximately 38% of the land surface. Between 1985 and 2005, the acreage increased about 7%, generating large increases in food production [10].

This pressure on agricultural activities promotes an increasing fertilizer consumption per each year that goes around the globe (Figure 1). Studies show that in the early 60's used to be about 146 million tons of fertilizer, increasing in 2010 to 176 million of tons, demonstrating a significant increase in the consumption of these inputs [11]. Also according to the International Fertilizer Association (IFA) [11], Brazil is the fourth largest consumer of mineral fertilizers ($N{:}P_2O_5{:}K_2O$) in the world, only behind China, India and the United States. In recent years the consumption of nutrients by Brazilian agriculture has increased considerably being that in 1961 Brazil was only the 25th world consumer, however in the 1990, the country was already appearing as the seventh in the rankings. Regarding to micronutrients, there is an increase in consumption of 13.3 times in the period 1990-2003 [12].

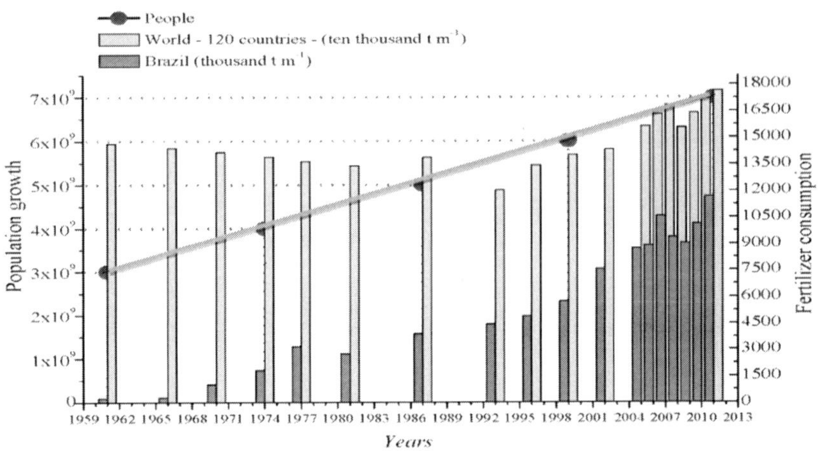

FIGURE 1. Population growth and consumption of fertilizers in Brazil and in the world [11]

However the use of these products is often not done correctly. The plan for fertilizer recommendation involves planning for the collection of soil samples, soil chemical analysis and recommendation by an appropriate professional, something that many producers do not perform, often using pre-determined recommendations without professional assistance. Another problem is the lack of crop rotation and the use of highly productive cultivar which, according to reference [13], require intense macronutrient application, resulting in micronutrient deficiencies in the soil.

Overuse of fertilizers can also provide soil contamination, it increases the concentration of a particular element or substance of environmental interest above its naturally occurring in the soil [7]. This contamination is related to eutrophication (excess nutrients), in this case the major nutrients to be considered are nitrogen (N) and phosphorus (P). Another concern is the availability of harmful elements such as toxic metals. These elements can express their pollutant potential directly on soil organisms, due to the availability to plants in phytotoxic levels, plus the ability to transfer to the food chain through the own plants, or by contamination of soil and water resources [14].

The presence of harmful elements in fertilizers is worrying because until the 70's, most fertilizers were obtained directly from ores found in nature, and although they also contained toxic metals, the beneficiation process of this material was less offensive from the environmental standpoint, because the contaminant levels were very low [6]. Recently, the shortage of mineral resources and modern extraction methods required large investments in equipment and technology, creating a technical difficulty and the pursuit of lower costs of production, which eventually encouraged companies to seek a cheaper solution such as using industrial waste, that may be feasible since it does not contaminate the environment. However, according to the Environmental Sanitation Technology Company of São Paulo State (CETESB) [15], in São Paulo, until the end of 2012, 4572 contaminated areas were found, and approximately 4% of these areas were contaminated with industrial waste, and among the contaminants, toxic metals are in 4th place, behind Polycyclic aromatic hydro carbons contaminants (PAHs), aromatic solvents and liquid fuels.

Between the sources of soil contamination from waste, there is the contamination of sewage sludge, that even though has a variable composition, with macro, micro nutrients and organic matter, their use may be limited by the presence of pathogens and toxic metals [16,17]. Studies evaluating the use of sewage sludge in sugarcane found that the doses used increased levels of chromium, nickel (Ni), Pb and zinc (Zn) in the soil [18].

Reference [19] studying the availability of Cd in dystrophic soil submitted to the application of sewage sludge, observed that at the first hour of contact between the sewage sludge and soil occurred transference of Cd to the solution. After this interval, the Cd concentration in the solution decreases, occurring stabilization in the following hours, due to adsorption and ion exchange which act providing less metal. According to the authors, in all observed time intervals, the levels of Cd were higher than reference values for CETESB (<0.5 mg kg^{-1}) [20], which characterize the contamination of the soil by sewage sludge.

The use of pig manure as fertilizer is also a practical often carried out mainly in south of Brazil, however, some studies have shown risks as to its overuse. Reference [21] evaluated the accumulation of metals in the soil under the systematic application of swine manure, and found an increase in the availability of copper (Cu), zinc and manganese (Mn). The authors also claim that in these cases there is a need to monitor these elements over time, so that the total contents do not exceed the critical values. In study conducted in China [22], the application of manure in soil for 16 years resulted in higher accumulation of Cd. One of the major sources of metals such as Pb, Ni, Cr and Cd in pig manure is mineral supplied by commercial feed [21].

The metals are also components of many pesticides, mainly Cu, Zn and Pb, which cause high soil contamination by these elements [23], another case is the use of Cu, that has been used since 1917 for the Bordeaux mixture [24]. In a study conducted in Caetés watershed in the county of Paty's Ensign, Rio de Janeiro, Brazil, it was aimed to evaluate the soil contamination by toxic metals, from the intensive use of agrochemicals, in which the authors found an increase in the total levels of metals in soil, stream water and dams. The soil metal concentration not reached critical established levels, however the result was worrying about the water stream and dam, which showed total contents of Cd, Mn and Pb above maximum standards established by the Brazilian legislation [23].

In England and Wales, reference [25] developed an inventory on the input of metals in agricultural soils. Forms of contamination are the most diverse, the main sources are generated by atmospheric deposition, sewage sludge, manure, inorganic fertilizers, lime, agrochemicals, irrigation water and industrial waste.

All these reports point to industrialization as the main cause for the large amount of metal dispersed in the environment, and human as the protagonist of all this pollution, because usually mankind seeks the easiest way supply their needs while generating countless volumes of waste. These actions put us in the position of predator. And thus there is a need of legislations that seek to limit the presence of these contaminants in the soil. In Table 1, the international maximum levels of metals allowed in soils of various places show what would be, in theory at least, a concern of many countries in the world about metal contamination.

TABLE 1. Maximum permitted levels of toxic metals in the soil in some countries

Countries	Toxic Metals (mg kg⁻¹)				
	As	Cd	Pb	Cr	Hg
(*) European Comunities	-	1-3	50-300	-	1-1,5
(**) United States	14	1.6	50-300	1500	0.5
(**) Poland (1977 – 1993)	30	1-3	70-150	50-80	5
(**) Germany (1984 – 1992)	20	3 – 1.5	100	100	2
(**) United Kingdon	10	3-15	500-2000	-	-
(**) Austria	50	5	100	100	5

Source: * [26], ** [27]

Currently in Brazil, the legislation that deals about soil contamination is Resolution n°. 420 of 2009 of the National Council of Environment (CONAMA) [28], which aims to establish the criteria and values that guide of soil quality as the presence of chemical substances, among them the toxic metals. The resolution establishes three categories of guiding values which are: The Reference Value Quality (RVQ) which is the concentration of a substance that will define the natural quality of the soil, based on previous studies for each Brazilian state, the Prevention Value (PV), which refers to the concentration limit of a substance in the soil, and the Investigation Value (IV) which is the concentration of a substance in the soil of which there are potential risks, direct or indirect, to human health. Table 2 shows the values set for preventing and detecting toxic metals in soil.

TABLE 2. Values guiding to the presence of toxic metals in Brazilian soils in accordance with Resolution n° 420 of 2009 of CONAMA [28]

Metal	PV	IV		
		Agricultural Area	Residential	Industrial
As (mg kg⁻¹)	15.0	35.0	55.0	150.0
Cd (mg kg⁻¹)	1.3	3.0	8.0	20.0
Pb (mg kg⁻¹)	72.0	180.0	300.0	900.0
Cr (mg kg⁻¹)	75.0	150.0	300.0	400.0
Hg (mg kg⁻¹)	0.5	12.0	36.0	70.0

PV: Prevention Value; IV: Investigation value

However, the Resolution n°. 420 is being severely questioned by authorities and researchers, since it present several misconceptions that violate the Federal Constitution regarding the "preservation, enhancement and restoration of environmental quality propitious to life" (this issue will be addressed with greater emphasis on section 5 of this chapter).

3. FERTILIZERS *VERSUS* CONTAMINANTS

There are three different categories of agricultural fertilizer: macronutrient fertilizer to supply major nutrients (N, P and K); macronutrient fertilizer to supply secondary nutrients [calcium (Ca), magnesium (Mg) and sulfur (S)] and micronutrients [Cu, Zn, Ni, Mn, iron (Fe), molybdenum (Mo), boron (B), chlorine (Cl), sodium (Na), cobalt (Co) and silicon (Si)]. All of these nutrients can be applied through simple fertilizer (only one nutrient) or compound form (formulations with more than one nutrient).There is a further classification based on the essentiality of nutrients, where metals may be classified as essentials, such as B, Cu, Fe, Mn, Mo and Zn; beneficial, such as Co, Ni and V; and toxic, such as Cd, Cr, Hg and Pb. Thus, it should be noted that the essential and beneficial elements can become toxic at high concentrations [30].

According to reports developed by the United States Environmental Protection Agency (USEPA) [31] and The Weinberg Group (requested by The Fertilizer Institute) [32], the main metal contaminants that are found in agricultural fertilizers are Cd, Cr, Co, Cu, Pb, Mo, Ni, Zn, arsenic (As), mercury (Hg) selenium (Se), vanadium (V) and radio 266 (Ra). It can be observed in this list, that among the metals commonly found as contaminants in fertilizers, occur the presence of micronutrients, required for plant development.

According to USEPA [31], among the fertilizers used to supply macronutrients, the phosphate shown the highest levels of toxic metals (especially Cd), followed by nitrogen and potassium fertilizers. With respect to micronutrients, the sources used to obtain it are varied, resulting in the occurrence of fertilizers from low to high levels of metals. In general, micronutrient sources have higher level of contaminant in comparison to sources of macronutrients, on the other hand, are applied larger amounts of fertilizer to supply the macronutrients, therefore, the possibility of soil and environment pollution with the use of any type of contaminated fertilizer is real and should be seriously studied.

The presence of contaminants in fertilizers occurs most of the time with the use of industrial waste or low cost sources for raw materials. These sources, when treated and used properly and rationally, present themselves as a good alternative to agriculture, however its improper use can cause serious damage to the soil, plants and the human beings.

This fact becomes even more important in the case of fertilizers to supply micronutrients, which often are obtained from raw materials originated from industrial wastes, which may show, in addition of the desirable elements, toxic metals. In this situation, the benefits are highly attractive for the fertilizer industries, once these companies acquire low quality raw material at low costs, without generating any type of waste, incorporating everything in their products.

One of the main cases regarding the presence of contaminants in fertilizers happened in the 90s in the United States, where many farmers and social actors reported the possibility of the presence of toxic metals in these inputs. The impact generated and population pressure led to the development of several research regarding the contamination of fertilizers and soil pollution, culminating with the publication by the U.S. government, of the first law regulating the content of toxic elements in the United States in 1998. From these studies and based on Canadian law, in 2002, the USEPA published the federal law that regulates the production of fertilizers to supply Zn from industrial waste [33].

Another case reported in the literature is the presence of contaminants in raw materials from China. In a publication from the Washington Department of Ecology (WDE) [34] it was reported the presence of Cd in raw materials for the manufacture of fertilizers, with samples showing more than 20% of Cd, i.e., more than 200,000 ppm. It is also reported that the final products produced with the use of these contaminated sources presented 10,600 ppm of Cd. The report also warns that Cd present in the analyzed material does not naturally occur in natural mineral sources, leading to the conclusion that the incorporation of Cd or the use of wastes contaminated with this metal was occurred.

In Brazil, there are frequent reports of companies selling fertilizer, mineral and organic, outside the standard quality. The most common cases are related to the presence of nutrients amounts below the specified labels, since this factor is more investigated. However, there are few studies evaluating the presence of toxic metals in these inputs, although the few that were conducted demonstrated the presence of these contaminants in fertilizers [6, 35, 36].

Currently, most grain-producing countries and other plant materials have laws regarding the limits of metals in fertilizer, which are presented in Table 3, where it can be observed a large variation between the allowed values. In general, the stipulated limits are for metals in phosphate fertilizer and micronutrient fertilizers, but there are some normatives that consider fertilizers in general, regardless of its composition or purpose. The limits of toxic metals can also be determined based on the amounts of nutrients or total amount in fertilizers.

TABLE 3. International maximum limits of contaminants allowable in fertilizers

Country	Fertilizer type	As	Cd	Pb	Cr
		mg kg^{-1}			
Switzerland[1]	Phosphate	NA	50.0	NA	NA
Finland[1]	Phosphate	NA	50.0	NA	NA
Sweden[1]	Phosphate	NA	100.0	NA	NA
Norway[1]	Phosphate	NA	100.0	NA	NA
Denmark[1]	Phosphate	NA	110.0	NA	NA
Belgium[1]	Phosphate	NA	210.0	NA	NA
Germany[1]	Phosphate	NA	210.0	NA	NA
Austria[1]	Phosphate	NA	275.0	NA	NA
U.S. (California)[2]	Phosphate	2.0*	4.0*	20.0*	NA
U.S. (Washington)[3]	Phosphate	13.0	165.0	61.0	1.0
Australia[1]	Phosphate	NA	300.0	NA	NA
Japan[1]	Phosphate	50.0	343.0	100.0	5.0
New Zealand[1]	Phosphate	NA	280.0	NA	NA
U.S. (Texas)[4]	Anyone	41.0	39.0	300.0	17.0
Canada[1]	Anyone	75.0	20.0	500.0	NA
China[4]	Anyone	50.0	8.0	100.0	NA
U.S. (California)	Micronutrient	13.0**	12.0**	140.0**	NA
U.S. (Washington)	Micronutrient	112.0**	83.0**	463.0**	6.0**
Australia	Micronutrient	NA	50.0	2000.0	5.0

1 [38], 2 [39], 3 [40], 4 [41]; * - Maximum allowed value per percentage point of P_2O_5 in the fertilizer; ** - Maximum allowed value per percentage point of the micronutrients sum in the fertilizer ; NA – not available

The Brazilian fertilizers law is based on the Normative Instruction (NI) 27/06 [37], which states like contaminating toxic metals, the elements As, Cd, Pb, Cr and Hg, determining the maximum allowable limits of these metals in four different categories of fertilizers (Table 4).

TABLE 4. Maximum allowable limits for metal contaminants in Brazilian fertilizers [37]

Fertilizer type	Maximum permitted quantity of the metal contaminant (mg kg^{-1})				
	As	Cd	Pb	Cr	Hg
Simple mineral fertilizer to suply P_2O_5	2.0*	4.0*	20.0*	40.0*	0.05*
Compound mineral fertilizer with micronutrient	500.0**	15.0**	750.0**	500.0**	10.0**
Compound mineral fertilizer with macronutrient and micronutrient	250.0	57.0	1000.0	NA	NA
Mineral fertilizer to suply micronutrient	4000.0	450.0	10000.0	NA	NA

* - Maximum allowed value per percentage point of P_2O_5 in the fertilizer; ** - Maximum allowed value per percentage point of the micronutrients sum in the fertilizer; NA – Not Available.

There are some doubts in relation to NI 27/06, which consider the legislation incipient, since that it does not determines which raw materials can be used as sources of nutrients in the fertilizers. Nutrient sources for macronutrients generally present good quality, however, for micronutrients, the use of low quality materials is frequently observed in Brazil, as mentioned previously, resulting in a possible contamination of the environment.

The legislation also does not tell how it will be the inspection of fertilizers industries, moreover tolerating the presence of contaminants in amounts up to 30% higher than the accepted values (Table 4). Another worth mentioning fact, is that the limits should be revised, since Article 5 affirms the need of a revision of the values predicted by NI 27/06 four years after its publication.

Nowadays, still does not exist an adequate legislation in Brazil regarding the use of toxic waste in the production of micronutrients. However, there are technical studies which classify these elements as hazardous materials, an example is the NBR 10004 of the Brazilian Association of Technical Standards (ABNT) [42], which stablish that the use of this wastes may present a risk to public health, causing or contributing to an higher incidence of diseases and mortality, also presenting adverse effects on the environment when handled or disposed inappropriately.

Another issue that is being discussed by the scientific community is that the determination of contaminants limits in fertilizers provides for the industries the possibility to include, as a source of nutrients, contaminated materials or with low quality, i.e., by the laws, the government provided the industries, exactly what they wanted, the possibility to incorporate low-quality raw materials and contaminated to supply nutrients [43]. It is observed so that there is a strong influence of the fertilizer industry in the studies and laws that limit the contaminants in these materials, showing that this problem is not only related to the environmental field, but also with the political and economic sector.

4. METALS: TOXICITY IN THE FOODCHAIN

Due to its mobility in aquatic systems and to its toxicity to life forms, the toxic metals or heavy metals, once in surface or in groundwater, are mentioned as one of the most problematic inorganic contaminants in the environment.

The toxic meals are strongly pollutants elements and present many harmful effects to ecosystems, causing physical-chemical alterations to water, what promote the decrease of water quality and mortality of the fauna and flora, prejudicing consequently, the human health [44].

The term heavy metal, is many times used with connotations of pollution and toxicity. However, "heavy" is conventionally used to imply the high density of the metal, referring to the pure element or to an alloy of metallic elements. The expression heavy metal apply to elements which have specific mass higher than 5 g cm^{-3} or which have atomic number higher than 20 [6].

Some metals, as the case of Cd, Pb, Ni and others, even when its presence is diluted in waters, many times in undetectable quantities, have recalcitrant effect, i.e., its persistence in the environment imply in process as biomagnification, which regards to the accumulation of these harmful substances in the many trophic levels of the ecological chains of ecosystems.

The contamination by toxic metals in animals can occur due to inhalation, skin penetration or ingestion of contaminated waters and foods, this last caused mainly by the capacity of plants in bioccumulate these elements in high proportions and later offer them to animals and human beings [45].

In animals, the toxic metals act in highly specific reactions, enzymatic in the most part and in alterations in systems that evolves these reactions, resulting in negative effects. One classic example is the metalloenzymes, in which dimensional structure is changed if one "intruder metal" binds to its active site, what prevents the enzymatic action [46].

Metal contaminations in the environment results from a sum of natural processes, between which also make part the natural sources of contamination, however, it is a common sense that the sources so called anthropogenic are more frequent when compared to the natural sources of contamination [47].

Significant increases of the metals concentrations are found in areas of intense industrial activity, where the accumulation can be many times higher than the average levels from non-contaminated areas. However, areas distant from the industrial centers in many cases also show metals contamination, due to the atmospheric transportation of these compounds [48] and in the case of agricultural soils, due to the use of fertilizes contaminated with these metals [6].

The contamination by metals is arising from the continuous disposal of industrial, domestic or livestock wastes in soils, rivers, lakes and seas. These pollutants include the toxic metals, which cause great concern to public health, because they are incorporated very easily in the tropic chain. These harmful elements cannot be destroyed by biological degradation, as in the case of the most part of organic pollutants. Incidences of the accumulation of metals in fish, oysters, mussels, sediment and other components from the aquatic systems have been reported in many countries around the world [49, 50].

Once the metals are present in cultivated soils, they are likely of radicular absorption by plants. It is important to point out that some of these elements present biological function whereas others are much known as toxic even in low concentrations. This way, the plants can be considered as constituents of the foundation of the trophic chain, causing great concern due to the possibility of the occurrence of toxic concentrations of certain metals which are transported from plants to higher levels of the food chain [51].

The toxic metals are not only subjected to radicular absorption by plants when presents in the soil solution, because may also occur the fixation of the metals in the soil particles, seepage/leaching to lower levels of soil or even the surface runoff, reaching the groundwater, rivers, lakes and seas. In general, there are two types of soils in Brazil, sandy and clayey, so that in sandy soils, due to the low ion exchange capacity, the metals are easily moved in the environment by leaching or surface runoff, reaching easily the hydric resources. In the other hand, in clayey soils, due to the higher ion exchange capacity, the metals tend to stand immobilized in the soil particle, becoming available for the plants in more than one cultivation season.

The toxic metals resulting from the anthropogenic activities, when disposed incorrectly in the environment, directly or indirectly reaches the top of the chain, i.e., humans will have contact with these elements by the consumption of food from vegetal or animal origin, contaminated by water, soil or air. Once in the human organism, these metals present cumulative character, i.e., are accumulated in the live tissues, i.e., hardly are eliminated by the organism. Down below are related some of the characteristics of some toxic metals and its respective collateral effects caused to human organisms.

4.1. Arsenic (As) – Cas Id 7440-38-2

In humans, the ingestion of high levels of arsenic can result in death. Exposure to lower levels can cause nausea and vomiting, decreased production of red and white blood cells, abnormal heart rhythm, damage to blood vessels, and a sensation of "pins and needles" in hands and feet. The ingestion of inorganic arsenic can increase the risk of skin cancer and cancer in the liver, bladder, and lungs. Inhalation of inorganic arsenic can cause increased risk of lung cancer. Exposure to higher than average levels of arsenic occur mostly in the workplace, near hazardous waste sites, or in areas with high natural levels. At high levels, inorganic arsenic can cause death [52].

Aquatic organisms accumulate arsenic mainly as inorganic forms, and some of the organisms such as phytoplankton, bacteria, etc. transform them into methylated and organic forms [53].

As is associated with ores containing metals, such as copper and lead. Arsenic may enter the environment during the mining and smelting of these ores. Small amounts of arsenic also may be released into the atmosphere from coal-fired power plants and incinerators because coal and waste products often contain some arsenic. It cannot be destroyed in the environment. It can only change its form, or become attached to or separated from particles. It may change its form by reacting with oxygen or other molecules present in air, water, or soil, or by the action of bacteria that live in soil or sediment [52].

Researching the effects of arsenic accumulation from plants cultivated near pressure-treated wood, Cao and Ma [54] found out that elevated As concentrations were observed in the soils adjacent to fences and poles, because As was frequently used for wood treatment, in order to increase the wood life time. Eating vegetables from these contaminated soils may pose a risk of As exposure since vegetables would readily take up As from these soils.

As, according to the Agency for Toxic Substances and Disease Registry (ATSDR) [55] is the first substance on the Priority List of Hazardous Substances.

4.2. Cadmium (Cd) – Cas Id 7440-43-9

At room temperature (25 °C/73.4 °F) the Cd is presented at solid state, with atomic number of 48 and atomic mass of 112.41 g. This element belong to the group of the metal of gray silver metallic color and it is available in many forms including sheet, granule, pellet, powder, rod, wire, among other. It is a soft metal, highly toxic which can be easily cut with a knife, and mostly produced as a byproduct of the extraction of Zn, Cu and Pb [56].

It is considered one of the most toxic metals, presenting important harmful effects for biological activity on soil, plant metabolism, human and animal health. The abundance of Cd in magmatic and sedimentary rocks, generally not more than 0.3 mg kg^{-1}, being that this metal tends to be concentrated in clayey deposits. The Cd is strongly associated to Zn in the geochemistry, however, have a strong affinity with S, presenting a higher mobility than Zn in acid environments [27].

Generally ore (CdS) is associated as a contaminant (0.25 to 0.5%) to ore Zn and Pb. According to the International Cadmium Association [57], Cd has been mainly used in manufacturing NI-Cd batteries, though, also has other applications as pigments, coatings, stabilizers for PVC (polyvinyl chloride) and alloys.

One of the main problems associated with Cd is its final destination, because it can reach the soil or the air through the burning of urban waste or the burning of fossil fuels, in this way polluting the environment and causing damage to the ecosystem [58].

In humans, Cd affects several organ systems, as the Cardiovascular (Heart and Blood Vessels), Developmental (effects during periods when organs are developing), Gastrointestinal (Digestive), Neurological (Nervous System), Renal (Urinary System or Kidneys), Reproductive and Respiratory (From the Nose to the Lungs), Cd is also known as a carcinogen agent [52]. According to the Agency for Toxic Substances and Disease Registry [55] Cd is the seventh substance on the Priority List of Hazardous Substances.

4.3. Lead (Pb) – Cas Id 7439-92-1

The Pb is rarely found in its natural state, but in combination with other elements, being the most important ores which present this metal are: Galena, cerussite, anglesite, perite and others.

Among the toxic metals, Pb is present in larger amounts in the earth's crust, with an average concentration between 10 and 20 mg kg^{-1} [59], therefore, is among the most used metals worldwide scale of Fe> Al> Cu> Zn> Pb [60].

To be quite versatile, Pb is much used in industrial products, and can be hardened by the addition of other metals and metal alloy being used as producing components, welds and other materials, in addition to being able to form various compounds employed in chemical industry [61].

According to unofficial information from the National Department of Mineral Production (DNPM), the main consuming sectors Pb are: accumulators (batteries) 80%, representing approximately 88 tons of metal, oxides, 12%, equivalent to 13 tons; and electronic (alloys, solders and various), 8%, with 9000 tons.

The Pb in soil is extremely stable and highly toxic to humans and animals, is currently ranked as the second most dangerous element in the Priority List of Hazardous Substances [55]. Most of the lead enters the human body through the respiratory and gastrointestinal routes, and after absorption, it can be found in the blood, soft and mineralized tissues [62].

In humans, the Pb poisoning may initially cause lack of appetite, metallic taste in mouth, muscle discomfort, malaise, headache and abdominal cramps strong. However, in childhood, symptoms often linked to the deposition of this metal in the brain are predominant. In its inorganic Pb^{2+} is an inhibitor of enzymes, further affecting the nervous system [63].

The organ systems affected by Pb are the Cardiovascular (Heart and Blood Vessels), Developmental (effects during periods when organs are developing), Gastrointestinal (Digestive), Hematological (Blood Forming), Musculoskeletal (Muscles and Skeleton), Neurological (Nervous System), Ocular (Eyes), Renal (Urinary System or Kidneys) and Reproductive [52].

4.4. Chromium (Cr) – Cas Id 7440-47-3

With regard to heavy metal Cr, this has turned into a serious pollutant of air, soil and water. About 70% of the production of Cr is used in alloys, including stainless steel and 15% in industrial chemical processes, mainly in leather tanning, pigments, electrochemical and cooling towers [64].

The ions of Cr are found in different oxidation states, however, the trivalent form (Cr^{3+}) is the most common, while the hexavalent form (Cr^{6+}) depends on the more specific conditions of pH and redox potential [65].

The toxicity of Cr depends on its oxidation state. While the Cr^{3+} is relatively stable and immobile, the Cr^{6+} moves easily through the soil and aquatic environments, with a strong oxidizing agent capable of being absorbed even by the skin in this way, the Cr presents itself as a carcinogenic species [66, 67].

In humans, the Cr affects the immune, renal (kidney or urinary system) and respiratory tract (the nose to the lungs), and is known as a powerful carcinogen [52].

According to reference [68], a lethal dose of Cr^{6+} is between 50 and 100 mg kg^{-1}, a value much lower when compared to Cr^{3+}, the values are between 1900 and 3300 mg kg^{-1}. Since the Cr^{6+} occupies the 17th position in the Priority List of Hazardous Substances [55].

4.5. Nickel (Ni) – Cas Id 7440-02-0

The Ni occurs in the earth's crust in the order of 0.01%, mainly as sulfide, oxide and silicate minerals. Geological activities as natural weathering and volcanoes led the natural distribution of this metal in natural environments at modest levels. The Ni metal and its compounds are used in multiple applications as stainless steel and other alloys, castings, catalysts, batteries, electronics, ceramics, pigments and even coins [69].

Among the main uses of this metal can still cite it use in coating parts, in a process called electroplating, widely used around the world by Electroplating industries [70].

Although not an element extensively released to the environment, this can present a risk to human health. Epidemiological investigations and experimental studies have shown that certain compounds of Ni are extremely carcinogenic after inhalation, however, the same studies show that these risks are limited by conditions specific occupational exposure to this metal [71].

Results show that this element can cross the placenta and present embryological and teratogenic properties. The main danger of Ni to the man beside the carcinogenicity is related to the ability to cause sensitivity reactions [71].

The most common harmful health effect in the general population is allergic contact dermatitis elicited by prolonged skin contact of sensitized individuals with Ni. Therefore, the majority of studies related to health effects of Ni to humans concern airborne Ni particulate matter. Strongest evidence for high cancer risk was found for sulfidic nickel species (NiS, NiS_2 and Ni_2S_3) in Ni refinery dust. On the molecular level the toxic Ni species responsible for severe health effects, as allergic contact dermatitis and respiratory tract cancer, has been suggested to be caused by Ni^{2+} [69].

Ni affects the following organ systems: Cardiovascular (Heart and Blood Vessels), Dermal (Skin), Immunological (Immune System), Respiratory (From the Nose to the Lungs) [52]. Actually, Ni ranks the 57th position on the Priority List of Hazardous Substances [55].

4.6. Copper (Cu)- Cas Id 7440-50-8

The Cu, despite being a heavy metal, meets the criteria of essentiality for plants and microorganisms, being classified as micronutrient [72]. This metal is essential to human health, and is present in all human tissues and many fluids and with amino acids, fatty acids and vitamins, is required for the metabolic processes [73].

Excess copper may be toxic to humans because of the affinity metal with S-H groups of many proteins and enzymes associated with diseases such as epilepsy, melanoma and rheumatoid arthritis, as well as the loss of taste [72]. Toxic levels of Cu occur naturally in some soils, whereas others may contain high levels of Cu as a result of the anthropogenic release of contaminated materials into the environment through mining, smelting, manufacturing, agriculture and waste disposal technologies [72].

In plants, either deficient or in excess, Cu can cause disorders in plant growth and development by adversely affecting important physiological process. In particular photosynthetic electron transport is altered under deficiency and excess of Cu. Thus, for healthy plant growth and development Cu must be acquired from the soil, transported throughout the plant, distributed and compartmentalized within different tissues and its content carefully regulated within different cells and organelles. For this purpose, plants, like all other organisms, have homeostatic mechanisms to maintain the correct concentrations of essential metal ions. Membrane transport systems are likely to play a central role in these processes [73].

Copper ranks the 125th position on the Priority List of Hazardous Substances [55].

4.7. Manganese (Mn)- Cas Id 007439-96-5

According to the International Manganese Institute (IMnI) [74], this metal is the fourth most used in terms of tonnage, being ranked behind Fe, Al and Cu, being the production of Mn alloys about 17.7 million metric tons at 2011. Mn the second major abundant metal and 12th most abundant element in the earth's crust, being its ores deposits usually of sedimentary origin, with oxide ore layers inter-bedded with iron-rich formations [75].

Steel is basically an alloy of iron and carbon, consisting of an iron phase and iron carbides. Crude steel produced from iron contains an undesirable amount of oxygen and some sulphur. Mn plays a key role because of two important properties: its ability to combine with sulphur and its powerful deoxidation capacity [76].

High grade Mn ores (> 40%) are typically processed into suitable metallic alloy forms by pyrometallurgical processes. Low grade manganese ores (< 40%) are conventionally processed by pyrometallurgical reductive roasting or melting followed by hydrometallurgical processing for production of chemical manganese dioxide (CMD), electrolytic manganese (EM) or electrolytic manganese dioxide (EMD) [77].

In the human body, Mn displays a somewhat unique behavior with regard to its toxicity. It is relatively non-toxic to the adult organism except to the brain where it causes Parkinson-like symptoms when inhaled even at moderate amounts over longer periods of time [78].

This element, when in large amounts, affect fertility in mammals and are toxic to the embryo and fetus, what implicates that pregnant women should not be exposed to Mn anytime [78].

Due to the substitution of Pb by Mn in gasoline, reference [79], researching the cultivation of beans and oats in soil next to roads, found not only Mn concentrations significantly higher in organic soil but also higher accumulation found in fruits and stems, suggesting that the addition of MMT to gasoline can be the cause for this increase in exchangeable Mn in organic soils.

According to the Agency for Toxic Substances and Disease Registry [55], Mn is the 140th substance on the Priority List of Hazardous Substances.

5. CASE STUDIES: CONTAMINATED FERTILIZERS IN BRAZILIAN CROPS

Based on information obtained from seizures made in Brazilian seaports, related to the entry of loads of industrial waste with the presence of toxic metals for use as source of micronutrients in fertilizers [6], three studies were conducted to evaluate the possibility of soil and plants contamination using different sources and Zn levels, since this is main micronutrient applied in the Brazilian agricultural crops. All studies were conducted at the State University of West Paraná - Brazil, being one for soybean [2], one for the wheat crop [4] and the last one for the corn crop [5].

In the work done with the soybean crop [2], an experiment was conducted in 2007, in Palotina-PR, with coordinates 24° 18' S and 53° 55' W (Figure 2) in a soil classified as Hapludox (RED). It were used four different fertilizers for the Zn supply sold in Brazil, in three doses (0.0; 0.9 and 1.8 kg ha^{-1}). The results of this study demonstrated the presence of toxic metals (Cd, Pb and Cr) in the soybean leaf tissue (Table 5) at levels above of the tolerable in agricultural crops [27]. We also observed the occurrence of these three metals in the soil after harvest (Table 6), which compared to the Brazilian legislation, that establishes the maximum concentrations of contaminants in soils [28], classify the obtained values as above the maximum allowed for the soil be agriculturable (IV - Agricultural area) in the case of Cd. For Pb and Cr the values obtained in the study were below of the prevention value (PV), which is the maximum allowable value of the contaminant in soil to that it is able to sustain its biological processes.

The work done with the wheat crop [4] was conducted on the residual fertilization of the work with the soybean crop, since the concentrations of the metals Cd, Pb and Cr in the soil of the experiment were significant, especially in the case of Cd. The results of this study showed no accumulation of Cd in the leaves of wheat at the level of detection by the analytical method (flame atomic absorption spectrometry - FAAS), however, we found concentrations of Pb and Cr. For Pb the maximum concentration found was under the tolerable limit, and for the Cr, the maximum concentration found was above the tolerable level for crops [27]. One of the main conclusions of this work was the occurrence of availability of Cr and Pb to wheat plants with fertilization performed in previous crop (soybean), demonstrating the persistence of contaminants in soil and their occurrence in more than one crop agriculture.

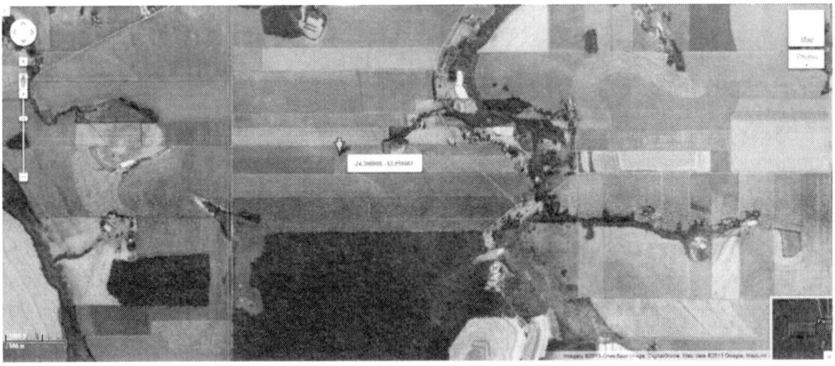

FIGURE 2. Local where the experiment involving fertilization with Zn in soybean and wheat culture were performed [80]

The work conducted with maize crop was conducted in Mercedes, Paraná State-Brazil, with geographic coordinates of 24° 25' S and 54° 15' W (Figure 3) in a Rhodic Eutrudox. The fertilization of this experiment was performed with eight different sources of Zn in four different doses (0.0; 2.0; 4.0 and 6.0 kg ha^{-1}). In the chemical analysis performed in the fertilizers used in the work, it was observed that one of the sources presented 11039.46 mg of Pb kg^{-1}, being the maximum limit permitted by Brazilian law (NI 27/06) [37] 10000.00 mg of Pb kg^{-1} (Table 3). However, as previously mentioned in the section 3, the NI 27/06 tolerates the presence of contaminants up to 30% of the maximum permitted value, in this way, the fertilizer would not be considered as irregular, and can be freely traded.

FIGURE 3. Local where the experiment involving fertilization with Zn in corn crop was performed [80]

Regarding the leaf concentration of Cd, Pb and Cr (Table 5), it was observed that only the Pb concentrations was found at detectable levels by the used method (FAAS), where the leaf content of Pb in corn plants can be considered inside the tolerable limits [27]. For the soil, it was observed the occurrence of Cd, Pb and Cr (Table 6), so that the concentrations of Cd were below the maximum allowable limit so that the soil is agriculturable (IV - Agricultural area), and the Pb and Cr were below the PV, which is the maximum allowable value of the contaminant in the soil, to that it be able to sustain its biological processes.

Based on the results obtained in the three presented studies it can be inferred that the presence of metal contaminants in fertilizers is real and should be studied more carefully by regulatory authorities in Brazil. The foliar and soil levels of the metals Cd, Pb and Cr found cause concern from the environmental point of view, due the possibility of entry of these contaminants in the food chain, either by consumption of contaminated materials or pollution of water resources from contaminants in the soil. It should be noted that the values presented in these works were obtained with only one fertilization, so that with successive fertilizations may occur a greater availability of these contaminants to the environment.

TABLE 5. Foliar levels of Cd, Pb and Cr in experiments conducted with mineral fertilizers to supply Zn in different crops in Brazil

Metal	Agricultural crop			Tolerable foliar level****
	Soybean*	Wheat**	Corn***	
Cd (mg kg^{-1})	0.73	ND	ND	0.05 a 0.50
Pb (mg kg^{-1})	25.53	1.95	8.67	0.50 a 10.00
Cr (mg kg^{-1})	3.36	9.28	ND	2.00

ND – not detected by the method used (FAAS) with a quantification limit of 0.01 mg L^{-1}; * - results from reference [2]; ** - results from reference [4]; *** - results from reference [5]; **** - Tolerable foliar level according to reference [27].

TABLE 6. Levels of toxic metals Cd, Pb and Cr in soil after cultivation of various crops fertilized with Zn in Brazil

Metal	Agricultural crop		Classification of the metal value in the soil*	
	Soybean*	Corn**	PV	IV – Agricultural area
Cd (mg dm^{-3})	5.52	2.52	1.30	3.00
Pb (mg dm^{-3})	24.25	51.48	72.00	180.00
Cr (mg dm^{-3})	22.19	18.43	75.00	150.00

* - results from reference [2]; ** - results from reference [5]; *** - Classification of the metal value in the soil according to reference [28]; PV–Prevention Value; IV – Investigation Value for agricultural area.

It is worth affirm also that the laws related to the presence of contaminants [37], and levels of toxic heavy metals allowed in Brazilian soils [28], are being severely questioned by Brazilian researchers and government agencies. Among the key questions are the lack of criteria for determining and monitoring the presence of contaminants in fertilizers, passivity with the addition of toxic metals in brazilians soils by the minimum quality and lack of discussion with researchers and entities responsible for the drafting of legislation.

It should furthermore be highlighted that before approval of Resolution No. 420 of CONAMA [28], there was serious questioning on the part of Brazilian authorities and researchers in relation to the criteria used for its formulation by means of Proceeding 02000.002955/2004-69 [81-83], thus confirming the need for more societal discussion (especially among the specialized technical and scientific segments in a multidisciplinary manner) before this is submitted for approval of possible resolution. The resolution, moreover, violates the Brazilian Federal Constitution with respect to the "preservation, improvement and recovery of the environmental quality favorable to life" because it permits the introduction of contaminants in the soil up to the maximum limit permitted by the resolution, thus allowing the use of contaminated fertilizers and soil conditioners in agriculture until they reach and exceed these values and thereby degrade environmental quality. Therefore, the need for more in-depth field studies in relation to environmental legislation before it is drafted and approved becomes clear. These studies must be undertaken together with environmental regulatory bodies and research entities in the country. Based on the results obtained in this study, it is evident that some of the Zn fertilizer sold by sources in Brazil significantly affect the quantity of Cd, Pb, and Cr in soil. If their use is continued, there may be contamination of soil, surface and underground waters, plants, animals, and humans, leading to severe damage to health as well as the environment. This fact raises concern because according to references [84-86], even low levels of heavy metals in soil may be available to plants, an effect that is difficult to ascertain because the symptoms they case occur only over the long term, thus increasing even more the hazards associated with the use of contaminated materials in agricultural activities.

In a review about the environmental risks of adding phosphate fertilizers contaminated with trace elements in long-term, reference [87] warn about the possibility of increasing concentrations of As, Cd and Pb in agricultural soils using contaminated fertilizers. The authors also emphasize that to prevent the entry of potentially toxic elements in agriculture, it is necessary a strict control on the sources of macro and micronutrients used in agriculture. This demonstrates the importance and need for further studies related to the use of fertilizers contaminated with toxic metals in crops and the environmental impacts that these sources of nutrients can cause.

6. SUSTAINABLE STRATEGIES FOR SOIL REMEDIATION

The introduction of toxic chemicals by man in various environmental compartments (soil, water and atmosphere), the of controlled or uncontrolled elimination of wastes (accidental spills, mining processes and smelting of metallic ores, fertilizer application, sewage sludge and pig manure in agricultural soils) puts the ability to self-cleaning of contaminated ecosystems in doubt. Consequently, the pollutants that accumulate toxic metals are of concern in relation to human exposure of ecosystems and the potential impact [88].

Today there are several techniques for soil decontamination. It is worth mentioning that the choice of one or other technique, among other factors, also depends on the characteristics of the contaminant in question, between the different techniques is possible enumerate:

1. Thermal treatment: Use of heat to remove, stabilize or destroy the contaminants;
2. Physical treatment: Physical processes used to separate toxic substances from the environment;
3. Chemical treatment: Uses chemical reactions to remove, destroy or modify toxic substances;
4. Biological treatment: Makes use of natural metabolic means, such a microorganisms and other biological agents to remove, destroy or modify the contaminants;
5. Stabilization / solidification: This step the contaminants are stabilized chemically and / or modified to reduce the potential for contamination.

These technologies can in many cases be applied "in situ", i.e., at the site of origin of the contamination, or "ex situ" outside the place of contamination, which results in the removal and transport of the contaminated soil.

Among the physical-chemical treatments, these mostly are based on soil washing. These methods are based on the technological principle of the transfer of a contaminant from the soil for a liquid or gaseous phase. The main products are treated soil and the contaminants concentrated. The specific process of treatment depends on the type of contaminant, in particular as regards the type of connection established with the soil particles.

It is observed by the description above, that such treatments become process more expensive the decontamination because are many cases in which the soil treatment requires the removal of several layers of the same and specific treatment in laboratories.

Thus, many approaches have been developed, evaluated and performed to deal with soil pollution. Current technologies for decontamination soil are often expensive and of high energy consumption, and in many cases, the soil cannot be reused after treatment [89]. Thus, several studies have been developed to promote proper techniques and low cost of remediation in order to prevent the spread of these contaminants in the food chain [88, 90].

As an alternative, an ecological approach technology that involves the use of plants to clean up or remediate soils contaminated with toxic metals, called phytoremediation, has been developed and encouraged by the fact of being a simple and inexpensive option for remediation of contaminated soils [90-92].

Phytoremediation can be divided into phytoextraction (hiperacumulation) phytostabilization, rhizofiltration, fitotransformation and fitovolatilization (Figure 4).

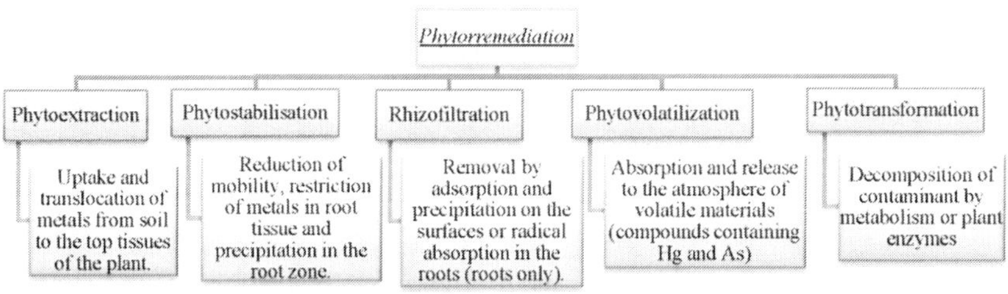

FIGURE 4. Types and characteristics of phytoremediation [92,93]

However for the phytoremediation to be effective it depends on the selection of plant species, preferably hyperaccumulators, which are able to grow in soils of low fertility, and degraded by the presence of contaminants [94].

In studies carried out in Iran using *Euphorbia macroclada* and *Centaurea virgata*, demonstrate that the sampled species not only were able to grow in soils highly contaminated by metals, as well as were capable of accumulating high concentrations of Zn, Mn, Cu, Pb and Fe, in this way these plants can be classified as hyperaccumulators, in other words, have adequate potential for phytoremediation of contaminated soils [95].

However, the metal accumulation by the plants is only efficient after the contaminant be removed from the soil, for example, by harvesting the plant matter. If most of the toxic metals are located on the shoots, the harvest can be performed using traditional farming methods. In general, it is necessary to harvest the plants before the fall of the leaves, or before his death and decomposition, so that contaminants do not disperse or return to the soil.

After harvesting, the biomass must be processed for extraction and recollection of the majority of metals. Alternatively, the volume and the weight of the biomass can be reduced by thermal processes, physical, chemical or microbial. In the case of burning plant, for example, the energy produced represents an economic value of the process and the ash may be treated as an ore, which may still be extracted the metallic contaminant (especially if the ashes are enriched in one or two metals).

Even Brazil showing great potential species for bioremediation and phytoremediation, the recovery of contaminated areas, due to the great biodiversity and climate, favoring biological processes in pollution treatment, knowledge about these species of plants and microbial communities in phytoremediation potential are still scarce [96].

In research carried out by reference [97], it was reported that some plants were able to remediate soils contaminated with metals such as *Brassica juncea* for phytoextraction of Cd and Cr^{6+}, Cu, Ni and Pb,*Avena sativa* for Zn, *Pisum sativum* L. and *Zea mays* L. for removal of Pb.

However, from an economic point of view, the use of these species of accumulator plants generally do not generate financial returns to the farmers, because these plants only remove the contaminant from the soil, being later discarded after the remediation. Thus if the farmer in question require to decontaminate a particular area, the phytoremediation only would generate costs, especially with seeds, agricultural machinery, irrigation, etc.

Nevertheless, currently there is a widespread idea that the species used for phytoremediation of soil can also generate economic gains beyond decontamination. The use of oil crops, for example, *Crambe abyssinica H., Moringa oleifera L.* and *Jatropha curcas*, in which the product marketing is not consumed as food, but as biofuel (biodiesel), is a possibility to decrease the cost of recovery of the contaminated area. Thus the fact that the plant has accumulated large amounts of heavy metal will have little influence on the end product of the culture, generating profit for the property.

Moreover, oil crops like *C. abyssinica, M. oleifera* and *J. curcas*, have been the subject of research in the removal of toxic metals such as Cd, Pb, Cr and water through the adsorption process, in which the pie their seeds are used as adsorbent [98-101]. The results of these studies demonstrated great potential for removal of these metals, and that these adsorbents are obtained at minimal charges, because they are wastes generated by the extraction of vegetable oil.

Thus, the use of cultures for vegetable oil production and recovery of contaminated areas can be an interesting alternative, which will generate the money from oil production and recovery / stabilization of the contaminated area with the development of these cultures, preventing them be lost through leaching or soil surface runoff and reach the groundwater and surface water and groundwater. However, more researches are still needed In order to evaluate this possibility, focusing on the development of oil-producing plants and cycles of addition of the adsorbent material and the biodegradability thereof. Between this and other issues to which scientific research is driven, many studies still need to be developed so that, in one way or another, the quest for sustainability will one day be achieved.

7. CONCLUSION

Based on information existing in the Brazilian and international literature, we can make some important observations:

The fertilization of plants is one of the major reasons of the increase in food production across the globe. Consequently, the use of fertilizers in agricultural crops is extremely high and growing, however, for the maintenance of this production capacity, we should be very careful with the fertilizers sources that are being used in agriculture.

The presence of toxic metals in fertilizers has been reported in the major food producing countries, like United States, China and Brazil. This demonstrates that this issue deserves further detailed studies of long-term about the successive addition and the effects of these contaminants to crops, soil, supply chain and the wider environment. Thus, the institutions of research and supervision, have to evaluate the products that are used for the manufacture of fertilizers on the market, as this is an activity that generate large amounts of money and that will always be a target of people with questionable character.

The current legislation, not only in Brazil, but internationally, with respect to the use of wastes as sources of nutrients, presence of contaminants in fertilizers and presence of toxic metals in soil, need higher accompaniments and answers for some of the community questions, because the legislation must ensure that the agricultural soils are not being degraded and becoming unproductive. In Brazil, it is strongly recommended that the laws must be harder and restrictive, until we have reliable and conclusive results related to the addition of contaminated fertilizer in long-term. In this case, it should be emphasized that these laws should be developed based on results from institutional researches and reliable groups, without the influence of political and industrial groups. Meanwhile, the damage caused by man in air, soil and waters, must be repaired, by remediation methods (such as phytoremediation) aiming not only the decontamination, but also ensuring a small profit, as an incentive to the farmer.

Brazil is considered as one of the countries with largest biodiversity and most promising in food production for the globe, however, for that prediction become a reality, it is very important that the productive areas are not degraded. The technologies developed by our civilization provide ways to avoid the degradation of our environment, but often they are not used because they prevent higher industrial profits, this should not occur, because the balance of the ecosystem and the quality of human life should be above of any monetary profit.

REFERENCES

1. Nacke H, Gonçalves Jr AC, Stangarlin JR,Schwantes D, Strey L, Nava I. Productivity and yield components of maize fertilized with different sources and doses of zinc. Spanish Journal of Rural Development 2011;2(1) 47-55.
2. Nava I, Gonçalves Jr AC, Nacke H, Guerini VL, Schwantes D. Availability of cadmium, lead and chromium toxic heavy metals in soil and soybean leaf tissue fertilized with different sources of NPK+Zn. Ciência e Agrotecnologia 2011;35(1), 884-892.
3. Gonçalves Jr AC, Nacke H, Marengoni, NG, Carvalho, EA, Coelho GF. Yield and production components of soybean fertilized with different doses of phosphorus, potassium and zinc. Ciência e Agrotecnologia 2010;34(3) 660-666.
4. Gonçalves Jr AC, Nacke H, Schwantes D, Nava IA, Strey L. Phytoavailability of toxic heavy metals and productivity in wheat cultivated under residual effect of fertilization in soybean culture. Water, Air and Soil Pollution 2011;220(1) 205-211.
5. Nacke H, GonçalvesJr AC, Schwantes D, Nava I, Strey L, Coelho GF. Availability of heavy metals (Cd, Pb and Cr) in agriculture from commercial fertilizers. Archives of Environmental Contamination and Toxicology 2013;64(1), 371-379.
6. Gonçalves Jr AC, Luchese EB, Lenzi E. Evaluation of phytoavailability of the cadmium, lead and chromium in soybean cultivated in the Latossolo vermelho escuro, treated with commercial fertilizers. Química Nova 2000;23(2) 173-177.
7. Manzatto CV, Freitas Junior E, Peres JRR. Agricultural use of Brazilian soils. Rio de Janeiro: Embrapa Solos; 2002.

8. FAO - Food and Agriculture Organization of the United Nations. The state of food and agriculture: Women in agriculture closing the gender gap for development. Rome: FAO; 2011.

9. Nelson GC, Rosegrant MW, Palazzo A, Gray I, Ingersoll C, Robertson R, Tokgoz S, Zhu T, Sulser TB, Ringler C, Msangi S, You L. Food Security, Farming, and Climate Change to 2050: scenarios, results, policy options. Washington: International Food Policy Research Institute; 2010.

10. Foley JA, Ramankutty N, Brauman KA, Cassidy ES, Gerber JS, Johnston M, Mueller ND, O'Connell C, Ray DK, West PC, Balzer C, Bennett EM, Carpenter SR, Hill J, Monfreda C, Polasky S, Rockstrom J, Sheehan J, Siebert S, Tilman D, Zaks DPM. Londons: Solutions for a cultivated planet. Nature 2011;478(1) 337-342.

11. International Fertilizer Industry Association. http://www.fertilizer.org/ifa/ifadata/search (accessed 15 June 2013).

12. Yamada T. Deficiências de micronutrientes, ocorrência, detecção e correção: o sucesso da experiência brasileira. Piracicaba: Potafos; 2004.

13. Cakmark I. Plant nutrition research: Priorities to meet human needs for food in sustainable ways. Plant and Soil 2002;247(1) 03-24.

14. Gonçalves Jr AC, Pessoa ACS. Phytoavailability of the cadmium, lead and chromium in soybean cultivated in argissolo vermelho eutrófico, treated with commercial fertilizers. Scientia Agrária2002;3(1-2) 19-23.

15. CETESB - Environmental Sanitation Technology Company of São Paulo State. Relation of contamined areas in São Paulo 2012. http://www.cetesb.sp.gov.br/userfiles/file/areas-contaminadas/2012/texto-explicativo.pdf (accessed 15 June 2013).

16. Lopes JC, Ribeiro LG, Araújo MG, Beraldo MRBS. Lettuce production using doses of sewage sludge. HorticulturaBrasileira 2005;23(1), 143-147

17. Pegorini ES, Andreoli CV. Introduction. In: Andreoli, CV. (ed.) Alternative use of sanitary waste. São Paulo:ABES; 2006. p01-06.

18. Marques MO, Nogueira TAR, Fonseca IM, Marques TA. Cr, Ni, Pb e Zn contents in sewage sludge added Red Argisol and sugarcane cropped. Revista de Biologia e Ciência da Terra2007;27(1) 133-143.

19. Alamino RCJ, Polivanov H, Campos TMP, Silva VHG, Santos LV, Mendes JC. Bioavailability of Cadmium on a Latosol Increased of Sewage Sludge. Anuário do Instituto de Geociência 2007;30(2) 45-54.

20. CETESB - Environmental Sanitation Technology Company of São Paulo State. Report establishment of guiding values for soils and groundwater in the State of São Paulo. São Paulo: CETESB; 2001.

21. Mattias JL, Ceretta CA, Nesi CN, Girotto E, Trentin EE, Lourenzi CR, Vieira RCB. Copper, zinc and manganese in soils of two watersheds in Santa Catarina with intensive use of pig slurry. Revista Brasileira de Ciência do Solo 2010;34(4) 1445-1454.

22. Wu L, Tan C, Liu L, Zhu P, Peng C, Luo Y, Christie P. Cadmium bioavailability in surface soils receiving long-term applications of inorganic fertilizers and pig manure. Geoderma 2012;173-174(1) 224–230.

23. Ramalho JFGP, Sobrinho NMB, Velloso ACX. Heavy metals contamination of a watershed in Caetés by the use of agrochemicals. Pesquisa Agropecuária Brasileira 2000;35(7) 1289-1303.

24. Baker DE. Copper. In: Alloway, BJ. (ed.) Heavy metals in soils. Glasgow: Blackie and Son; 1990. p.51-74.

25. Nicholson FA, Smith SR, Alloway BJ, Carlton-Smith C, Chambers BJ. An inventory of heavy metals inputs to agricultural soils in England and Wales. The Science of the Total Environment 2003;311(1) 205–219.

26. Council Directive of 12 June 1986 on the protection of the environment, and in particular of the soil, when sewage sludge is used in agriculture (86/278/EEC). http://eur-lex.europa.eu/LexUriServ/LexUriServ.do?uri=CELEX:31986L0278:EN:HTML (accessed 15 July 2013).

27. Kabata-Pendias A, Pendias H. Trace elements in soils and plants. 3rd ed. Boca Raton: CRC Press; 2001.

28. Brazil. Resolution 420. Brasília: Ministry of the Environment; 2009.

29. Novais RF, Alvarez VVH, Barros NF, Fontes RLF, Cantarutti RB, Neves JCL. editors. Soil Fertilty: Viçosa: SBCS; 2007.

30. USEPA – United States Environmental Protection Agency. Background Report on Fertilizer Use, Contaminants and Regulations. Columbus: USEPA; 1999.

31. The Weinberg Group. Scientific Basis for Risk-Based Acceptable Concentrations of Metals in Fertilizers and Their Applicability as Standards. Washington: The Weinberg Group; 2001.

32. USEPA – United States Environmental Protection Agency. Zinc fertilizers made from recycled hazardous secondary materials. Washington: USEPA; 2002.

33. WDE – Washington Department of Ecology. Focus: Imported Cadmium-Contaminated Zinc Sulfate Used in Fertilizer and Other Products. Washington: WDA; 2000.

34. Gabe U, Rodella AA. Trace elements in Brazilian agricultural limestones and mineral fertilizers. Communications in Soil Science and Plant Analysis 1999;30(5-6) 605-620.

35. Martins ALC, bataglia OC, Camargo AO, Cantarella H. Corn yield and uptake of Cu, Fe, Mn and Zn from sewage sludge-amended soil with and without liming.Revista Brasileira de Ciência do Solo 2003;27(3) 563-574.

36. Brazil. NormativeInstruction 27. Brasília: Ministry of Agriculture, LivestockandSupply; 2006.

37. Hooda SP. Trace Elements in Soils. London: Wiley; 2010.

38. CCR – California Code of Regulations. Standards and labeling. California: CCR; 2007.

39. AAPFCO – Association of American Plant Food Control. AAPFCO's Statement of Uniform Interpretation and Policy (SUIP) #25: the heavy metal rule. http://www.aapfco.org/rules.html (accessed 15 June 2013).

40. AFPC – Association of Fertilizer and Phosphate Chemists. Latest Metal Regulations in Fertilizers. http://afpc.net/Metal%20Regulations.html (accessed 15 June 2013).

41. ABNT – Brazilian National Standards Organization. NBR 10004. Rio de Janeiro: ABNT; 2004.

42. Rodella AA. Regulation of contaminant contents in fertilizers: A Case study. Revista Brasileira de ciência do solo 2005; 29(1) 797-801.

43. Pino GAH. Biosorption of heavy metals using *Cocos nucifera*. Master's Dissertation. Pontifical Catholic University; 2005.

44. Morillo J, Usero J, Gracia I. Partitioning of metals in sediments from the Odiel river (Spain). Environment International 2002;28(4) 263-271.

45. Bueno BYM. Removal of Pb, Cr and Cu by a Combined Biosorption / Bioflotation process using a *Rhodococcusopacus* straint. Doctorate Thesis. Pontifical Catholic University; 2007.

46. Nriagu JO, Pacyna JM. Quantitative assessment of worldwide contamination of air, water and soils by trace metals. Nature 1988;333(1) 134–139.

47. Loska K, Wiechula D, Korus I. Metal contamination of farming soils affected by industry. Environmental International 2004;30(1) 159-165.

48. Naimo TJ. A review of the effects of heavy metals on freshwater mussels. Ecotoxicology 1995;4(6) 341-362.

49. Sayler GS, Nelson Jr JD, Colwell RR. Role of bacteria in bioaccumulation of mercury in the oyster *Crassostrea virginica*. Applied Microbiology 1975;30(1) 91–96.

50. Peralta-Videa JR, Lopez ML, Narayan M, Saupe G, Gardea-Torresdey J. The biochemistry of environmental heavy metal uptake by plants: Implications for the food chain. The international Journal of biochemistry & Cell Biology 2009;41(8-9) 1665-1677.

51. ATSDR – Agency for Toxic Substances and Disease Registry. Toxic Substances Index. http://www.atsdr.cdc.gov/az/a.html (accessed 15 June 2013).

52. Rahman MA, Hasegawa H, Lim PR. Bioaccumulation, biotransformation and trophic transfer of arsenic in the aquatic food chain. Environmental Research 2012;116(1) 118-135.

53. Cao X, MA LQ. Effects of compost and phosphate on plant arsenic accumulation from soils near pressure-treated wood. Environmental Pollution 2004;132(3) 435-442.

54. ATSDR – Agency for Toxic Substances and Disease Registry. The Priority List of Hazardous Substances. http://www.atsdr.cdc.gov/SPL/index.html (accessed 15 June 2013).

55. Brady JE, Humiston GE. General Chemistry. 2th ed. Rio de Janeiro: Livros Técnicos e Científicos; 1986.

56. ICdA. INTERNACIONAL Cadmium ASSOCIATION, Revision of the Battery Directive: Public Consultation, 2001.

57. PINO G.A.H., Biossorption of Heavy metals on green coconut powder. Master Dissertation, PUC-RJ, March, 113p. 2005.

58. Paoliello MMB, Chasin AAM. Ecotoxicology of lead and its compounds. Salvador: CRA; 2001.

59. ILZSG – International Lead and Zinc Study Group. Statistic of Lead and Zinc. 2010. http://www.ilzsg.org/static/home.aspx (accessed 15 June 2013).

60. Kreush MA. Assessment with proposals for improving the industrial process of recycling lead and indicating applicability to slag generated. Master's Dissertation. Federal University of Parana; 2005.

61. ATSDR – Agency for Toxic Substances and Disease Registry. Case studies in environmental medicine: lead toxicity. Atlanta: ATSDR; 1999.

62. Capitani EM. Diagnosis and treatment of lead poisoning in children and adults. Medicina 2009;42(3) 319-329.

63. Cervantes C, Campos-García J, Devars S, Gutiérrez-Corona F, Loza-Tavera H, Torres'Guzmán JC, Moreno-Sánchez R. Interactions of chromium with microorganisms and plants. FEMS Microbiology Reviews 2001;25(3) 335-347.

64. WHO - World Health organization. Environmental health criteria 61 - Chromium. Geneva: WHO; 1988.

65. Smith WLE, Gadd GM. Reduction and precipitation of chromate by mixed culture sulphate-reducing bacterial biofilms. Journal of Applied Microbiology 2000;88(6) 983-991.

66. Kumral E. Speciation of chromium inwaters via sol–gel preconcentration prior to atomic spectrometric determination. Doctorate Thesis. The Graduate School of Engineering and Sciences of Izmir Institute of Technology; 2007.

67. Frois SR, Grassi MT, Fernandes TC, Barreto RAS, Abate G. Preconcentration of Cr(III) and speciation analysis of chromium employing montmorillonite saturated with potassium ions. Química Nova 2011;34(3) 462-467.

68. Schaumlöffel D. Nickel species: analysis and toxic effects. Journal of Trace Elements in Medicine and Biology 2012;26(1) 1-6.

69. Veit MT, Silva EA, Fagundes-Klen MR, Tavares CRG, Gonçalves GC. Biosorption of nickel and chromium from a galvanization effluent using seaweed pre-treated on a fixed-bed column. Acta Scientiarum. Technology 2009; 31(2) 175-183.

70. Leonard A, Gerber P, Jacquet P. Carcinogenicity, mutagenicity and teratogenicity of nickel.Mutation. Research/Reviews in Genetic Toxicology 1981;87(1) 1-15.

71. Sodré FF, Lenzi E, Costa ACS. Applicability of adsorption models to the study of copper behaviour in clayey soils. Química Nova 2001;24(3) 324-330.

72. Lima AJB, Cardoso, MG, Guerreiro MC, Pimentel FA. Using activated carbon to remove copper from sugar cane spirit. Química Nova 2006;29(2) 247-250.

73. Yruela I. Copper in plants. Brazilian Journal of Plant Physiology 2205;17(1) 145-156.

74. International Manganese Institute. Applications of Mn. http://www.manganese.org/about_mn/applications (accessed 15 June 1013).

75. Das AP, Sukla LB, Pradhan N, Nayak S. Manganese biomining: A review. Bioresource Technology 2011;102(16) 7381-7387.

76. Zhang W, Cheng, CY. Manganese metallurgy review. Part I: Leaching of ores/secondary materials and recovery of electrolytic/chemical manganese dioxide. Hydrometallurgy 2007;89(3-4) 137-159.

77. Gerber GB, Leonard A, Hantson PH. Carcinogenicity, mutagenicity and teratogenicity of manganese compounds. Critical Reviews in Oncology/Hematology 2002;42(1) 25-43.

78. Brault N, Loranger S, Courchesne F, Kennedy G, Zayed J. Bioaccumulation of manganese by plants: influence of MMT as a gasoline additive. Science of the Total Environment 1994;153(1-2) 77-84.

79. Google. Google Maps. https://maps.google.com (accessed 15 June 2013).

80. Paula Jr D. Office 2147/09-MP-PJCS-UMA. http://www.mma.gov.br/port/conama/processos/FE4582B1/Of2147-09-MP-PJCS-U MA_7maio09.pdf. 2009. (accessed 15 June 2013).

81. Varjabedian R, Leite CBB. Office 1029/2009 - PGJ/CAO. http://www.mma.gov.br/port/conama/processos/FE4582B1/ConsideracoesMinisPublSaoPaulo.pdf. 2009 (accessed 15 june 2013).

82. Brazilian Ministry of the Environment. Process 02000.002955/2004-69. 2011. http://www.mma.gov.br/port/conama/processo.cfm?processo=02000.002955/2004-69 (accessed 15 June 2013).

83. Gimeno-García E, Andreu V, Boluda R. Heavy metals incidence in the application of inorganic fertilizers and pesticides to rice farming soils. Environmental Pollution 1996;92(1) 19–25.

84. Ryser P, Sauder WR. Effects of heavy-metal-contaminated soil on growth, phenology and biomass turnover of *Hieraciumpiloselloides*. Environmental Pollution 2006;140(1) 52–61.

85. Jan FA, Ishaq M, Khan S, Ihsanullah I, Ahmad I, Shakirullah M. A comparative study of human health risks via consumption of food crops grown on wastewater irrigated soil (Peshawar) and relatively clean water irrigated soil. Journal of Hazardous Materials 2010;179(1-3) 612–621.

86. Jiao W, Chen W, Chang AC, Page AL. Environmental risks of trace elements associated with long-term phosphate fertilizers applications: A review. Environmental Pollution 2012;168(1) 44-53.

87. Ghosh M, Singh SP. A Review on Phytiremediation of heavy metals and utilization of its byproducts. Applied ecology and environmental research 2005;3(1) 1-18.

88. Lorestani B, Cheraghi M, Yousefi N. Phytoremediation Potencial of Native Plants Growing on a Heavy Metals Contaminated soil of Copper mine in Iran. World Academy of Science, Engineering and Technology 2011;53(1) 377-382.

89. Susarla S, Medina VF, McCutcheon SC. Phytoremediation: An ecological solution to organic chemical contamination Ecological Engineering 2002;18(1) 647–658.

90. Gratão PL, Prasad MNV, Cardoso PF, Lea PJ, Azevedo RA. Phytoremediation: Green technology for the clean up of toxic metals in the environment. Brazilian Journal of Plant Physiology 2005;17(1) 53-64.

91. Lorestani B, Cheraghi M, Yousefi N. Accumulation of Pb, Fe, Mn, Cu and Zn in plantsand choice of hyperaccumulator plant in the industrial town of viean, Iran. Archives of Biological Science Belgrade 2011; 63(3) 739-745.

92. Jing Y, He Z, Yang X. Role of soil rhizobacteria in phytoremediation of heavy metal contaminated soils Journal of Zhejiang University Science B 2007;8(3) 192-207.

93. Mkumbo S, Mwegoha W, Renman G. Assessment of the phytoremediation potential for Pb, Zn and Cu of indigenous plants growing in a gold mining area in Tanzania. International Journal of Environmental Sciences 2012;2(4) 2425-2434.

94. Cheraghi, M,Lorestani, B,Yousefi, N. Introduction of Hyperaccumulator Plants with Phytoremediation Potential of a Lead- Zinc Mine in Iran. World Academy of Science, Engineering and Technology 2011, 53(1) 163-168.

95. Marques M, Aguiar CRC, Silva JJLS. Technical challenges and social, economic and regulatory barriers to phytoremediation of contaminated soils. Revista Brasileira de Ciência do Solo 2011;35(1) 1-11.

96. Prasad MNV, Freitas HMO. Metal hyperaccumulation in plants – Biodiversity prospecting for phytoremediation technology.Eletronic Journal of Biotechnology 2003;6(3) 285-321.

97. Gonçalves Jr. AC, Rubio F, Meneghel AP, Coelho GF, Dragunski DC, Strey L. The use of *Crambe abyssinica* seeds as adsorbent in the removal of metals from waters. Revista Brasileira de Engenharia Agrícola e Ambiental 2013;17(2) 306-311.

98. Rubio F, Gonçalves Jr. AC, Meneghel AP, Tarley CRT, Schwantes D, Coelho GF. Removal of cadmium from water using by-product *Crambe abyssinca* Hochst seeds as biosorbent material. Water Science and Technology 2013;68(1) 227-233.

99. Rubio F, Gonçalves Jr. AC, Strey L, Meneghel AP, Coelho, GF, Nacke, H. Applicability of *Crambe abyssinica* Hochst byproduct as biosorbent in the removal of chromium from water. Spanish Journal of Rural Development. 2013;4(1) 25-40.

100. Gonçalves Jr. AC, Meneghel AP, Rubio F, Strey L, Dragunski DC, Coelho GF. Applicability of *Moringa oleifera* Lam. pie as an adsorbent for removal of heavy metals from waters. Revista Brasileira de Engenharia Agrícola e Ambiental 2013;17(1): 94-99.

101. Nacke H, Gonçalves Jr. AC, Coelho GF, Strey L, Schwantes D, Laufer A. Removal of cadmium from aqueous solutions by adsorption on *Jatropha biomass* In: Bartolo H. (ed.) Green Design, Materials and Manufacturing Processes. Boca Raton: CRC Press Taylor & Francis Group; 2013. p367-372.

7
CHAPTER

Urban Air Pollution Modeling

Anjali Srivastava[1] and B. Padma S. Rao

[1] National Environmental Engineering Research Institute, Kolkata Zonal Centre, India

1. INTRODUCTION

All life form on this planet depends on clean air. Air quality not only affects human health but also components of environment such as water, soil, and forests, which are the vital resources for human development.

Urbanization is a process of relative growth in a country's urban population accompanied by an even faster increase in the economic, political, and cultural importance of cities relative to rural areas. Urbanization is the integral part of economic development. It brings in its wake number of challenges like increase in population of urban settlement, high population density, increase in industrial activities (medium and small scale within the urban limits and large scale in the vicinity), high rise buildings and increased vehicular movement. All these activities contribute to air pollution. The shape of a city and the land use distribution determine the location of emission sources and the pattern of urban traffic, affecting urban air quality (World Bank Reports 2002). The dispersion and distribution of air pollutants and thus the major factor affecting urban air quality are geographical setting, climatological and meteorological factors, city planning and design and human activities.

Cities in the developing countries are characterized by old city and new development. The old cities have higher population density, narrow lanes and fortified structures.

In order to ensure clean air in urban settlements urban planning and urban air quality management play an important role. New legislations, public awareness, growth of urban areas, increases in power consumption and traffic pose continuous challenges to urban air quality management. UNEP (2005) has identified niche areas Urban planning need to primarily focus on as:

- Promotion of efficient provision of urban infrastructure and allocation of land use, thereby contributing to economic growth,
- managing spatial extension while minimizing infrastructure costs,
- improving and maintaining the quality of the urban environment and
- Prerecording the natural environment immediately outside the urban area.

Air quality modelling provides a useful support to decision making processes incorporating environmental policies and management process. They generate information that can be used in the decision making process. The main objectives of models are: to integrate observations, to predict the response of the system to the future changes, to make provision for future development without compromising with quality.

2. URBAN AIR QUALITY

The urban air is a complex mixture of toxic gases and particulates, the major source is combustion of fossil fuels.Emissions from fossil fuel combustion are reactive and govern local atmospheric chemistry.Urban air pollution thus in turn affect global troposphere chemistry and climate (e.g. tropospheric O_3 and NO_X budgets, radiative forcing by O_3 and aerosols).

Sources of air pollutants in urban area, their effect and area of concern are summarized in Table 1.

TABLE 1. Urban sources of air pollutants, their effect and area of concern.

Source	Pollutants	Effects	Area of concern
Large number of vehicles	Particulate matters (PM_{10}, $PM_{2.5}$), Lead (Pb), Sulphur dioxide (SO_2), Oxides of nitrogen (NO_x), Ozone (O_3), Hydro carbons (HCs), Carbon monoxide (CO), Hydrogen fluoride (HF), Heavy metals (e.g. Pb, Hg, Cd etc.)	Human Health (acute and chronic)	Local, Regional and Global
Use of diesel powered vehicle in large number			
Use of obsolete vehicles in large number			
Large number of motor cycles/three wheelers (2 stroke and three stroke)		Ecosystem (acute and chronic)	Local, Regional and Global
Unpaved and/or poorly maintained street			
Open burning			
Inadequate		Greenhouse	Global

infrastructure		gas emission	
Low quality of fuel/fuel adulteration			
Little emission control & technology in industry			
Presence of industries (e.g. ceramic, brick works, agrochemical factory)		Acid rain	Global
Waste incineration		Stratospheric ozone depletion	Global
Limited dry deposition of pollutants		Long-range transport	Global

Urban air pollution involves physical and chemical process ranging over a wide scale of time and space. The urban scale modeling systems should consider variations of local scale effects, for example, the influence of buildings and obstacles, downwash phenomena and plume rise, together with chemical transformation and deposition. Atmospheric boundary layer, over 10 to 30 km distances, governs the dispersion of pollutants from near ground level sources. Vehicular emissions are one the major pollution source in urban areas. Ultrafine particles are formed at the tailpipe due to mixing process between exhaust gas and the atmosphere. Processes at urban scale provide momentum sink, heat and pollutant source thereby influencing the larger regional scale (up to 200 km). Typical domain lengths for different scale models is given in table 2.

TABLE 2. Typical domain length for different scale model

Model	Typical Domain Scale	Typical resoluti on	Motion Example
Micro scale	200x200x100m	5m	Molecular diffusion, Molecular viscosity
Mesoscale (urban)	100x100x5km	2km	Eddies, small plumes, car exhaust, cumulus clouds
Regional	1000x1000x10k m	36km	Gravity waves, thunderstor ms, tornados, cloud clusters, local winds, urban air pollution
Synoptic(contine ntal)	3000x3000x20k m	80 km	High and low pressure system, weather

			fronts, tropical storms, Hurricanes Antarctic ozone hole,
Global	65000x65000x20km	4° x 5°	Global wind speed, rossby (planetary) waves stratospheric ozone reduction Global worming

Piringer et al., 2007, have demonstrated that atmospheric flow and microclimate are influenced by urban features, and they enhance atmospheric turbulence, and modify turbulent transport, dispersion, and deposition of atmospheric pollutants. Any urban scale modeling systems should consider effects of the various local scales, for example, the influence of buildings and obstacles, downwash phenomena and plume rise, chemical transformation and deposition. The modelling systems also require information on emissions from various sources including urban mobile pollution sources. Simple dispersion air quality pollution transport models and complex numerical simulation model require wind, turbulence profiles, surface heat flux and mixing height as inputs. In urban areas mixing height is mainly influenced by the structure heights and construction materials, in terms of heat flux. Oke (1987,1988, 1994), Tennekes (1973), Garrat (1978, 1980), Raupach et al (1980) and Rotach (1993, 1995) divided the Atmospheric Boundary Layer within the urban structures into four sub layers (Figure 1).

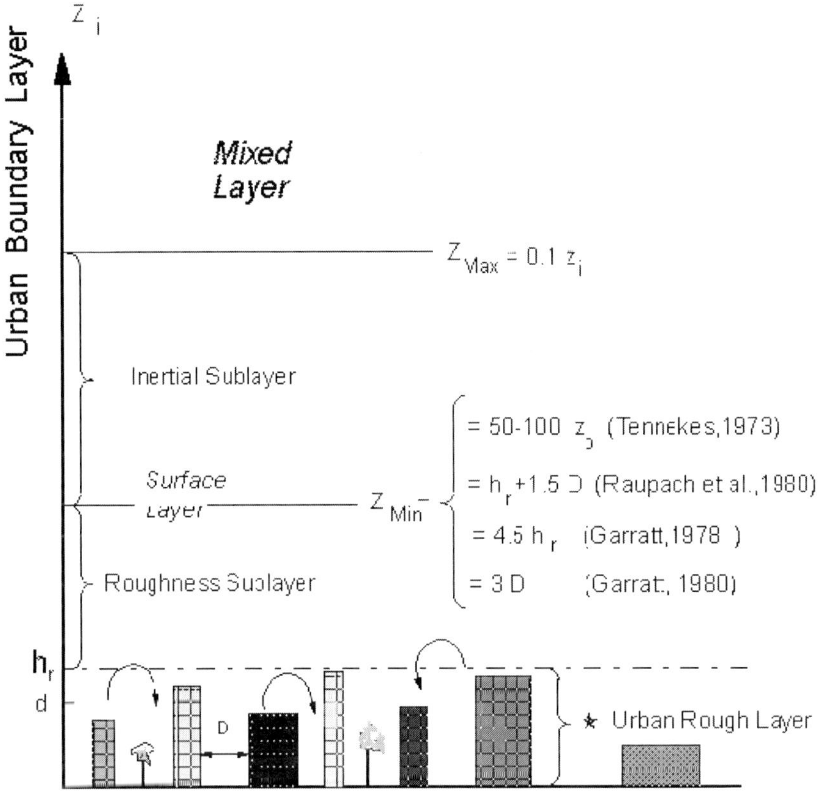

FIGURE 1. Boundary- layer structure over a rough urban built- up area A daytime situation is displayed where Z I denotes the mixed layer height. Modefied after Oke (1988) and Rotach (1993).

In urban establishments anthropogenic activities take place between the top of highest building and the ground. People also live in this area. The layer of atmosphere in this volume is termed as Urban Canopy. The thermal exchanges and presence of structures in urban canopy modify the air flows significantly and this makes the atmospheric circulations in urban canopy highly complex. The heterogeneity of urban canopies poses a challenge for air quality modeling in urban areas. The importance of various parameters in different models for urban atmosphere study is given in Table 3.Figure 2 shows the flow and scale lengths within an urban boundary layer, UBL.

TABLE 3. Ranking of parameters in different applications for urban air environment

Parameter	Air Quality	Urban Climatology	Urban Planning
Wind speed	Very important	Important	Very Important
Wind Direction	Very important	Important	Very Important
Temperature Humidity	Important	Extremely important	Very Important
Pollutant Concentration	Extremely important	Very important	Very important
Turbulent Fluxes	Very important	Very important	Very important

Vehicles are one of the important pollution sources in urban areas. Maximum exposure to local public is from this source and thus they form important receptor group. Pollutant dispersion of vehicular pollution is at street scale and is the smallest scale in urban environment. Hosker (1985) showed that flows in street canyon are like recirculating eddy driven by the wind flow at the top with a shear layer which separates the above canyon flows from those within it. In deep street canyons the primary vortex does not extend to the ground but a weak contra rotating vortex is formed near the ground and is relatively shallow (Figure 3). Pavageau et al (2001) demonstrated that wind directions which are not normal to the street axis cause variations in the flow. The real geometry of the street canyon and the mean flow and turbulence generated by vehicles within the canyon also affect the recirculating flow.

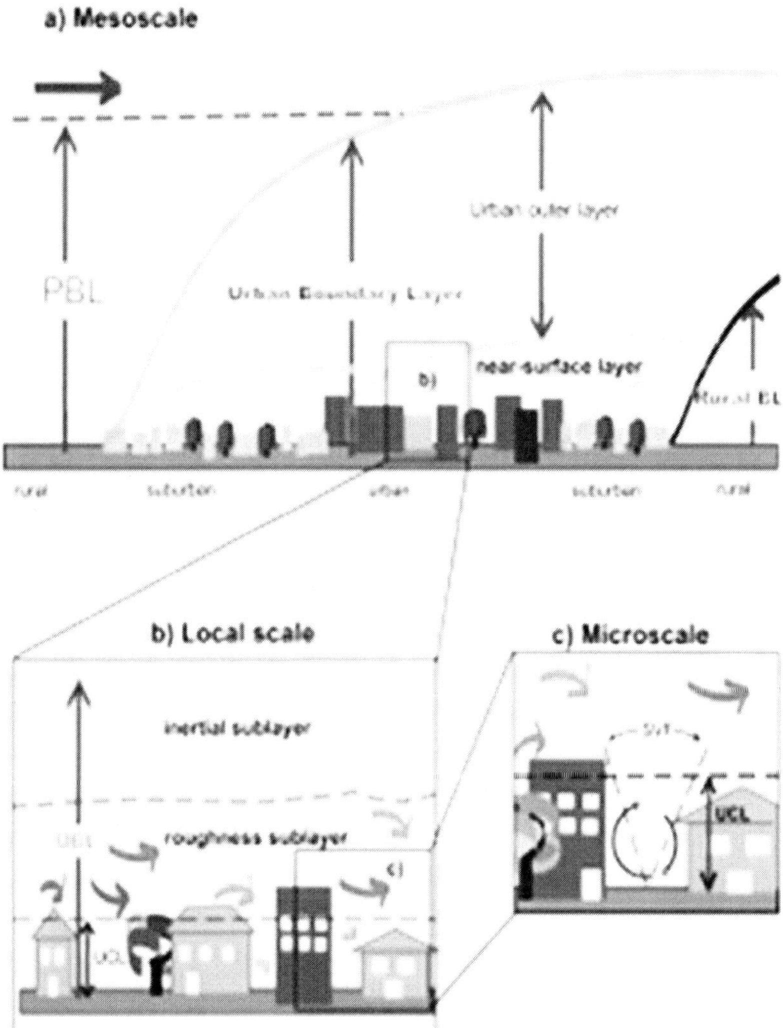

FIGURE 2. Schematic diagram showing processes, flow and scale lengths within an urban boundary layer, UBL. This is set in the context of the planetary boundary layer, PBL, the urban canopy layer, UCL, and the sky view factor, SVF, a measure of the degree to which the sky is obscured by surrounding buildings at a given point which characterises the geometry of the urban canopy. Ref:. Meteorology applied to urban pollution problems-Final report COST Action 715. Dementra Ltd Publishers

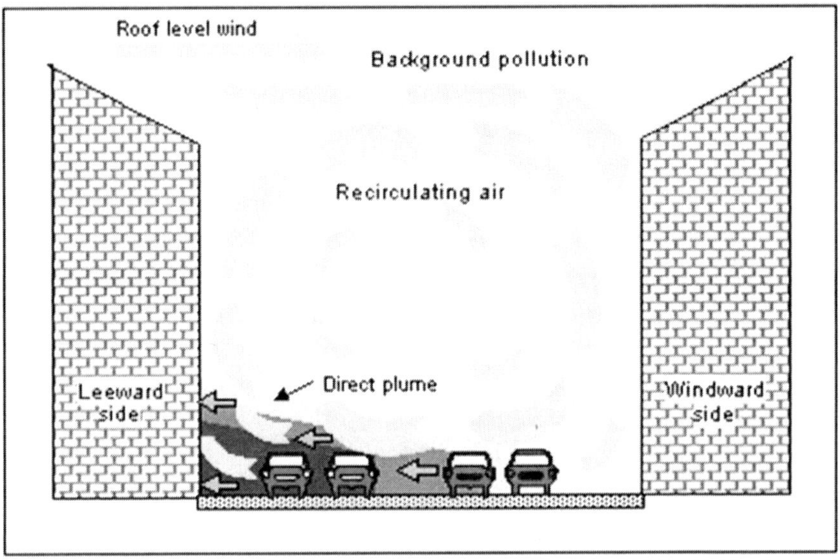

FIGURE 3. Air flow pattern in a Street Canyon

Concentrations of pollutants at a receptor are governed by advection, dispersion and deposition. Air pollutants can be divided into two main categories namely conventional air pollutants and Hazardous Air Pollutants (HAPs). Conventional air pollutants include particulate matters, sulphur dioxide, nitrogen dioxide, carbon monoxide, particles, lead and the secondary pollutant ozone. HAPs include Volatile Organic Compounds, toxic metals and biological agents of many types. All pollutants are not emitted in significant quantities. Secondary pollutants like some VOCs, carbonyls and ozone are formed due to chemical transformation in air. These reactions are often photochemical.

The important components of air quality modelling are thus,

- Knowledge of sources and emissions
- Transport, diffusion and parametrisation
- Chemical transformations
- Removal process
- Meteorology

Understanding contribution from various sources to air quality is the key for effective management of the air quality. Air quality models offer a useful tool in comprehending these issues. These models evaluate the relationship between air pollutant emissions and their resulting concentration in the ambient air. Commonly used air quality models are: 1) Conceptual Models 2) Emission Models 3) Meteorological Models, 4) Chemical Models, 5) Source Oriented Models and 6) Receptor Models.

3. AIR QUALITY MODEL CLASSIFICATION

Air quality models cover either separately or together atmospheric phenomena at various temporal and spatial scales. Urban air models generally focus from local (micro- tens of meters to tens of kilometers) to regional (meso) scale. Models can be broadly divided into two types namely physical and mathematical.

Physical models involve reproducing urban area in the wind tunnel. Scale reduction in the replica and producing scaling down actual flows of atmospheric motion result in limited utility of such models. Moreover these are economically undesirable.

Mathematical models use either use statistics to analyse the available data or mathematical representation of all the process of concern. The second type of mathematical models is constrained by the ability to represent physical and chemical processes in equations without assumptions.

Statistical model are simple but they do not explicitly describe causal relationships and they cannot be extrapolated beyond limits of data used in their derivation. Thus dependence on past data becomes their major weakness. These cannot be used for planning as they cannot predict effect of changes in emissions.

3.1. Eulerian And Lagrangian Models

Eulerian approach has been used to predict air pollutant concentrations in urban areas. The space domain (geographical area or air volume), are divided into "small" squares (two-dimensional) or volumes (three-dimensional), i.e. grid cells. Thus Eulerian models are sometimes called "grid models". Equidistant grids are normally used in air pollution modeling. Then the spatial derivatives involved in the system of Partial Differential Equations are discretized on the grid chosen. The transport, diffusion, transformation, and deposition of pollutant emissions in each cell are described by a set of mathematical expressions in a fixed coordinate system. Chemical transformations can also be included. Long range transport, air quality over entire air shed, that is, large scale simulations are mostly done using Eulerian models. Reynolds (1973), Shir and Shieh (1974) applied Eulerian model for ozone and for SO_2 concentration simulation in urban areas, and Egan (1976) and Carmichael (1979) for regional scale sulfur. Holmes and Morawska (2006) used Eulerian model to calculate the transport and dispersion over long distances. The modeling studies by Reynolds (1973) on the Los Angeles basin formed the basis of the, the well-known Urban Air shed Model-UAM. Examples of Eulerian models are CALGRID model and ARIA Regional model or the Danish Eulerian Hemispheric Model (DEHM).

Lagrangian Model approach is based on calculation of wind trajectories and on the transportation of air parcels along these trajectories. In the source oriented models the trajectories are calculated forward in time from the release of a pollutant-containing air parcel by a source (forward trajectories from a fixed source) until it reaches a receptor site. And in receptor oriented models the trajectories are calculated backward in time from the arrival of an air parcel at a receptor of interest (backward trajectories from a fixed receptor). Numerical treatment of both backward and forward trajectories is the same. The choice of use of either method depends on specific case. As the air parcel moves it receives the emissions from ground sources, chemical transformations, dry and wet depositions take place. If the models provide average time-varying concentration estimates along the box trajectory then Lagrangian box models have been used for photochemical modeling. The major shortcoming of the approach is the assumption that wind speed and direction are constant throughout the Physical Boundary Layer. As compared to the Eulerian box models the Lagrangian box models can save computational cost as they perform computations of chemical and photochemical reactions on a smaller number of moving cells instead of at each fixed grid cell of Eulerian models. Versions of EMEP (European Monitoring and Evaluation Programme) are examples of Lagrangian models. These models assume pollutants to be evenly distributed within the boundary layer and simplified exchange within the troposphere is considered.

3.2. Box Models

Box models are based on the conservation of mass. The receptor is considered as a box into which pollutants are emitted and undergo chemical and physical processes. Input to the model is simple meteorology. Emissions and the movement of pollutants in and out of the box is allowed. The air mass is considered as well mixed and concentrations to be uniform throughout. Advantage of the box model is simple meteorology input and detailed chemical reaction schemes, detailed aerosol dynamics treatment. However, following inputs of the initial conditions a box model simulates the formation of pollutants within the box without providing any information on the local concentrations of the pollutants. Box models are not suitable to model the particle concentrations within a local environment, as it does not provide any information on the local concentrations, where concentrations and particle dynamics are highly influenced by local changes to the wind field and emissions.

3.3. Receptor Models

Receptor modeling approach is the apportionment of the contribution of each source, or group of sources, to the measured concentrations without considering the dispersion pattern of the pollutants. The starting point of Receptor models is the observed ambient concentrations at receptors and it aims to apportion the observed concentrations among various source types based on the known source profile (i.e. chemical fractions) of source emissions. Mathematically, the receptor model can be generally expressed in terms of the contribution from 'n' independent sources to 'p' chemical species in 'm' samples as follows:

$$C_{ik}=\sum_{j=1}^{n} a_{ij} f_{jk} \quad C_{ik}=\sum_{j=1}^{n} a_{ij} f_{jk} \qquad (1)$$

Where C_{ik} is the measured concentration of the k_{th} species in the ith sample, a_{ik} is the concentration from the jth source contributing to the ith sample, and f_{jk} is the kth species fraction from the jth source. Receptor models can be grouped into Chemical mass balance (CMB), Principal Component Analysis (PCA) or Factor analysis, and Multiple Linear Regression Analysis (MLR) and multivariate receptor models.

The Chemical Mass Balance (CMB) Receptor Model used by Friedlander, 1973 uses the chemical and physical characteristics of gases and particulate at source receptor to both identify the presence of and to quantify source contributions of pollutants measured at the receptor. Hopke (1973, 1985) christened this approach as receptor modelling. The CMB model obtains a least square solution to a set of linear equation, expressing each receptor concentration of a chemical species as a linear sum product of source profile species and source contributions. The output to the model consists of the amount contributed by each source type to each chemical species. The model calculates the contribution from each source and uncertainties of those values. CMB model applied to the VOC emissions in the city of Delhi and Mumbai (Figure 4) shows that emissions from petrol pumps and vehicles at traffic intersection dominate.

PCA and MLR are statistical models and both PMF and UNMIX are advanced multivariate receptor models that determine the number of sources and their chemical compositions and contributions without source profiles. The data in PMF are weighted by the inverse of the measurement errors for each observation. Factors in PMF are constrained to be nonnegative. PMF incorporates error estimates of the data to solve matrix factorization as a constrained, weighted least-squares problem (Miller et al., 2002; Paatero, 2004).

Geometrical approach is used in UNMIX to identify contributing sources. If the data consist of 'm' observations of 'p' species, then the data can be plotted in a p-dimensional data space, where the coordinates of a data point are the observed concentrations of the species during a sampling period. If n sources exist, the data space can be reduced to a (n-1) dimensional space. An assumption that for each source, some data points termed as edge points exist for which the contribution of the source is not present or small compared to the other sources.

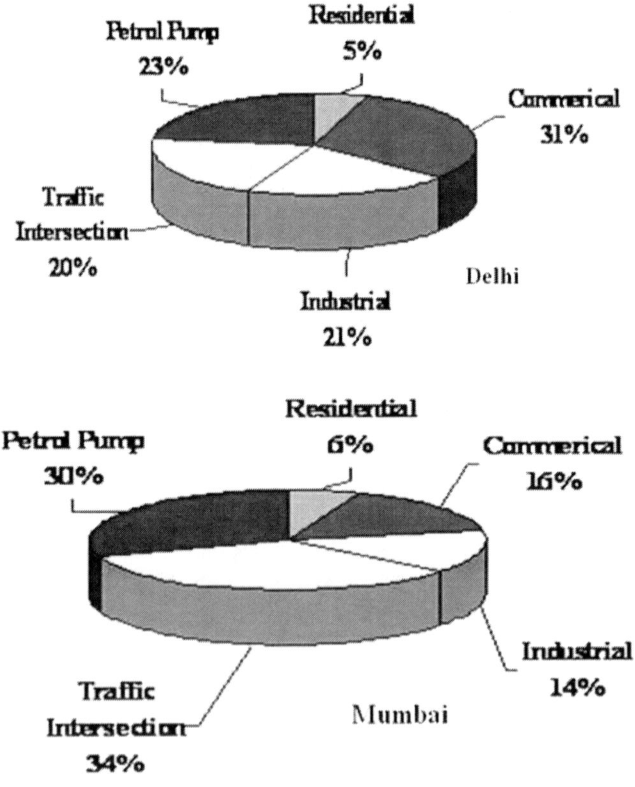

FIGURE 4. Category wise Contribution to Total VOCs at Mumbai and Delhi based on CMB results(Ref: Anjali Srivastava 2004, 2005)

UNMIX algorithm identifies these points and fits a hyperplane through them; this hyperplane is called an edge. If n sources exist, then the intersection of n-1 of these edges defines a point that has only one contributing source. Thus, this point gives the source composition. In this way, compositions of the n sources are determined which are used to calculate the source contributions (Henry, 2003).

3.4. Computational Fluid Dynamic Models

Resolving the Navier-Stokes equation using finite difference and finite volume methods in three dimensions provides a solution to conservation of mass and momentum. Computational fluid dynamic (CFD) models use this approach to analyse flows in urban areas. In numerous situation of planning and assessment and for the near-sources region, obstacle-resolved modeling approaches are required. Large Eddy Simulations (LES) models explicitly resolve the largest eddies, and parameterize the effect of the sub grid features. Reynolds Averaged Navier Stokes (RANS) models parameterize all the turbulence, and resolve only the mean motions. CFD (large eddy simulation [LES] or Reynolds-averaged Navier-Stokes [RANS]) model can be used to explicitly resolve the urban infrastructure. Galmarini et al., 2008and Martilli and Santiago,2008, used CFD models to estimate spatial averages required for Urban Canopy Parameters. Using CFD models good agreement in overall wind flow was reported by fieldGidhagen et al. (2004).They also reported large differences in velocities and turbulence levels for identical inputs

3.5. The Gaussian Steady-State Dispersion Model

The Gaussian Plume Model is one of the earliest models still widely used to calculate the maximum ground level impact of plumes and the distance of maximum impact from the source. These models are extensively used to assess the impacts of existing and proposed sources of air pollution on local and urban air quality. An advantage of Gaussian modeling systems is that they can treat a large number of emission sources, dispersion situations, and a receptor grid network, which is sufficiently dense spatially (of the order of tens of meters). Figure 5 shows a buoyant Gaussian air pollutant dispersion plume. The width of the plume is determined by σ_y and σ_z, which are defined by stability classes(Pasquill 1961; Gifford Jr. 1976)

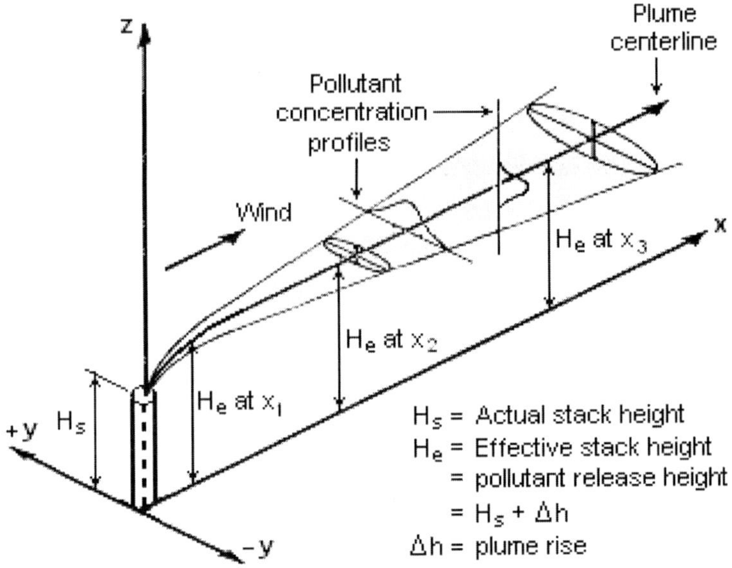

FIGURE 5. A buoyant Gaussian air pollutant dispersion plume

The assumptions of basic Gaussian diffusion equations are:

- that atmospheric stability and all other meteorological parameters are uniform and constant throughout the layer into which the pollutants are discharged, and in particular that wind speed and direction are uniform and constant in the domain;
- that turbulent diffusion is a random activity and therefore the dilution of the pollutant can be described in both horizontal and vertical directions by the Gaussian or normal distribution;
- that the pollutant is released at a height above the ground that is given by the physical stack height and the rise of the plume due to its momentum and buoyancy (together forming the effective stack height);
- that the degree of dilution is inversely proportional to the wind speed;
- that pollutant material reaching the ground level is reflected back into the atmosphere;
- that the pollutant is conservative, i.e., not undergoing any chemical reactions, transformation or decay.

The spatial dynamics of pollution dispersion is described by the following type of equation in a Gaussian model:

$$C(x,y,z;He)=Q2\pi u\sigma y\sigma z\cdot \quad \cdot[\exp(-y22\sigma 2y)\cdot\{\exp(-(z-He)22\sigma 2z)+\exp(-(z+He)22\sigma 2z)\}]C(x,y,z;He)=Q2\pi u\sigma y\sigma z\bullet \bullet[\exp(-y22\sigma y2)\bullet\{\exp(-(z-He)22\sigma z2)+\exp(-(z+He)22\sigma z2)\}] \quad (2)$$

Where

C(x, y, z) : pollutant concentration at. point (x, y, z);

U: wind speed (in the x "downwind" direction, m/s)

Σ: represents the standard deviation of the concentration in the x and y direction, i.e., in the wind direction and cross-wind, in meters;

Q: is the emission strength (g/s)

He: is the effective stack height, see below.

From the above equation, the concentration in any point (x, y, z) in the model domain, from a constant emission rate source, in steady state can be calculated.

Plume rise equations have been developed by Briggs (1975). The effective stack height (physical stack height plus plume rise) depends on exit velocity of gas, stack diameter, average ambient velocity, stack gas temperature and stability of atmosphere

$$He=H+\Delta H,\Delta H=1.4Q14/H(d\theta dz)-38/,QH=\rho QCP(TG-15)3600/He=H+\Delta H,\Delta H=1.4QH14(d\theta dz)-38,QH=\rho QCP(TG-15)3600 \quad (3)$$

Where

H: height of stackT_G : Temperature of exit gasQ: Volume of exit gas dθ/dz : Temperature Gradient ρ: Density of exit gas C_P: Specific heat at constant pressure

Some major air pollution dispersion models in current use

- ADMS 3: Developed in the United Kingdom (www.cerc.co.uk)
- AERMOD: Developed in the United States, (www.epa.gov/scram001/dispersion_prefrec.htm)
- AUSPLUME: Developed in Australia, (http://www.epa.vic.gov.au/air/epa)

- CALPUFF: Developed in the United States, (www.src.com/calpuff/calpuff1.htm)
- DISPERSION2:Developed in Sweden,(www.smhi.se/foretag/m/dispersion_eng.htm)
- ISC3: Developed in the United States, (www.epa.gov/ttn/scram/dispersion_alt.htm)
- LADM: Developed in Australia, (Physick, W.L,et al, 1994)
- NAME: Developed in the United Kingdom,(www.metoffice.gov.uk/research/modelling-systems/dispersion-model)
- MERCURE: Developed in France, (www.edf.com)
- RIMPUFF: Developed in Denmark, (http://www.risoe.dtu.dk)

8 regional air quality modeling leading to setting up of air quality index for an industrial area in India is given in Fig 2. This study has resulted in estimating the air assimilative capacity of the region and delineating developmental plans accordingly

3.6. Urban Pollution And Climate Integrated Modeling

Integrated air quality modelling systems are tools that help in understanding impacts from aerosols and gas-phase compounds emitted from urban sources on the urban, regional, and global climate. Piringer et al., 2007 have demonstrated that urban features essentially influence atmospheric flow and microclimate, strongly enhance atmospheric turbulence, and modify turbulent transport, dispersion, and deposition of atmospheric pollutants. Numerical weather prediction (NWP) models with increased resolution helps to visualize a more realistic reproduction of urban air flows and air pollution processes.

Integrated models thus link urban air pollution, tropospheric chemistry, and climate. Integration time required is $\geq$ 10 years for tropospheric chemistry studies in order to consider CH_4 and O_3 simulation and aerosol forcing assessment. Tropospheric chemistry and climate interaction studies extend the integration time to $\geq$ 100 years.

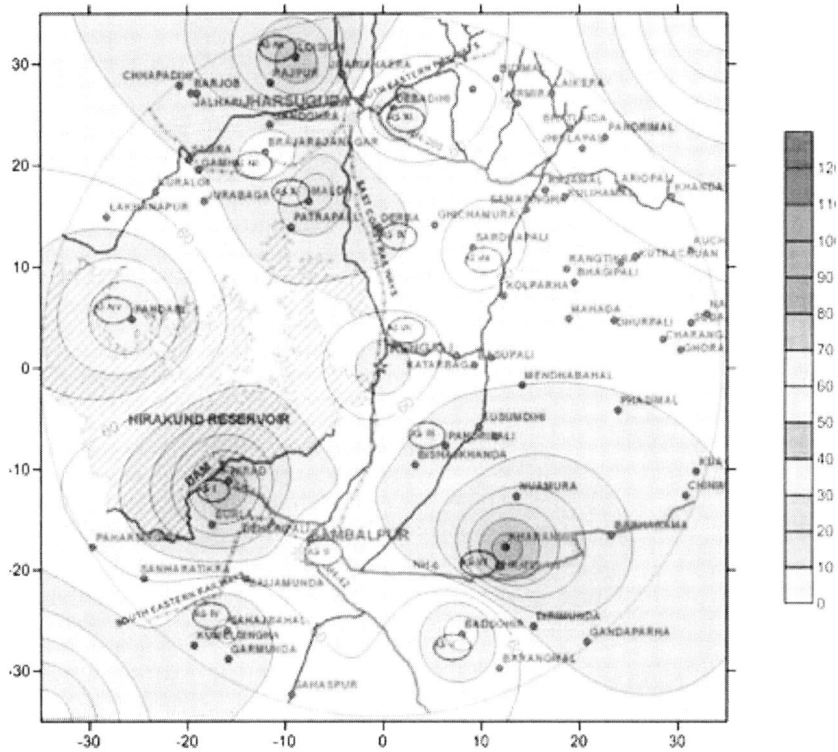

AQI of ambient air	Description of air quality
Below 20	Excellent
Between 20 and 39	Good
Between 40 and 59	Fair
Between 60 and 79	Poor
Between 80 and 99	Bad
Beyond 100	Dangerous

FIGURE 6. Air Quality Index of an Industrial Area: Orissa, India

Urban air quality and population exposure in the context of global to regional to urban transport and climate change is proposed to be assessed by integrating urbanized NWP and Atmospheric Chemistry (ACT) models (Baklanov et al., 2008; Korsholm et al., 2008). A. A. Baklanov and R. B. Nuterman (2009) sugested a multi-scale modelling system which comprised of downscaling from regional to city-scale with the Environment

–HIgh Resolution Limited Area Model (Enviro-HIRLAM) and to micro-scale with the obstacle-resolved Microscale Model for Urban Environment (M2UE). Meteorology governs the transport and transformations of anthropogenic and biogenic pollutants, drives urban air quality and emergency preparedness models; meteorological and pollution components have complex and combined effects on human health (e.g., hot spots, heat stresses); and pollutants, especially urban aerosols, influence climate forcing and meteorological events (precipitation, thunderstorms, etc.), thus this approach is closer to real life scenario. Examples of integrated models are Enviro-HIRLAM:Baklanov and Korsholm, 2007, WRF-Chem: Grell et al., 2005; EMS-FUMAPEX: Forecasting Urban Meteorology, Air Pollution and Population Exposure; CFD (large eddy simulation [LES] or Reynolds-averaged Navier-Stokes [RANS]) models: Galmarini et al., 2008 and Martilli and Santiago., 2008;MIT Integrated Global System Model Version 2 (IGSM2): A.P. Sokolov, C.A. Schlosser, S. Dutkiewicz, S. Paltsev, D.W. Kicklighter,H.D. Jacoby, R.G. Prinn, C.E. Forest, J. Reilly, C. Wang, B. Felzer,M.C. Sarofim, J. Scott, P.H. Stone, J.M. Melillo and J. Cohen., 2005; US EPA and NCAR communities for MM5 (Dupont et al., 2004; Bornstein et al., 2006; Taha et al., 2008), WRF models (Chen et al., 2006); THOR - an Integrated Air Pollution Forecasting and Scenario Management System: National Environmental Research Institute (NERI), Denmark.

The outline of overall methodology of FUMAPEX and MIT interactive chemistry model is shown inFigure 6 and 7. Schematic of couplings between atmospheric model and the land model components of the MIT IGSM2 is given in Figure 8.

Need of integrated models

All of these models have uncertainties associated with them. Chemical transport models, such as Gaussian plume models and gridded photochemical models, begin with pollutant emissions estimates and meteorological observations and use chemical and physical principles to predict ambient pollutant concentrations. Since these models require temporally and spatially resolved data and can be computationally intensive, they can only be used for well-characterized regions and over select time periods. Eulerian grid models are not suitable to assess individual source impacts, unless the emissions from the individual source are a significant fraction of the domain total emissions. This limitation arises from the assumption that emissions are uniformly mixed within the grid cell, and thus do not properly address the initial growth and dispersion of the pollutants.

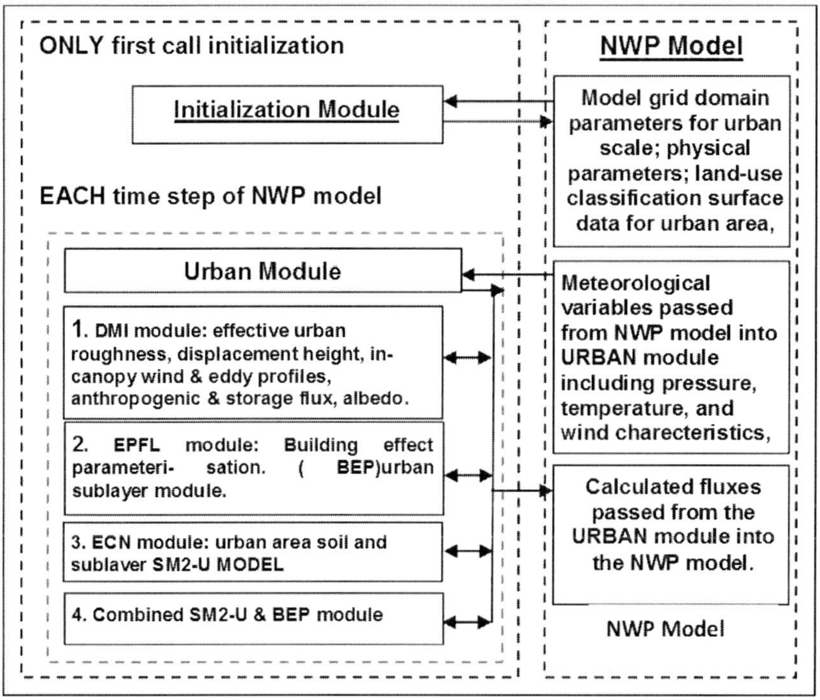

FIGURE 7. General scheme of the FUMAPEX urban module for NWP models.

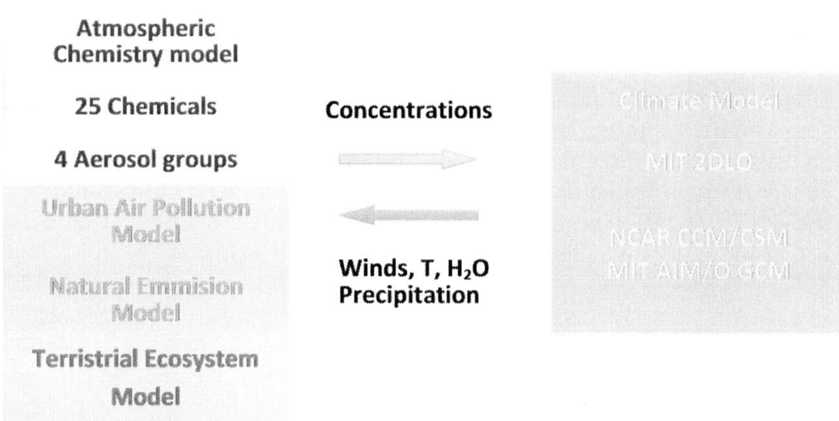

FIGURE 8. Overall Scheme MIT Interactive Chemistry-Climate Model

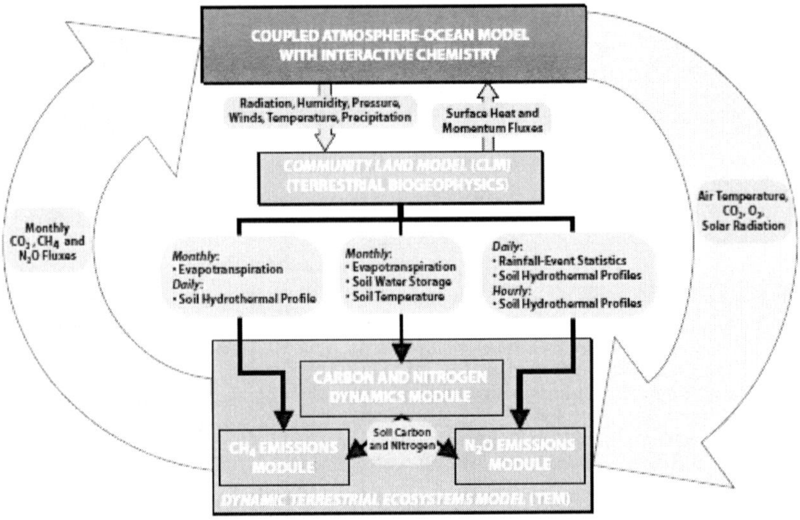

FIGURE 9. Schematic of coupling between the atmospheric model (which also includes linkages to the air chemistry and ocean models) and the land model components of the IGSM2, also shown are the linkages between the biogeophysical (CLM) and biogeochemical (TEM) subcomponents. All green shaded boxes indicate fluxes/storage that are explicitly calculated/tracked by this Global Land System (GLS). The blue shaded boxes indicate those quantities that are calculated by the atmospheric model of the IGSM2.

Lagrangian plume and puff models account for chemical processes by simple linear transformations in time. These models can track individual source impacts, thus enabling user to outline source specific air pollution control strategies. Considerable differences are observed when concentrations are compared in time and space because of uncertainties in the characterization of the direction of transport that are of the order of the actual plume width. The observed and simulated concentrations for fixed receptors, give estimates of maximum concentration values within a factor of two or three of those observed. These differences are an order of magnitude larger than those observed for estimates of secondary pollutants. Both Eulerian and Lagrangian, models are not suitable to handle inert pollutants and secondary pollutants whose concentrations depend on reaction rates and are photochemical in nature.

Receptor models, such as Positive Matrix Factorization and Chemical Mass Balance (CMB), source apportionment addresses the problem by statistical inference of source contributions to total pollution from observations of ambient air chemical composition. Mass balance methods of source apportionment use linear models with chemical composition vectors of sources as covariates. Knowledge of meteorological variables is not required but may be used to refine the analysis. Knowledge of emission sources is useful for the interpretation of results from statistical-based receptor models and is required by receptor models that use a mass balance approach. Less data and computational resource requirement by Receptor models as compared to chemical transport models, make them more convenient tool for evaluation of ambient pollutant concentrations and pollutant emission inventory. However, their utility for reactive air pollutants is uncertain and questionable. The disadvantage of CMB model arises from its assumptions. such as constant compositions of source emissions over the period of ambient and source sampling; linear additive and unreactive chemical species; identification of all sources contributing to the receptor and knowledge of their emission profile, linearly independent emission profiles.

The urban air quality models requires

- Good net work ambient air concentrations of pollutants of concern: Geography of the urbanarea, constructed clusters, road network, location of bluidings etc play a major role in dispersion of pollutants. Thus to understand the ambient status of pollutants it is necessary to have sufficient number of monitoring locations to cover the urban sprawl of concern.
- Micro metereology data: The wind patterns, temperature, humidity alter in urban areas according to anthropogenic activity and architecture
- Bluilding details: To account for the effect of anthropogenic architecture falling in path of plume, its geometry is required to be known.
- Knowledge of all sources: All sources and their emission profiles are required to be known to plan for further development in urban area and control of pollutant emission
- Atmospheric Chemistry: All transformations of emitted chemical species, their reaction rates pathways must be known to account for observed concentration of pollutants.
- Healthy Impacts: Models need to incorporate health effect of pollutants

None of the models available can handle all the requirements of urban air quality management. Each one focuses of one aspect and thus coupling of different models are required.

4. FURTHER ISSUES TO BE ADDRESSED

COST an intergovernmental framework for European Cooperation in Science and Technology, Europe, addressed issues related to urban air quality models in its action programmes. Cost 728 focussed on enhancing mesoscale meteorological Modeling capabilities for air pollution and Dispersion applications under larger programme of urbanization of meteorological and air quality models. The issues identified for improvements to the state of urbanization of models can be summarized as

- Systematic evaluation of urban land surface schemes
- Increasing the range of variables observed to ensure as complete a range of evaluation as possible
- evaluation over a broad spectrum of conditions (meteorological, morphological, geographical setting, etc.
- Testbeds and observatories with different objectives and dataset richness.
- A deeper understanding of urban PBL dynamics i.e development of long-term urban test beds in a variety of geographic regions (e.g., inland, coastal, complex terrain) and in many climate regimes, with a variety of urban core types (e.g., deep versus shallow, homogeneous versus heterogeneous).
- A framework to address conceptual issue of evaluation of model prediction of the flow within the canopy
- User friendly and multifaceted urban databases and enabling technology
- Developing core capabilities for advancing urban modeling and boundary layer research
- An open database to address issues of availability and sources of high-resolution data sets easily to all with mechanism for its maintenance, upgrading, updating, and archiving.
- www.unep.org/urban_environment/pdfs/handbook.pdf

REFERENCES

1. A. Baklanov, U. Korsholm, 2007 On-line integrated meteorological and chemical transport modelling: advantages and prospective. In: Preprints ITM 2007: 29th NATO/SPS International Technical Meeting on Air Pollution. Modelling and its Application, 24-28.09.2007, University of Aveiro, Portugal, 2134 .
2. A. Baklanov, U. Korsholm, A. Mahura, C. Petersen, A. Gross, E. N. V. I. R. O. -H, I. R. L. A. M. on-line, modelling. coupled, urban. of, meteorology, pollution. air, Adv, Sci. Res., 2 41-46, 2008.
3. A. A. Baklanov, R. B. Nuterman, atmospheric. Multi-scale, modeling. environment, urban. for, Advances. areas, Science. in, . Research, 53 EOF57 EOF 5357 2009
4. R. Bornstein, R. Balmori, H. Taha, D. Byun, B. Cheng, J. Nielsen-Gammon, S. Burian, S. Stetson, M. Estes, D. Nowak, P. Smith, 2006 Modeling the effects of land-use land cover modifications on the urban heat island phenomena in Houston, Texas. SJSU Final Report to Houston Advanced Research Center for Project R-04 -0055, 127
5. G. A. Briggs, 1975 Plume Rise Predictions. Lectures on Air Pollution and Environmental Impact Analysis. D. A. Haugen. Boston, MA, American Meteorology Society: 59111 .
6. G. R. Carmichael, L. K. Peters, 1979 Numerical simulation of the regional transport of SO2 and sulfate in the eastern United States, Proc. 4 th Symp. on turbulence, diffusion and air pollution, AMS 337.
7. F. Chen, M. Tewari, H. Kusaka, T. L. Warner, 2006 Current status of urban modeling in the community Weather Research and Forecast (WRF) model. Sixth AMS Symposium on the Urban Environment,Atlanta GA, January 2006.
8. S. Dupont, T. L. Otte, J. K. S. Ching, 2004 Simulation of meteorological fields within and above urban and rural canopies with a mesoscale model (MM5) Boundary-Layer Meteor., 2004, 113:111-158.
9. B. A. Egan, K. S. Rao, A. Bass, 1976 A three dimensional advective-diffusive model for long-range sulfate transport and transformation 7 th ITM, 697, Airlie House.
10. S. K. Friedlander, 1973 Chemical element balances and identification of air pollution sources, Environmental Science and Technology 7 235240

11. S. Galmarini, J. , F. Vinuesa, A. Martilli, 2008 Relating small-scale emission and concentration variability in air quality models, Chapter 1.2, URBANIZATION OF METEOROLOGICAL AND AIR QUALITY MODELS, COST Action 728, 15 May 2008, http://www.cost728.org

12. J. R. Garratt, 1978 Transfer Characteristics for a Heterogeneous Surface of Large Aerodynamic Roughness. Quart. J. R. Meteorol. Soc., 104: 491502 .

13. Garratt J.R. 1980 Surface influence upon vertical profiles in the nocturnal boundary layer Boundary-Layer Meteorology 26 1 6980 , DOI: 10.1007/BF00164331

14. Garratt J.R. 1994 Review: the atmospheric boundary layer, Earth-Science Reviews, 37, 89134 .

15. L. Gidhagen, C. Johansson, J. Langner, G. Olivares, 2004 "Simulation of NOx and ultrafine particles in a street canyon in Stockholm, Sweden." Atmospheric Environment 38 14 20292044 .

16. Gifford Jr, F. A. , 1976 "Consequences of Effluent Releases." Nuclear Safety 17 1 6886 .

17. G. A. Grell, S. E. Peckham, R. Schmitz, S. A. Mc Keen, G. Frost, W. C. Skamarock, B. Eder, 2005 Fully coupled "online" chemistry within the WRF model, Atmos. Environ., 39(37), 6957-6975.

18. R. C. Henry, 2003 Multivariate receptor modeling by N-dimensional edge detection. Chemometrics and Intelligent Laboratory Systems 65 179-189.

19. N. S. Holmes, L. Morawska, et al. 2005 "Spatial distribution of submicrometre particles and CO in an urban microscale environment." Atmospheric Environment 39 22 39773988 .

20. N. S. Holmes, L. Morawska, 2006 A review of dispersion modelling and its application to the dispersion of particles: An overview of different dispersion models available. Atmospheric Environment, 40 5902-5928.

21. P. K. Hopke, 1985 Receptor Modeling in Environmental Chemistry, Wiley, New York.

22. P. K. Hopke, ed, 1991 Receptor Modeling for Air Quality Management, Elsevier, Amsterdam.

23. G. Hosker, 1985 1985 Clin. Phys. Physiol. Meas. 6 131 A uni-directional urethral force gauge Pavageau, M., Rafailidis, S. and Schatzmann, M. (2001) 'A comprehensive experimental databank

for the verification of urban car emission dispersion models', International Journal of Environment and Pollution, 15, 417425

24. H. D. Jacoby, R. G. Prinn, C. E. Forest, J. Reilly, C. Wang, B. Felzer, M. C. Sarofim, J. Scott, P. H. Stone, J. M. Melillo, J. Cohen, Report, 124 July 2005

25. U. S. Korsholm, A. Baklanov, A. Gross, J. H. Sørensen, the. On, of. importance, meteorological. the, interval. coupling, dispersion. in, during. E. T. E. modeling, 1 Atmos. Environ., doi:10.1016/j.atmosenv.2008.11.017, 2008.

26. A. Martilli, J. L. Santiago, 2008 How to use computational fluid dynamics models for urban canopy parameterizations. Chapter 2.1, URBANIZATION OF METEOROLOGICAL AND AIR QUALITY MODELS, COST Action 728, 15 May 2008, http://www.cost728.org

27. S. L. Miller, M. J. Anderson, E. P. Daly, J. B. Milford, 2002 Source apportionment of exposures to volatile organic compounds. I. Evaluation of receptor models using simulated exposure data. Atmospheric Environment 36 3629-3641.

28. D. Mumovic, J. Crowther, Z. Stevanovic, 2006 Integrated Air Quality Modelling for a Designated Air Quality Management Area in Glasgow, Building and Environment, 41 12 17031712 , doi:10.1016/j.buildenv.2005.07.006

29. D. Mumovic, J. Crowther, 2006 Assessing Urban Air Quality using Microscale CFD Modeling, PHOENICS CFD Newsletter Spring 2006 1010

30. T. R. Oke, 1988 "The urban energy balance," Progress in Physical Geography, 12 471508 .

31. P. Paatero, 2004 User's Guide for Positive Matrix Factorization Programs PMF2 and PMF3, Part 1: Tutorial. University of Helsinki, Finland.

32. F. Pasquill, 1961 "The Estimation of the Dispersion of Windborne Material." Meteorology Magazine 90(1063): 3340 .

33. M. Pavageau, S. Rafailidis, M. Schatzmann, 2001 'A comprehensive experimental databank for the verification of urban car emission dispersion models', International Journal of Environment and Pollution, 15 417425

34. W. L. Physick, J. A. Noonan, Gregor. J. L. Mac, 1994 LADM: a Lagrangian Atmospheric Dispersion Model [Monograph], CSIRO Division of Atmospheric Research technical paper.

35. M. Piringer, E. Petz, I. Groehn, G. Schauberger, 2007 A sensitivity study of separation distances calculated with the Austrian Odour Dispersion Model (AODM), Atmospheric Environment, 41 725-1735.

36. M. R. Raupach, A. S. Thom, I. Edwards, 1980 "A Wind Tunnel Study of Turbulent Flow Close to Regularly Arrayed Rough Surfaces", Boundary-Layer Meteorol. 18 373397 .

37. S. Reynolds, P. Roth, J. Seinfeld, 1973 Mathematical modeling of photochemical air pollution Atm.Env 7.

38. M. W. Rotach, 1993, close. Turbulence, a. to, urban. rough, Part. I. I. surface, Variances, Boundary. gradients, Meteorol. Layer, 66 75-92,

39. M. Rotach, 1995 (1995) 'Profiles of turbulence statistics in and above an urban street canyon', Atmospheric Environment, 29 14731486

40. C. C. Shir, L. J. Shieh, 1974 A generalized urban air pollution model and its application to the study of SO2-distribution in the St. Louis Metropolitan area, J. Appl. Met. 19 185204 .

41. 2002: South Asia Urban Air Quality Management Briefing. Note 4 , What Do We Know About Air Pollution?-India Case Study,http://www.worldbank.org/sarurbanair

42. 2002: South Asia Urban Air Quality Management Briefing Note 5 , Impact of Traffic Management, http://www.worldbank.org/sarurbanair.

43. Srivastava, Anjali ; Sengupta B., Dutta S.A., 'Source apportionment of ambient VOCs in Delhi City Science of The Total Environment', Volume 343 1-3 , 1 May2005,207220

44. Srivastava, Anjali ; 'Source apportionment of ambient VOCS in Mumbai city Atmospheric Environment', Volume 38, Issuey39, December2004, pp. 6829-6843

45. D. Stanners, P. . Bourdeau, Eds., 1995 Europe's Environment: The Dobris Assessment, European Environment Agency.Office for Publications of the European Communities, Luxemburg.

46. A. P. Sokolov, C. A. Schlosser, S. Dutkiewicz, S. Paltsev, D. W. Kicklighter, 2005Th :The MIT Integrated Global System Model (IGSM),Version 2: Model Description and Baseline Evaluation,

47. H. Taha, 2008 Sensitivity of the urbanized MM5 (uMM5) to perturbations in surface properties in Houston Texas. Boundary-Layer Meteorology, 127 193-218

48. H. Tennekes, 1973 A model for the dynamics of the inversion above a convective boundary layer. Journal of Atmospheric Science 30, 550567

49. THOR- an Integrated Air Pollution Forecasting and Scenario Management System. Available at thor.dmu.dk National Environmental Research Institute (NERI), Denmark

50. Urban Environnent Management, Tool book United Nations Environment Programme, United Nations Human Settlements Programme. Available at www.unep.org/urban_environment/pdfs/handbook.pdf

8

CHAPTER

Economic evaluation of the air pollution effect on public health in China's 74 cities

Li Li, Yalin Lei, Dongyan Pan, Chen Yu and Chunyan Si

School of Humanities and Economic Management, China University of Geosciences, Beijing 100083, China

ABSTRACT

Air deterioration caused by pollution has harmed public health. The existing studies on the economic loss caused by a variety of air pollutants in multiple cities are lacking. To understand the effect of different pollutants on public health and to provide the basis of the environmental governance for governments, based on the dose–response relation and the willingness to pay, this paper used the latest available data of the inhalable particulate matter (PM_{10}) and sulphur dioxide (SO_2) from January 2015 to June 2015 in 74 cities by establishing the lowest and the highest limit scenarios. The results show that (1) in the lowest and highest limit scenario, the health-related economic loss caused by PM_{10} and SO_2 represented 1.63 and 2.32 % of the GDP, respectively; (2) For a single city, in the lowest and the highest limit scenarios, the highest economic loss of the public health effect caused by PM_{10} and SO_2 was observed in Chongqing; the highest economic loss of the public health effect per capita occurred in Hebei Baoding. The highest proportion of the health-related economic loss accounting for GDP was found in Hebei Xingtai. The main reason is that the terrain conditions are not conducive to the spread of air pollutants in Chongqing, Baoding and

Xingtai, and the three cities are typical heavy industrial cities that are based on coal resources. Therefore, this paper proposes to improve the energy structure, use the advanced production process, reasonably control the urban population growth, and adopt the emissions trading system in order to reduce the economic loss caused by the effects of air pollution on public health.

Keywords: Air pollution The public health effect The economic loss 74 cities China

BACKGROUND

Air pollution mainly refers to human activities or the natural processes that cause a certain substance to continuously enter the atmosphere at a sufficient concentration to endanger the health and cause environmental pollution. There are many types of air pollutants, the primary of which are total suspended particulates (TSP), inhalable particulate matter (PM_{10}), fine particulate matter ($PM_{2.5}$), SO_2 and NOx, among others (Yu et al. 2008). After inhalation of harmful pollutions, humans may develop respiratory disease and can suffer from serious diseases, such as tracheitis, bronchitis, asthma, lung disease and lung cancer, for many years. The energy consumption structure in China is mainly based on coal resources, and the rapid growth of motor vehicles in cities results in increasingly more serious air pollution in China's large cities; in addition, the questions regarding public health and air pollution have garnered widespread attention (Wilkinson and Smith 2007).

The main air pollutions in China are PM_{10}, SO_2 and NOx (Chen et al. 2001). In the 12th Five-Year period, China clearly proffered the target to reduce the total SO_2 emissions by 8 % and increase the ratio of the urban air quality to achieve the second level of 8 % (Ministry of Environmental Protection of the People's Republic of China 2011). According to an environmental analysis reported by the Asian Development Bank in 2013, the Chinese government was taking measures to control air pollution; however, of the world's 10 most seriously polluted cities, 7 cities were in China. Of China's 500 major cities, less than 1 % met the standards of the World Health Organization (Zhang2012). In 2015, the World Health Organization released a report stating that at least one in eight people died of air pollution globally. Air pollution has become the world's largest environmental health risk (Huanqiunet 2015).

Since January 1, 2013, the Ministry of Environmental Protection has monitored the air quality index of Beijing, Tianjin, Hebei, the Yangtze River Delta, the Pearl River Delta region, the municipality directly under the central government, the provincial capital cities and the cities specifically designated in the state plan, which are collectively called the 74 cities, in brief (China's National Environmental Monitoring Centre 2013a). The concentrations of pollutants such as PM_{10} and SO_2 have been monitored since November 2014 (China's National Environmental Monitoring Centre 2013b). As World Bank (1997) didn't identify the health effects of NOx and NOx wasn't included in the dose–response relation of Ho and Jorgenson (2007), thus to quantitatively evaluate the economic loss due to the effects of air pollution on public health in China, this paper analyzes PM_{10} and SO_2 based on the latest available data from January 2015 to June 2015, uses the method of foreign study on China's economic loss due to air pollution effects on public health for reference (Wang and Smith 1999; Ho and Nielsen2007) and estimates the economic loss caused by the effects of air pollution on public health in the 74 cities. An evaluation of the health-related economic loss can provide a basis for the government to develop and initiate preventative measure for controlling air pollution. At the same time, these findings can also improve the environmental protection awareness of the local government and the public.

LITERATURE REVIEW

In recent decades, industrialization and urbanization have experienced rapid development, which has resulted in increasing air pollution. According to the World Bank, there is a close relationship between air pollution and public health. There is a positive relation between the concentration of air pollutants and respiratory diseases, lung function loss, chronic bronchitis and premature death (World Bank SEPA 2007). The evaluation of health-related economic loss caused by air pollution has become a hot topic for scholars and institutions.

Research progress on the economic loss regarding to public health impacts caused by air pollution

Ridker (1967) calculated the economic loss associated with different diseases which caused by air pollution in the USA in 1958 by using the human capital method. The results showed that the economic loss related to the effects on public health was 80.2 billion dollars in the USA. This study hailed the beginning of the calculation of health-related economic loss caused by air pollution.

Employing a survival analysis and the data from a 14- to 16-year mortality follow-up of 8111 adults in the six cities in the U.S., Dockery et al. (1993)

estimated the associations between particulate air pollution and daily mortality rates. Their results confirmed that the mortality rate was associated with the level of air pollution. Using data from 1994 to 1995 in Hong Kong, Wong et al. (1999) determined that adverse health effects were evident at the current ambient concentrations of air pollutants. Samet et al. (2000) recognized an association between daily changes in the concentration of ambient particulate matter and the daily number of deaths (mortality) in the United States. Wong et al. (2002) used Poisson regression to estimate the associations between daily admissions and the levels of $PM_{10}$and SO_2 in Hong Kong and London. The results confirmed that air pollution caused detrimental short-term health effects.

Using the collective data regarding to PM_{10} and SO_2 from January 1999 to September 2000, Kaushik et al. (2006) assessed the ambient air quality status in the fast growing urban centres of Haryana state, India. Adopting the daily data during 2008 and 2009 in Beijing, Xu et al. (2014b) confirmed that short-term exposure to particulate air pollution was associated with increased ischemic heart disease (IHD) mortality.

In 1981, the concept, theory and method of environmental pollution economic loss assessment were put forward and discussed in the congress of the National Symposium on Environmental Economics (Xia 1998). Thereafter, the economic loss associated with environmental pollution was of interest to scholars. Gao et al. (1993) adopted the GEE (Generalized Estimation Equation) to study the relationship between TSP in Haidian District, Beijing and low air pollution. Using the two methods (ecology and time series) and the data of 1992 in Shenyang, Xu et al. (1996) determined that total mortality, chronic obstructive pulmonary disease (COPD), cardiovascular disease and pollution levels were significantly correlated. Jing and Ren (2000) conducted an epidemiological survey on adults who were older than 25 years using the multiple logistic regression analysis. The results showed that 6 types of respiratory system diseases or symptoms appeared with an increasing frequency as air pollution levels increased. Chen and Hong (2002) quantitatively evaluated the air pollution in Shanghai based on the risk evaluation method and found that the health effects caused by SO_2 exhibited a gradually declining trend. Chen et al. (2010) evaluated the health impacts of particulate air pollution on urban populations in 113 Chinese cities, and it was estimated that the total economic cost of the health impact was approximately 341.4 billion Yuan, 87.79 % of which was attributable to premature deaths. Chen et al. (2015) employed a Poisson regression model to estimate residents' health benefits in two scenarios: environmentally controlled scenario 1 and environmentally controlled scenario 2. Scenario 2 showed a potentially higher reduction of emissions and greater health benefits than scenario 1. Xu et al. (2014a) used the established model between PM_{10} and thermal environmental indicators to evaluate the PM_{10}—related health risk in Beijing.

Certain scientific institutions also focused more on the economic loss associated with public health effects caused by air pollution. The World Health Organization estimated that the total loss globally caused by air pollution-related disease was 0.5 % (Murray and Lopea 1997) in 1997. In the same year, the World Bank systematically studied the health effects caused by air pollution in China (World Bank 1997). The U.S. Environmental Protection Agency estimated that the economic benefits of health and ecological improvement in the United States from 1990 to 2010 were as high as $6–50 trillion, most of which could be attributed to the decrease in the number of deaths caused by air pollution (U.S. EPA 1999). The World Health Organization reported that 80 % of the world's cases of heart disease and stroke deaths were due to air pollution, and a total of 7 million people in the world died of air pollution in 2014 (Huanqiunet 2014). In 2015, the World Health Organization released data that at least 1 in every 8 people died of air pollution throughout the world. Air pollution has become the world's largest environmental health risk (Huanqiunet 2015).

Research progress on the method used to evaluate the economic loss associated with public health effects caused by air pollution

The previous studies regarding to the economic loss caused by the effects of air pollution on public health generally included the determination of economic loss using the contents of the environmental pollution assessment, the public health impact assessment and a choice of methods. Generally, the methods used were as follows.

Modified human capital method

Ridker (1967), Dockery et al. (1993), Wang et al. (2005), Jia et al. (2004), Wan et al. (2005), Han et al. (2006), Zhang et al. (2008), Shang et al. (2010), Han (2011), and Shen et al. (2014) quantitatively estimated the economic loss in different regions and obtained different results.

Illness cost method

Air pollution led to changes in the disposable income of people, particularly, an increase in medical expenses. Medical expenses became a recognized fact, and they also became a very heavy burden on civilians. Based on the above views, certain scholars obtained conclusions by analysing the illness costs caused by air pollution. These scholars include Chen et al. (2010), Zmirou et al. (1999), Hedley et al. (2008), Patankar and Trivedi (2011), Brandt et al. (2014), and Yan (2012).

Willingness to pay

Willingness to pay is an indirect evaluation method which constructs a simulated market to reveal people's willingness to pay for certain environmental goods, in order to evaluate the value of environmental quality. Researchers included Carlsson and Martinsson (2001), Wang and John (2006), Koop and Tole (2004), Pascal et al. (2013), Yaduma et al. (2013), Ami et al. (2014), Istamto et al. (2014), Cai and Yang (2003), Peng and Tian (2003), Cai et al. (2007), Zhou et al. (2010), and Zeng et al. (2015).

Literature summary

1. In the studies of the economic loss caused by air pollution, domestic and foreign researchers studied the qualitative relationship between and quantitative analysis of air pollution and its health effect. However, generally, previous studies were solely based on a country, a city or a type of air pollutant; the health effects of many types of air pollutants in a city of a typical city are moderate.

2. Regarding to the method for evaluating the economic loss caused by the effects of air pollution on public health, the deficiencies of the modified human capital method were that the life prediction of the society may not be reasonable, and the different choices of the discount rate would have a large impact on the evaluation results. The disadvantage of the illness cost method is that it may underestimate the illness value. Additionally, the method's other disadvantage was that the individual may have a willingness to pay.

From the economic perspective, the willingness to pay method is the most reasonable method because it can reveal the value of all goods and utilities, and it can completely evaluate the economic values of environmental resources, which is currently being widely recognized and accepted. Thus, this paper utilizes the willing to pay method to evaluate the economic loss associated with public health effects caused by air pollution in 74 cities.

METHODS AND DATA

The dose–response relationship and the willingness to pay

To study the economic loss related to public health effects caused by air pollution, it is necessary to consider the types of public health effects and to establish the relation between the concentration of air pollutants and the effect on public health, which is called a dose–response. In different studies, the dose–response relationship is different. The indexes of public health effects caused by air pollution, which the World Bank put forward, included premature deaths, hospitalization and emergency caused by respiratory diseases, the number of restriction days caused by health problems related to the inhalation of particulate matter, lower respiratory tract infections, childhood asthma, asthma, chronic bronchitis, respiratory symptoms, and chest discomfort. This paper used the dose–response relationship of Ho and Jorgenson (2007) for reference and assumed that all the people in the 74 cities were exposed to the same concentrations of PM_{10} and SO_2. The dose–response relationship is shown in the Eq. (1).

$$HE_{xrh} = DR_{xh} \times C_{rx} \times POP_r \quad (1)$$

where HExrh is the h-th type of public health effect caused by the air pollutant x (including PM_{10} and SO_2) in the region r. DRxhDRxh is the dose–response coefficient of the air pollutant x (unit: the number of the people suffer with the concentration of the air pollutant increasing by 1 μg/m^3) and the h-th type of public health. CrxCrx is the concentration of the air pollutant x in the region r. POPrPOPr is the number of the people in the region r. Ho and Jorgenson (2007) used the survey data of Beijing and Anqing in 1997 to estimate the economic loss caused by the health effects of Chinese residents using the willingness to pay method, the population of was were 6.53 million and 0.35 million respectively. Based on the loss value estimation of Ho and Jorgenson (2007), this paper modified it, as shown in Table 1. The total economic loss of the 74 cities is obtained by adding up all of the economic loss types relating to the health effects, as shown in the Eqs. (2) and (3):

$$HEV_{xrh} = V_{xh} \times HE_{xrh} \quad (2)$$

$$\text{THEV} = \sum_r \sum_x \sum_h HEV_{xrh} \quad (3)$$

where HEVxrhHEVxrh is the economic loss of the h-th type of public health effect caused by the air pollutant x (including PM_{10} and SO_2) in the region

r. VxhVxh is the economic loss value of the h-th type of public health effect caused by the air pollutant x (including PM_{10} and SO_2). THEVTHEV is the total economic loss related to public health effects caused by the air pollutant. In calculating total health-related economic loss, the paper adds up eight effects of PM_{10} on public health and three effects of SO_2 on public health together, which may appear double counting. As there is little literature of this issue, the paper hasn't analyzed it.

Table 1: The dose–response relationship and the loss value estimation of the public health effects

	Coefficient of the dose–response relationship		Loss value estimation (yuan, the price in 2002)	Loss value estimation (yuan, the mean price from January 2015 to June 2015)
	The lower limit scenario	The highest limit scenario	Ho and Jorgenson's estimation	Modified estimation
Effects of PM_{10} on public health				
Premature death	1.3	2.6	370,000	528,370.2
Hospitalization caused by respiratory disease	12	12	1751	2500.5
Emergency	235	235	142	202.8
Number of restriction days	18,400	57,500	14	20
Lower respiratory tract infections and childhood asthma	23	23	80	114.2
Asthma	1770	2608	2.5	3.57
Chronic bronchitis	61	61	48,000	68,545.33
Respiratory symptoms	49,820	183,000	3.7	5.28
Effects of SO_2 on public health				
Premature death	1	2.6	370,000	528,370.2
Chest discomfort	10,000	10,000	6.2	8.85
Lower respiratory tract infections and childhood asthma	5	5	6.2	8.85

Source: Ho and Jorgenson (2007)

Data

The environmental data of 74 cities during the period from January 2015 to June 2015 were reported by China's National Environmental Monitoring Centre (2015a, b, c, d, e, f), which mainly contained the monthly mean concentrations of PM10 and SO_2 (Figs. 1, 2). The number of the population and GDP in 74 cities were obtained from Askcinet (2015).

Fig. 1 Monthly mean concentration of SO_2 in China's 74 cities

Fig. 2 Monthly mean concentration of PM_{10} in China's 74 cities

From Figs. 1 and 2, Taiyuan, Shenyang and Yinchuan are the top three cities with the highest monthly mean concentration of SO_2. Baoding, Zhengzhou and Xingtai are the top three cities with the highest monthly mean concentration of PM_{10}. According to the ambient air quality standard GB3095-2012, the paper compared Chongqing, Baoding and Xingtai with Beijing and found that: (1) For SO_2, Beijing and Chongqing achieved the first level of national standards from January 2015 to June 2015. Baoding and Xingtai achieved the first level of national standards from April to June 2015, and they achieved the second level of national standards from January to March 2015. (2) For PM_{10}, Beijing achieved the second level of national standards from January 2015 to June 2015. Chongqing achieved the second level of national standards from February 2015 to June 2015, which didn't achieve the national standards in January 2015. Baoding and Xingtai achieved the second level of national standards from April to June 2015, but they didn't achieve the national standards from January to March 2015. Overall, the monthly mean concentration of SO_2 and PM_{10} in the 4 cities appeared a downward trend.

RESULTS AND DISCUSSIONS

The total economic loss associated with public health effects caused by air pollution in 74 cities

The dose–response relationship and the loss value estimation of the public health effects in different cities vary; therefore, this paper establishes different scenario parameters for the lowest limit scenario and the highest limit scenario in order to evaluate the total economic loss related to the effects of air pollution on public health in 74 cities.

1. The lowest limit scenario

As shown in Table 2, there were 84,917 premature deaths caused by PM_{10} and SO_2. There were 646,282 hospitalizations caused by respiratory disease, 12.66 million emergencies, more than 990 million restriction days and 1.23 million lowest respiratory tract infections and occurrence of childhood asthma due to PM_{10}. However, the economic loss caused by the effects of SO_2 on public health was less than that of PM_{10}.

Table 2 The total lowest economic loss caused by the effects of air pollutants on public health in 74 cities

	Coefficient of the dose–response relation	Cases	Loss value estimation (yuan, the mean price from January 2015 to June 2015)	Economic loss (million yuan)
Effects of PM_{10} on public health				
Premature death	1.3	70,014	528,370.2	36,993.31
Hospitalization caused by respiratory disease	12	646,282	2500.5	1616.03
Emergency	235	12,656,359	202.8	2566.71
Number of restriction days	18,400	990,965,979	20	19,819.32
The lowest respiratory tract infections and childhood asthma	23	1,238,707	114.2	141.46
Asthma	1770	95,326,619	3.57	340.32
Chronic bronchitis	61	3,285,268	68,545.33	225,189.78
Respiratory symptoms	49,820	2,683,148,101	5.28	14,167.02
Total economic loss of the public health effect caused by PM_{10}	300,833.95			
Effects of SO_2 on public health				
Premature death	1	14,903	528,370.2	7874.30
Chest discomfort	10,000	149,030,000	8.85	1318.92
The lowest respiratory tract infections and childhood asthma	5	74,515	8.85	0.66
Total economic loss caused by the effects of SO_2 on public health effect	9193.88			
Total economic loss related to public health effects	310,027.82			
Economic loss related to public health effects per capita	598 yuan			
GDP in the 74 cities	18,987,854			
Total economic loss of the public health effects accounting for the GDP	1.63 %			

This paper calculated that the total health-related economic loss caused by the air pollutant in 74 cities was approximately 310 billion yuan, explaining approximately 1.63 % of the 74 cities' GDP, which was higher than the result of Wei et al. (2012). The total economic loss of the public health effect caused by PM_{10} was 300.8 billion yuan, explaining approximately 97.03 % of the total economic loss, which was the major economic loss and was consistent with the result of Zhang (2012). The economic loss caused by chronic bronchitis, which was approximately 225.2 billion yuan. It was the largest in the total economic loss, explaining approximately 72.64 % of the total economic loss. The result was different from that of Chen et al. (2010), who determined that the economic loss caused by premature death was the largest.

2. The highest limit scenario

Table 3 showed there were 178,776 premature deaths caused by PM_{10} and SO_2. There were 646,282 hospitalizations caused by respiratory disease, 12.66 million emergencies, more than 3090 million restriction days and 1.23 million lower respiratory tract infections and childhood asthma caused by PM_{10}. Similarly, the economic loss of the public health effect caused by SO_2 was also less than that of PM_{10}.

Table 3 The total highest economic loss of the public health effect caused by the air pollutant in 74 cities

	The coefficient of the dose–response relation	Cases	the loss value estimation (yuan, the mean price from January 2015 to June 2015)	economic loss (million yuan)
PM_{10}'s public health effect				
Premature death	2.6	140,028	528,370.2	73,986.62
Hospitalization caused by respiratory disease	12	646,282	2500.5	1616.03
Emergency	235	12,656,359	202.8	2566.71
The number of restriction days	57,500	3,096,768,683	20	61,935.37
Lower respiratory tract infections and childhood asthma	23	1,238,707	114.2	141.46
Asthma	2608	140,458,656	3.57	501.44
Chronic bronchitis	61	3,285,268	68,545.33	225,189.78
Respiratory symptoms	183,000	9,855,802,940	5.28	52,038.64
The total economic loss of the public health effect caused by PM_{10}	417,976.05			
SO_2's public health effect				
Premature death	2.6	38,748	528,370.2	20,473.29
Chest discomfort	10,000	149,030,000	8.85	1318.92
Lower respiratory tract infections and childhood asthma	5	74,515	8.85	0.66
The total economic loss of the public health effect caused by SO_2	21,792.86			
The total economic loss of the public health effect	439,768.91			
Economic loss related to public health effects per capita	848 yuan			
The GDP in 74 cities	18,987,854			
The total economic loss of the public health effect accounts for the GDP	2.32 %			

The total economic loss of the public health effect caused by the air pollutant in 74 cities was approximately 439.8 billion yuan in the highest limit scenario, representing approximately 2.32 % of the GDP in 74 cities, a slight difference from the result of Wei et al. (2012). The total highest economic loss of the public health effect caused by PM_{10} was 418 billion yuan, explaining approximately 95.04 % of the total economic loss, which was the major economic loss and was also consistent with the result of Zhang (2012). The economic loss caused by chronic bronchitis was also the largest in the total economic loss, approximately 225.2 billion yuan, explaining approximately 51.24 %. The result was also different from that of Chen et al. (2010).

The economic loss caused by effect on the public health effect in the major cities

To further understand the health-related economic loss caused by the air pollutant, this paper estimated the health effects of air pollutants and the economic loss in the major cities from January 2015 to June 2015.

1. The lowest limit scenario

From Fig. 3, Chongqing's public health economic loss was 17 billion yuan, ranking the first among the 74 cities, followed by Beijing, Baoding, and Tianjin. From the regional perspective, there were 4 municipalities and 4 cities in Hebei Province in the top 10 cities with the highest economic loss. There were 7 cities in North China excluding Chongqing, Shanghai and Chengdu in the top 10 cities with the highest economic loss.

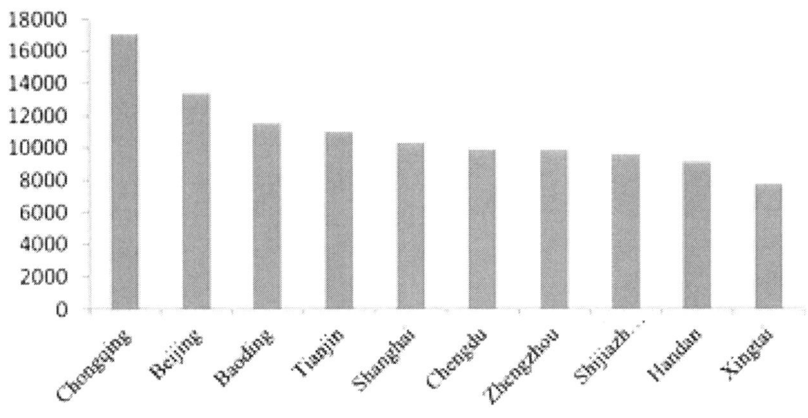

Fig. 3 The 10 cities with the highest economic loss (million yuan)

As shown in Fig. 4, there were 7 cities in Hebei Province in the top 10 with the highest economic loss. Zhengzhou and Ji'nan also ranked in the top 10 cities due to their poor air quality. Urumqi's health-related economic loss was not high; however, because of its low population, the resulting health economic loss per capita was higher.

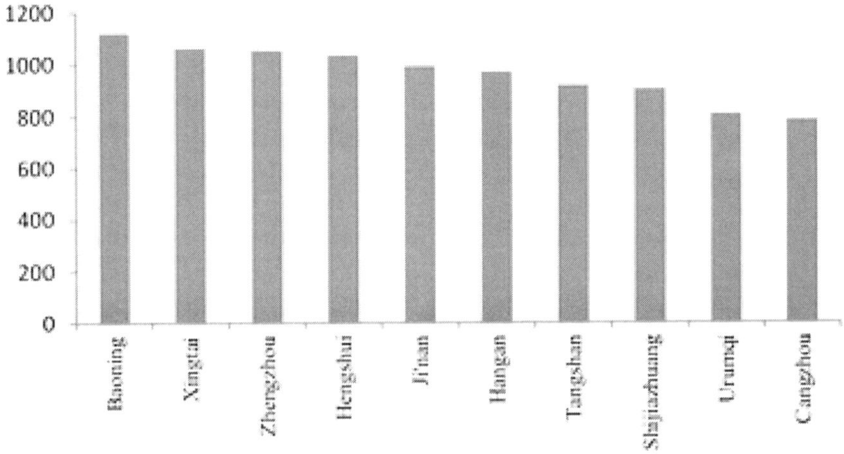

Fig. 4 The 10 cities with the highest health-related economic loss per capita (yuan/per person)

Figure 5 shows that the proportion of the health-related economic loss accounting for GDP in Xingtai was the highest, which was 10.15 %. In the 10 cities with the highest proportion of health-related economic loss accounting for GDP, there were 9 cities in the Hebei province. Suqian in Jiangsu province had a higher health-related economic loss and a lower GDP, ranking the tenth; therefore, the proportion of its health-related economic loss accounting for GDP was higher, ranking in the top 10. The economic losses of the public health effect of 4 municipalities ranked in the top 10. Although their GDPs were higher, the proportions of their health-related economic loss accounting for GDP were relatively lower, which did not rank in the top 10.

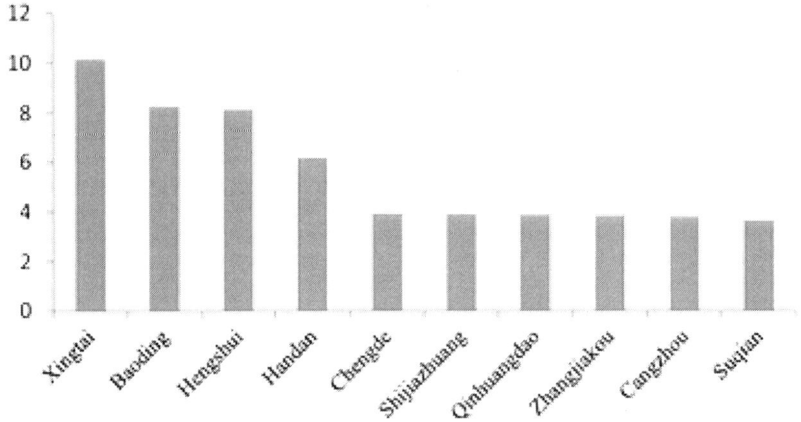

Fig. 5 The 10 cities with the highest proportion of health-relate economic loss accounting for GDP (%)

2. The highest limit scenario

In the highest limit scenario, the cities ranking in the top 10 with the highest economic loss in Fig. 6 were the same as those in the lowest limit scenario in Fig. 3. However, the rankings of Zhengzhou and Chengdu were different in Figs. 4 and 3.

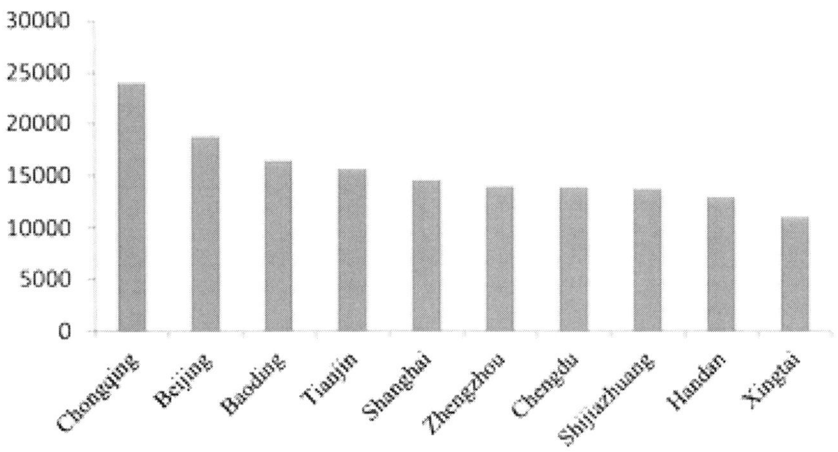

Fig. 6 The 10 cities with the highest economic loss (million yuan)

The cities ranking in the top 10 with the highest economic loss per capita in Fig. 5 in the highest limit scenario were the same as in the lowest limit scenario in Fig. 4. However, as shown in the Fig. 7 in the following, the economic loss per capita in the top 10 cities in the highest limit scenario was higher than that in the lowest limit scenario. For example, the economic loss per capita in Baoding was 1599.9 yuan/per person, higher by 477.2 yuan than that in the lowest limit scenario.

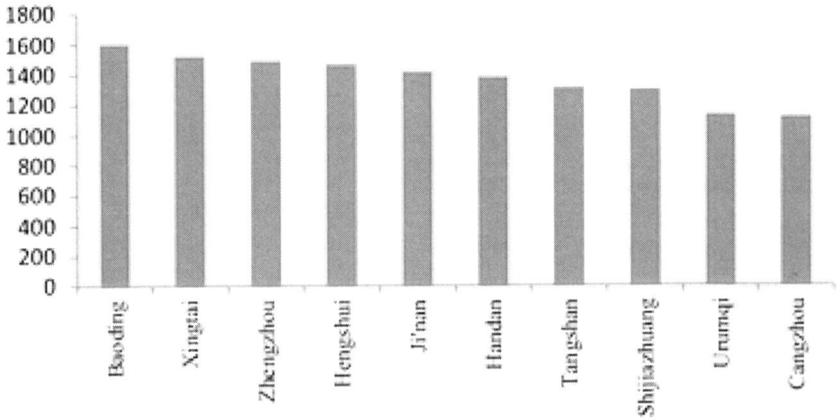

Fig. 7 The 10 cities with the highest health-related economic loss per capita (yuan/per person)

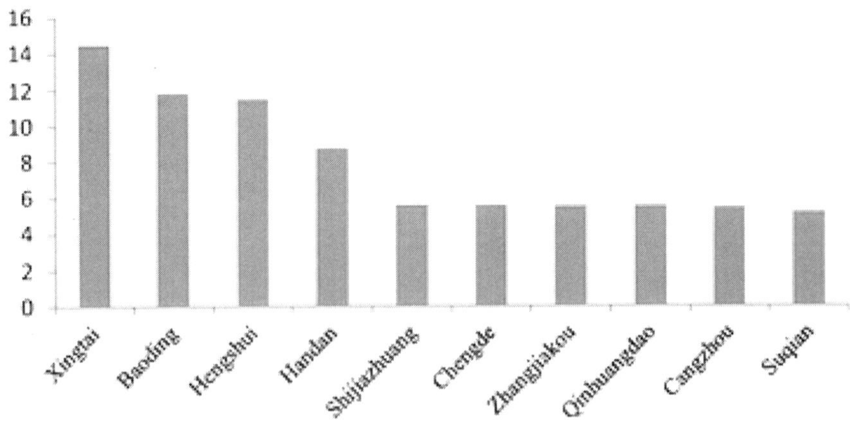

Fig. 8 The 10 cities with the highest proportion of the health-related economic loss accounting for GDP (%)

The 10 cities with the highest proportion of health-related economic loss accounting for GDP in Fig. 8 in the highest limit scenario were the same as in the lowest limit scenario in Fig. 3. However, the rankings of Shijiazhuang, Chengde, Zhengjiukou and Qinhuangdao and the proportions of the health-related economic loss accounting for GDP were different in Figs. 8 and 5.

CONCLUSIONS AND POLICY IMPLICATIONS

Based on the dose–response relationship and the willingness to pay method, this paper evaluated the health-related economic loss caused by air pollution in China's 74 cities using the latest available data regarding to $PM_{10}$and SO_2 from January 2015 to June 2015, by establishing lowest and highest limit scenarios. The conclusions and policy implications are as follows.

Conclusions

1. For the whole 74 cities

The health-related economic loss caused by PM_{10} was larger than that caused by SO_2 in the lowest and highest limit scenarios, and the economic loss associated with chronic bronchitis caused by PM_{10} was the largest in all the losses. Thus, PM_{10} has become the main air pollutant in 74 cities, and it is necessary to focus on the issue of chronic bronchitis caused by PM_{10}.

2. For the major cities

In the lowest and highest limit scenarios, the health-related economic loss in Chongqing, Beijing, Baoding, Tianjin and other major cities was larger than in other cities. The health-related economic loss per capita in Baoding, Xingtai and Zhengzhou was higher than in other cities. Regarding to the proportion of the health-related economic loss accounting for GDP, there were 9 cities in the Hebei Province included in the top 10 cities with the highest loss. It was evident that the air pollution was serious in North China, particularly in the Hebei Province, except in Shanghai in East China and in Chongqing and Chengdu in the southwest of China.

Policy implications

1. According to the results of this paper, Hebei Province is a typical polluted area. Energy consumption structure in Hebei Province is mainly composed of coal resource. Main pollutants from coal combustion are PM_{10} and SO_2, etc. Thus, to reduce air pollution, the coal-based energy structure needs to be improved, and new energies and advanced production processes need to be utilized. Furthermore, strengthening the development of technology and equipment; improving combustion technology, combustion devices and the fuel utilization rate; and reducing the additional pollutants generated by fuel burning will help reduce air pollution.

2. Pollutant concentration is an important factor affecting the loss of health effects. The government should adopt the emissions trading system, limit the pollutant discharges of enterprises, grasp the influence of the enterprises on environmental pollution, strengthen the management of enterprises, and develop a number of administrative regulations conducive to environmental protection to ensure the implementation of environmental protection measures.

3. From the calculation results, the urban population is also an important factor affecting the loss of health effects. The government should reasonably control the urban population and improve people's awareness of environmental protection. Accelerating the transfer of industries and upgrading may reduce the non-household population in order to reduce the side-effects caused by population growth and improve the urban environment and public health.

Here the paper uses the data from January to June 2015 to calculate the health-related economic loss in 74 cities. If the paper uses a long term data, for example, which covers from January to December 2015, there may be a larger health-related economic loss (assuming the population in 74 cities is constant). However, the health-related economic loss from January to December 2015 may not be twice as much as that from January to June 2015. As it can been seen from Figs. 1 and 2, the monthly mean concentration data of SO_2 and PM_{10} from January to June 2015 appeared a downward trend. In the future, data (if complete and available) can be combined using a geographic information system and other new tools to determine the economic loss of caused by the effects of air pollution on public health in the typical resource-based regions or cities and to provide references for environmental management and sustainable development.

DECLARATIONS

Authors' contributions

LL and YL designed the research and methodology; DP, CY collected the data and compiled all the data and literature; LL and CS finished the experiment and calculation; YL and LL analyzed the results and put forward the policies; YL and LL revised the manuscripts and approved the manuscripts; YL will be responsible for the future questions from readers as the corresponding authors. All authors read and approved the final manuscript.

Acknowledgements

The authors express their sincere thanks for the support from the National Natural Science Foundation of China under Grant No. 71173200 and the support from the Development and Research Center of China Geological Survey under Grant Nos. 1212011220302 and 12120114056601, Key Laboratory of Carrying Capacity Assessment for Resource and Environment, Ministry of Land and Resources (Chinese Academy of Land and Resource Economics, China University of Geosciences Beijing) under Grant No. CCA2015.08.

Competing interests

The authors declare that they have no competing interests.

REFERENCES

1. Ami D, Aprahamian F, Chanel O (2014) Willingness to pay of committed citizens: a field experiment. Ecol Econ 105:31–39

2. Askcinet (2015) The people's number and GDP ranking in 74 cities in the first half of 2015.http://www.askci.com/data/2015/07/23/15652fovg.shtml. Accessed on 23 Jul 2015

3. Brandt S, Perez L, Künzli N (2014) Cost of near-roadway and regional air pollution–attributable childhood asthma in Los Angeles County. J Allergy Clin Immunol 5:1028–1035

4. Cai YP, Yang ZF (2003) Health loss estimation of air pollution in the township industrial enterprises in Tianjin. J Saf Environ 1:33–36

5. Cai CG, Chen G, Qiao XC (2007) Comparison of the value evaluation method on the two fractional conditions of the single boundary and two boundary conditions—taking the investigation on health hazard caused by air pollution in Beijing city as an example. China Environ Sci 1:39–43

6. Carlsson F, Martinsson P (2001) Willingness to pay for reduction in air pollution: a multilevel analysis. Environ Econ Policy Stud 1:17–27

7. Chen BH, Hong CJ (2002) Quantitative assessment on the health effect of SO_2 pollution in Shanghai City. J Environ Health 1:11–13

8. Chen BH, Hong CJ, Kan HD (2001) Methodological research on the health-based risk assessment on air pollution. J Environ Health 2:67–69

9. Chen RJ, Chen BH, Kan HD (2010) A health-based economic assessment of particulate air pollution in 113 Chinese cities. China Environ Sci 3:410–415

10. Chen J, Li W, Cheng HG (2015) Evaluation of emission reduction potentials of key air pollutants and health benefits for residents of Beijing. Res Environ Sci 7:1114–1121

11. China's National Environmental Monitoring Centre (2013a) Monthly air quality report in January 2013 in 74 cities.http://www.cnemc.cn/publish/106/news/news_33883.html . Accessed on 07 Feb 2013

12. China's National Environmental Monitoring Centre (2013b) Monthly air quality report in November 2014 in 74

cities.http://www.cnemc.cn/publish/106/news/news_43865.html . Accessed on 21 Dec 2013

13. China's National Environmental Monitoring Centre (2015a) Monthly air quality report in January 2015 in 74 cities.http://www.cnemc.cn/publish/106/news/news_44259.html . Accessed on 05 Feb 2015

14. China's National Environmental Monitoring Centre (2015b) Monthly air quality report in Febuary 2015 in 74 cities.http://www.cnemc.cn/publish/106/news/news_44345.html . Accessed on 17 Mar 2015

15. China's National Environmental Monitoring Centre (2015c) Monthly air quality report in March 2015 in 74 cities.http://www.cnemc.cn/publish/106/news/news_44415.html . Accessed on 14 Apr 2015

16. China's National Environmental Monitoring Centre (2015d) Monthly air quality report in April 2015 in 74 cities.http://www.cnemc.cn/publish/106/news/news_44867.html . Accessed on 25 May 2015

17. China's National Environmental Monitoring Centre (2015e) Monthly air quality report in May 2015 in 74 cities.http://www.cnemc.cn/publish/106/news/news_45103.html . Accessed on 16 May 2015

18. China's National Environmental Monitoring Centre (2015f) Monthly air quality report in June 2015 in 74 cities.http://www.cnemc.cn/publish/106/news/news_45341.html . Accessed on 14 Jul 2015

19. Dockery DW, Pope CA, Xu X (1993) An association between air pollution and mortality in six US cities. N Engl J Med 24:1753–1759

20. Gao J, Xu XP, Li BL (1993) Investigation on the relationship between air pollution and death in Haidian District, Beijing. Chin J Prev Control Chronic Non-Commun Dis 5:207–210

21. Han Q (2011) The health damage caused by the particulate matters in air pollutants in Beijing—human capital method. North Environ 11:150–152

22. Han MX, Guo XM, Zhang YS (2006) Human capital loss of urban air pollution. China Environ Sci 4:509–512

23. Hedley AJ, McGhee SM, Barron B (2008) Air pollution: costs and paths to a solution in Hong Kong—understanding the connections among visibility, air pollution, and health costs in pursuit of

accountability, environmental justice, and health protection. J Toxicol Environ Health Part A 9–10:544–554

24. Ho MS, Jorgenson DW (2007) Sector allocation of emissions and damages in clearing the air: the health and economic damages of air pollution in China. The MIT Press, Cambridge

25. Ho MS, Nielsen CP (2007) Clearing the air: the health and economic damages of air pollution in China. The MIT Press, Cambridge

26. Huanqiunet (2014) The World Health Organization reported that a total of 7 million people died from air pollution in 2012.http://world.huanqiu.com/exclusive/2014-03/4929416.html. Accessed on 25 Mar 2014

27. Huanqiunet (2015) WHO: air pollution has become the world's largest environmental health risk.http://health.huanqiu.com/health_news/2015-06/6581336.html?referer=huanqiu. Accessed on 02 Jun 2015

28. Istamto T, Houthuijs D, Lebret E (2014) Willingness to pay to avoid health risks from road-traffic-related air pollution and noise across five countries. Sci Total Environ 497:420–429

29. Jia L, Guttikunda SK, Carmichael GR (2004) Quantifying the human health benefits of curbing air pollution in Shanghai. J Environ Manage 1:49–62

30. Jing LB, Ren CX (2000) Relationship between air pollution and acute and chronic respiratory diseases in Benxi City. J Environ Health 5:268–270

31. Kaushik CP, Ravindra K, Yadav K (2006) Assessment of ambient air quality in urban centres of Haryana (India) in relation to different anthropogenic activities and health risks. Environ Monit Assess 1–3:27–40

32. Koop G, Tole L (2004) Measuring the health effects of air pollution: to what extent can we really say that people are dying from bad air? J Environ Econ Manage 1:30–54

33. Ministry of Environmental Protection of the People's Republic of China (2011) National environmental protection "12th Five-Year Plan". http://gcs.mep.gov.cn/hjgh/shierwu/201112/t20111221_221595.htm. Accessed on 21 Dec 2011

34. Murray CJ, Lopea AD (1997) Global mortality, disability, and the contribution of risk factors: global burden of disease study. Lancet 349:1436–1442

35. Pascal M, Corso M, Chanel O (2013) Assessing the public health impacts of urban air pollution in 25 European cities: results of the Aphekom project. Sci Total Environ 449:390–400

36. Patankar AM, Trivedi PL (2011) Monetary burden of health impacts of air pollution in Mumbai, India: implications for public health policy. Public Health 3:157–164

37. Peng XZ, Tian WH (2003) Study on the willingness to pay for the economic loss of air pollution in Shanghai City. World Econ Pap 2:32–44

38. Ridker RG (1967) Economic costs of air pollution: studies in measurement. Praeger, New York

39. Samet JM, Zeger SL, Dominici F (2000) The national morbidity, mortality, and air pollution study. Part II: morbidity and mortality from air pollution in the United States. Res Rep Health Eff Inst 2:5–79

40. Shang YH, Zhou DJ, Yang J (2010) Study on the economic loss of the human health caused by air pollution. Ecol Econ 1:178–179

41. Shen XW, Wang YH, Zhang WX (2014) The economic loss of the human health caused by air pollution in Kunming. China Market 46:124–126

42. U.S. EPA (1999) Benefits and costs of the clean air act. U.S. EPA Office of Air and Radiation, Washington, DC

43. Wan Y, Yang HW, Masui T (2005) Health and economic impacts of air pollution in China: a comparison of the general equilibrium approach and human capital approach. Biomed Environ Sci 6:427

44. Wang H, John M (2006) Willingness to pay for reducing fatal risk by improving air quality: a contingent valuation study in Chongqing, China. Sci Total Environ 1:50–57

45. Wang X, Smith K (1999) Near-term health benefits of greenhouse gas reductions: a proposed assessment method and application to two energy sectors of China. World Health Organization, Geneva

46. Wang Y, Zhao XL, Xu Y, Li Y (2005) Estimation on the economic loss of air pollution in Shandong Province. Urban Environ Urban Ecol 2:30–33

47. Wei YM, Wu G, Liang QM (2012) China Energy Security Report (2012): Study on energy security. Science Press, Beijing

48. Wilkinson P, Smith KR (2007) A global perspective on energy: health effects and injustices. The Lancet 370(9591):965–978

49. Wong TW, Lau TS, Yu TS (1999) Air pollution and hospital admissions for respiratory and cardiovascular diseases in Hong Kong. Occup Environ Med 10:679–683

50. Wong CM, Atkinson RW, Anderson HR (2002) A tale of two cities: effects of air pollution on hospital admissions in Hong Kong and London compared. Environ Health Perspect 1:67

51. World Bank (1997) Clear water, blue skies: China's environment in the new century. World Bank, Washington, DC

52. World Bank SEPA (2007) Cost of pollution in China. World Bank, World Bank

53. Xia G (1998) Economic measurement and study on the loss of environmental pollution in China. China Environmental Sciences Press, Beijing

54. Xu ZY, Liu YQ, Yu DQ (1996) Impact of air pollution on mortality in Shenyang. Chin J Public Health 1:61–64

55. Xu LY, Yin H, Xie XD (2014a) Health risk assessment of inhalable particulate matter in Beijing based on the thermal environment. Int J Environ Res Public Health 12:12368–12388

56. Xu MM, Guo YM, Zhang YJ (2014b) Spatiotemporal analysis of particulate air pollution and ischemic heart disease mortality in Beijing, China. Environ Health 1:1–12

57. Yaduma N, Kortelainen M, Wossink A (2013) Estimating mortality and economic costs of particulate air pollution in developing countries: the case of Nigeria. Environ Resour Econ 3:361–387

58. Yan J (2012) Residents' breathing, the hospital cost of the disease of the circulatory system caused by air pollution in Lanzhou. In: Food safety and healthy life conference in Gansu in 2012, pp 61–64

59. Yu F, Guo XM, Zhang YS (2008) Evaluation on health economic loss caused by air pollution in 2004. J Environ Health 12:999–1003

60. Zeng XG, Xie F, Zong Q (2015) The behavior choice and willingness to pay to reduce the health risk of PM2. 5—taking the residents in Beijing as an example. China Popul Resour Environ 1:127–133

61. Zhang QF (2012) Toward an environmentally sustainable future country environmental analysis of the People's Republic of China. China Financial and Economic Publishing House, Beijing

62. Zhang GZ, Chun R, Nan ZR (2008) Study on the effects of air pollution on human health and economic loss in Lanzhou. J Arid Land Resour Environ 8:120–123

63. Zhou J, Wang Y, Ren L (2010) Willingness to pay for the improvement of air quality in Shandong typical city. J Environ Health 6:507–510

64. Zmirou D, Deloraine AMD, Balducci F, Boudet C (1999) Health effects costs of particulate air pollution. J Occup Environ Med 10:847–856

9

CHAPTER

Assessment of Nitrogen Losses Through Nitrous Oxide from Abattoir Wastewater-Irrigated Soils

Raghupathi Matheyarasu, Balaji Seshadri, Nanthi S. Bolan, Ravi Naidu

1. Future Industries InstituteUniversity of South AustraliaAdelaideAustralia

2.Global Centre for Environmental RemediationThe University of NewcastleAdelaideAustralia

3.Cooperative Research Centre for Contamination Assessment and Remediation of the EnvironmentSalisbury, AdelaideAustralia

ABSTRACT

The land disposal of waste and wastewater is a major source of N_2O emission. This is due to the presence of high concentrations of nitrogen (N) and carbon in the waste. Abattoir wastewater contains 186 mg/L of N and 30.4 mg/L of P. The equivalent of 3 kg of abattoir wastewater-irrigated soil was sieved and taken in a 4-L plastic container. Abattoir wastewater was used for irrigating the plants at the rates of 50 and 100 % field capacity (FC). Four crop species were used with no crop serving as a control. Nitrous oxide emission was monitored using a closed chamber technique. The chamber was placed inside the plastic container, and N_2O emission was measured for 7 days after the planting. A syringe and pre-evacuated vial were used for collecting the gas samples; a fresh and clean syringe was used each time to avoid cross-contamination. The collected gas samples were injected into a gas chromatography device immediately after each sampling to analyse the concentration of N_2O from different treatments. The overall N_2O emission was compared for all the crops under two different abattoir wastewater treatment rates (50 and 100 % FC). Under 100 % FC (wastewater irrigation), among the four species grown in the abattoir wastewater-irrigated soil, *Medicago sativa* (23 mg/pot), *Sinapis alba* (21 mg/pot), *Zea mays* (20 mg/pot) and *Helianthus annuus* (20 mg/pot) showed higher N_2O emission compared to the 50 % treatments—*M. sativa* (17 mg/pot), *S. alba*(17 mg/pot), *Z. mays* (18 mg/pot) and *H. annuus* (18 mg/pot). Similarly, pots with plants have shown 15 % less emission than the pots without plants. Similar trends of N_2O emission flux were observed between the irrigation period (4-week period) for 50 % FC and 100 % FC. Under the 100 % FC loading rate treatments, the highest N_2O emission was in the following order: week 1 > week 4 > week 3 > week 2. On the other hand, under the 50 % FC loading rate treatments, the highest N_2O emission was recorded in the first few weeks and in the following order: week 1 > week 2 > week 3 > week > 4. Since N_2O is a greenhouse gas with high global warming potential, its emission from wastewater irrigation is likely to impact global climate change. Therefore, it is important to examine the effects of abattoir wastewater irrigation on soil for N_2O emission potential.

Keywords: Abattoir Wastewater Irrigation Nitrous oxide Emission Greenhouse gas and global warming

INTRODUCTION

The emission of greenhouse gases (GHGs) increases with the rising global population (Preston et al. 2006). Human activities such as agriculture (e.g. chemical fertilisers), energy production (e.g. coal combustion), transport (e.g. fossil fuels) and other industrial activities are directly or indirectly contributing to the GHG emissions (Crutzen et al. 2008). Among the various sources that are responsible for GHG emissions, energy production is the major contributor followed by land use change for agriculture and industrial activities (Meinshausen et al. 2009; Cerri et al.2009). The major global greenhouse gases and their percentage of emission are illustrated in Fig. 1. At a global scale, nitrous oxide (N_2O) is a major GHG after carbon dioxide (CO_2) and methane (CH_4) (IPCC-2007). However, the global warming potential (GWP) varies between these

GHGs, for example, N_2O is 282 times more powerful than CO_2 (Ravishankara et al. 2009). Hence, the management of N_2O by reducing their emission is important in mitigating climate change (McCarl and Schneider 2000). Since N_2O is a highly potential GHG towards GWP, measurement and mitigation need to be done on a broader scale (Shine et al. 2005). N_2O emission contributes about 6 % of the overall global warming effect, but its contribution from the agricultural sector is about 16 %. Of that, almost 80 % of N_2O is emitted from Australian agricultural lands, originating from N fertilisers (32 %), soil disturbance (38 %) and animal waste (30 %) (Dalal et al. 2003).

Fig. 1 Global greenhouse gases emission (in %) (adapted from IPCC report 2007) (*F-gases* = fluorinated gas (hydrofluorocarbons (HFCs), perfluorocarbons (PFCs), sulphur hexafluoride (SF6) and nitrogen trifluoride (NF3)

Globally, denitrification is the primary process of N_2O production in temperate grassland soils and accounts for 60 % of the total N_2O emissions (Jha et al. 2012). A recent study at the European Union states that ruminants (cows, sheep and goats) have the highest carbon footprint (FAO 2006). Total net GHG emission of EU livestock production was estimated at 661 mt of carbon dioxide equivalent (CO_2-eq) which is about 9–13 % of the total GHG emission for the EU agricultural sector, comprising 23 % CH_4, 24 % N_2O, 21 % CO_2 (energy use) and 29 %

CO_2 (land use). A considerable amount of GHGs is emitted by the global animal industry, which is more than all the cars in the world put together, and a large part of that is 18 % N_2O and CH_4emissions; both of these gases have a far more powerful greenhouse gas effect than carbon dioxide (Garnaut et al. 2008).

The livestock sector accounts for 5–50 % of the total contribution, but it may vary from place to place (Gunnarsson et al. 2011). The overall contribution consists of pigs 0.4 %, sheep 3.4 %, cattle 2.7 % and beef cattle 11.2 %, which on average emits 554 kg CO_2-eq/tonne hot standard carcass weight (MLA 2010). Untreated abattoir wastewater is unsuitable for reuse or discharge into the receiving environment. It will cause serious environmental hazards in the receiving environment such as eutrophication, land degradation, nutrient leaching, groundwater contamination, greenhouse gas emission and effects on ecosystem value; hence, a proper reduction in pollutant levels in the prior stage is essential. In the recent years, meat production and consumption have been increasing considerably and predicted to peak in 2020 predominantly in Asia and the Pacific. Global per capita meat consumption is projected to increase from 32.9 kg/rwt in 2011 to 35.4 kg/rwt in 2020 (OECD-FAO 2011). Meat production is a considerable source of global GHG emission, emitting methane, nitrous oxide and carbon dioxide through various stages. GHGs are emitted by direct energy consumption and indirectly by feedstock production, herding, movement of animals, product transport, slaughtering, cleaning and dressing the animal product, waste and wastewater. The disposal of waste and wastewater to land is a major source of N_2O emission (Oene et al. 2005; Bolan et al. 2004). N_2O emission from wastewater-irrigated soils has been explored by many researchers (Bhandral et al. 2007; Saggar et al. 2004; Dalal et al. 2003).

Agricultural industry wastewater such as abattoir and dairy wastewaters are significant contributors towards N_2O emissions (Kampschreur et al. 2007; Russell et al. 1993). This is due to the presence of the high concentration of N and carbon; for example, abattoir wastewater (AWW) contains 200 to 400 mg/L of N and 545 mg/L of dissolved organic carbon (Longhurst et al. 2000; Cassidy and Belia 2005; Matheyarasu et al. 2012). Tsujimoto (1994) observed that N loss through leaching and gaseous emissions increases with an increasing level of animal waste and wastewater application in soil. Soils are the major source of the greenhouse gas nitrous oxide (N_2O) in our atmosphere.

The rate of denitrification in soils and the relative proportions of NO, N_2O and N_2 produced are controlled by various factors, such as soil microorganisms, climatic factors and management practices. The availability of mineral N (both NH_4^+ and NO_3^-) and labile C, together with processes that affect reaction rates, such as temperature, pH and redox potential in soil microsites (Saggar et al. 2013). N_2O can be produced by nitrifiers, denitrifiers and nitrifiers paradoxically denitrifying. Soil moisture conditions are sub-optimal for denitrification, nitrifier denitrification and the major contributor to N_2O emission (Kool et al. 2011). The majority of studies indicate that a significant amount of N in waste and wastewater is lost as N_2O emissions (Kampschreur et al. 2009; Czepiel et al. 1995; Bolan et al. 2004; Saggar et al. 2015). Since N_2O is a potential greenhouse gas with very high GWP, even low emissions can cause serious effects (Shine et al. 2005). In addition, the application of chemical fertilisers alone contributes 46 % of N_2O emission from the agriculture sector (Baumert et al. 2005) and therefore minimising the usage of these fertilisers will also reduce the emission rate.

There are increased concerns about denitrification associated with the loss of N in the environment. Denitrification can be both detrimental and beneficial to the environment (Bolan et al. 2004). For example, N_2O, one of the gaseous products from denitrification, has possible deleterious effects on global warming (Bateman and Baggs 2005). The primary consideration for mitigating gaseous N emissions from arable land is to match the supply of mineral N (from fertiliser application, legume-fixed N) to plant needs, although it is possible to achieve uniform application of N fertilisers. Mitigation approaches need to focus on ways to reduce the production of N_2O during denitrification and enhance the reduction of N_2O to N_2 thus lowering the N_2O:N_2 product ratio (Saggar et al. 2013). On the other hand, effective management practices help to minimise processes such as leaching, denitrification and NH_3^- volatilisation, all of which lead to the loss of plant-available N from the soil–plant system. These management practices include optimum N supply to pasture crops, proper animal residue management, controlled-release fertiliser and proper water management (Bolan et al. 2004).

Wastewater irrigation increased the concentration of major nutrients (N, P and K) in soil (Matheyarasu et al. 2014). Although there was an increase in soil fertility, the potential for N losses through nitrate leaching and N_2O emission is likely to be high. There is a great scope and need for reducing N_2O emissions from various sources to the atmosphere (Ravishankara et al.2009; McCarl and Schneider 2000). Currently, there are limited reports on N_2O emission from AWW-irrigated soils or AWW-irrigated cropping system in Australia. Therefore, it is important to examine the effects of AWW irrigation on the GHG emission potential. This paper examines the effect of different types of nutrient source (wastewater and urea) and irrigation intensity on biomass productivity of selected plant species and their N uptake efficiency and N_2O emission under greenhouse conditions. The overall objective of the study is to examine the effects of AWW irrigation on N uptake and N_2O emission in a calcareous soil. The specific objectives of the study included (a) to quantify N_2O emission from agricultural soil treated with various soil moisture conditions and N supplements (urea and AWW), (b) to investigate the effects of AWW irrigation on N_2O emission and (c) to study the rate of plant-induced denitrification in two different moisture gradients (50 and 100 % FC) with and without plants.

MATERIALS AND METHODS

Contaminated site assessment and soil sample collection

The study area (sampling site) is situated at 89.7 km north of Adelaide, South Australia. The latitude and longitude of the study area are 34° 8′ 26.60″ S and 138° 11′ 7.35″ E; the range is 749 m and the elevation of the treatment site is generally flat ranging from 13.5 m Australian height datum (AHD) to 14.5 m AHD. The region has mean annual rainfall of 287.3 mm and annual mean maximum temperature of 22.8 °C and minimum temperature of 10.7 °C. Abattoir wastewater and soil under abattoir wastewater irrigation were collected from land treatment. The AWW-irrigated soil were collected, air-dried and sieved to <2 mm for physiochemical characterisation. The site was under long-term wastewater irrigation to manage wastewater economically and was used for forage production, alternatively. The land treatment site (CI) has received around 385 mm of secondary treated effluent applied over the year at the rate of 32 mm per month.

The CI soil also received an additional 310 mm of water through rainfall, during the period (2012). In the study site, the rate of irrigation was not adjusted according to annual rainfall since it is intended for land treatment. The stored soil samples as collected from different locations and depths were analysed for pH, electrical conductivity (EC), nitrogen (N), phosphorus (P), carbon (C) and micro nutrients. Soil analyses were performed following standard methods as described in the *Soil Chemical Methods—Australasia* (Rayment and Lyons2011) manual. Soil pH was measured in water using glass electrodes at a 1:5 soil to water ratio. Soil EC was also measured at the same time using an EC meter. Soil total C and total N were estimated by dry combustion on air-dry soil using a LECO 2000 CNS analyser (Sparling et al.2006). Olsen P was estimated by soil extraction with sodium bicarbonate (0.5 M at pH 8.5) and measured by the molybdenum blue method (Olsen et al. 1954). Absorbance was measured at 882 nm in an Agilent UV–visible spectroscopy system (Germany), and the Olsen P concentration was calculated by preparing a calibration curve against the standards. The total P and micronutrients were determined using inductively coupled plasma-optical emission spectrometry (ICP-OES), with acid-digested soil samples (1:3 ratio of concentrated nitric–hydrochloric acid mixture/aqua regia) (Chen and Ma 2001). Similarly, available N (nitrate-N and ammonia-N) was measured using the SKALAR SANS system (analyser) with potassium chloride (2 M)-extracted soil samples (Luo et al. 2004).

Plant growth experiment

The plant growth experiment was conducted at the University of South Australia greenhouse using the contaminated soil collected from the land treatment sites. The wastewater used in this experiment was collected from the Primo abattoir at Port Wakefield, which was rich in major plant nutrients such as total nitrogen (TN) and total phosphorus (TP). Two sets of experiments were conducted to examine the effects of wastewater irrigation on N loss through gaseous emission (N_2O). In experiment 1, urea was used as a N source to study the gaseous emission from the soil. In experiment 2, AWW was used as N source and the effects of adding wastewater at different loading rates towards N_2O gaseous emission were studied.

Experiment 1—a study on N_2O emission without plants in a laboratory condition

Experiment 1 comprised seven treatments with three replicates to study the effects of urea addition on gaseous emission (N_2O). In this experiment, five different moisture levels were applied (e.g. 25, 50, 75, 100 and 120 % FC) with two levels of nitrogen loading (500 and 1000 mg/kg of soil) to examine the N loss through gaseous emission.

Experiment 2—gaseous emission with plants

The equivalent of 3 kg of soil was sieved and taken in a 4-L plastic container. AWW was used for irrigation of the plants at the rates of 50 and 100 % FC. Crop species including *Helianthus annuus*, *Sinapis alba*, *Medicago sativa* and *Zea mays* were used with no crop serving as a control. The entire experiment was carried out with three replications (Plate 1). Treatment details are as follows: 2 moisture levels*4 + 1 control (no plant)*3 replicates (2*5*3 = 30).

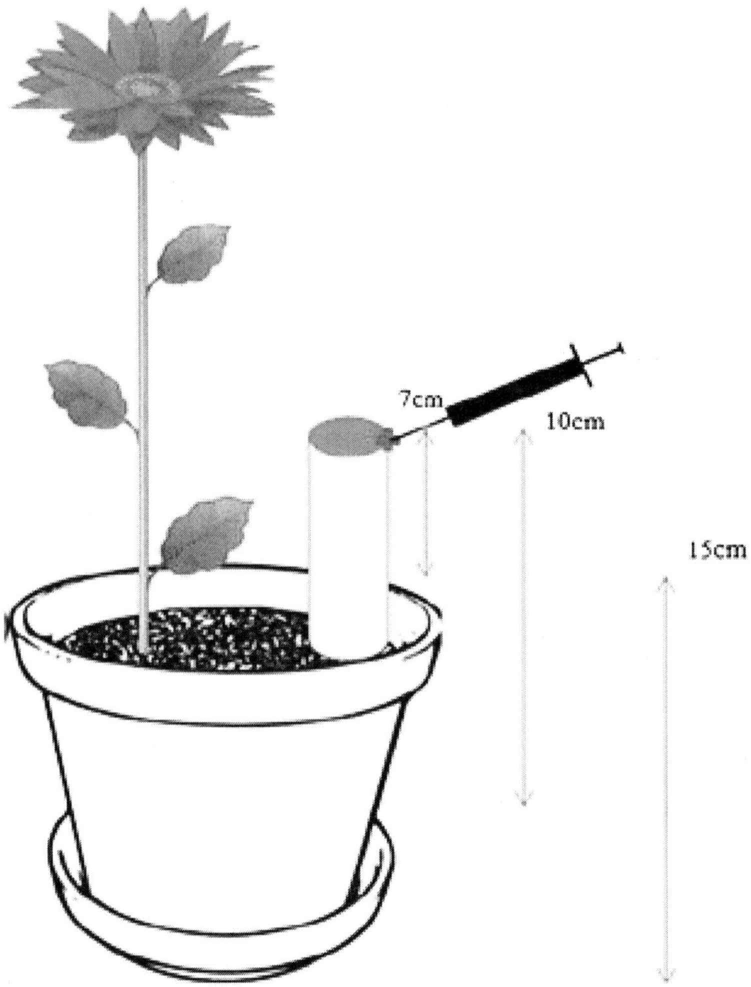

Plate 1 A study on plant-induced denitrification at a greenhouse located at the University of South Australia

Gas sampling unit

The N_2O emission from the treatment pot was measured weekly using a modified closed chamber technique. The modified chamber technique was developed during this study based on the principles described by Saggar et al. (2002; 2004) and Bhandral et al. (2007). This modified closed chamber technique involves sampling of N_2O emission close to the soil surface (10 cm), with the apparatus full description of the modified closed chamber presented in Fig. 2. The closed chamber was placed inside a pot (3 kg of soil). A total of 30 chambers were designed and used in this study with three replicates of each treatment. The dimensions of the chamber were 17 cm in height and 4 cm in diameter. The chamber was inserted into the soil with 7 cm exposed above the soil. Background N_2O emissions were measured for each pot on the first day after the initial setup to validate and check the efficiency of the modified closed chamber. After the application of treatments, measurements were made every week until 4 weeks after planting.

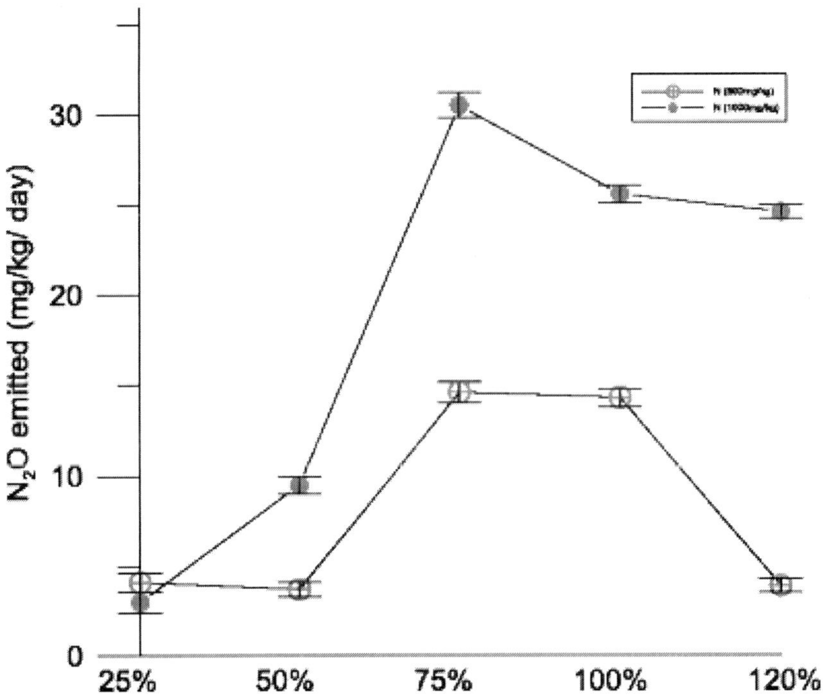

Fig. 2 Modified chamber techniques for the measurement of N_2O emission from the controlled environment

Gas sampling

Each gas sampling unit was placed inside the pot (Plate 2). Emissions of N_2O were measured for 7 days after the incubation in experiment 1 and planting in experiment 2 using the active efflux method. A syringe and pre-evacuated vials were used for collecting the gas samples; a fresh and clean syringe was used each time to avoid cross-contamination. The collected gas samples were injected to a gas chromatograph immediately after each sampling to analyse the emission of N_2O from different treatments. At the end of the greenhouse studies, a nutrient loss percentage as N_2O (%) was calculated to study the nutrient losses using nutrient input and loss ratio. Plants were harvested after 6 weeks of germination. The shoots and roots were separated and dried in a hot air oven at 70 °C to a constant weight. Plant samples were stored in a dry airtight container for further nutrient analysis to calculate nutrient uptake.

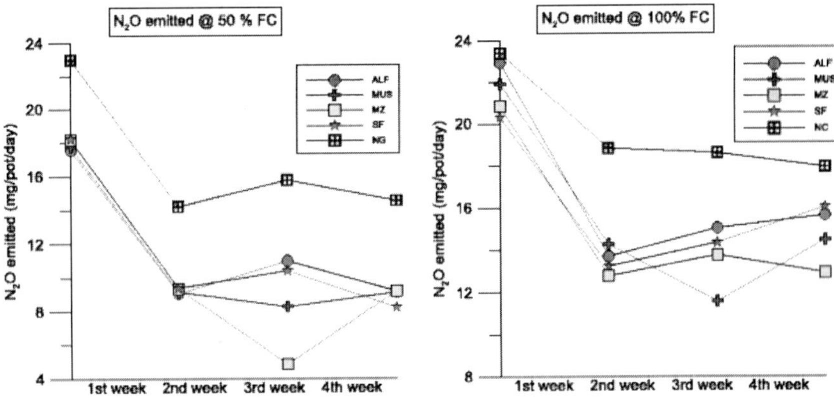

Plate 2 Gas sampling method for N_2O analysis

Statistical analysis

SPSS (Inc., 2001) was used to analyse the data. The differences in the replicates were determined using standard deviations for the gaseous emission (N_2O). Relationships between wastewater loading types (50 vs 100 % FC) and gaseous emission (N_2O) were analysed by Pearson correlation coefficients and the two-sample t test.

RESULTS AND DISCUSSION

Properties of soils and wastewater used in this experiment

The soils used in this study were collected from an abattoir wastewater-discharged landfill site at Port Wakefield, South Australia. The collected samples were air-dried, characterised for physicochemical properties and used for greenhouse plant growth experiment (assessment of nitrogen losses). Soil collected from the land treatment site was moderately alkaline; the pH of the CI soil was moderately acidic (6.3), with CNI and CTRL measuring 8 and 8.6, respectively. The electrical conductivity was very high ranging from 500 to 1109 μS/cm.

The AWW-irrigated soil was high in TN and TP at available nutrient concentrations. The nutrient contents in AWW-irrigated soil were up to 1165 mg/kg of nitrogen (N) and 223 mg/kg of phosphorus (P). The wastewater sample was characterised for its major nutrient concentration (N and P) immediately after collection. The AWW used in this experiment had a high TN and TP concentration (186 mg/L of N and 30.4 mg/L of P) (Table 1).

Effects of urea application and rate of application on nitrous oxide emission with varying soil moisture

The differences in the rate of N application (as urea) on N losses through N_2O were observed by the end of the experiment. The application of urea with two different rates (low = 500 mg/kg; high = 1000 mg/kg) showed significant impacts on the N_2O emission. The rate of emission was also significantly affected by the soil moisture content. The results showed that the high rate of N addition (1000 mg/kg) was most vulnerable in terms of N losses with varying moisture gradient. For both the levels of N application, the N_2O emission increased with the increasing soil moisture content (% FC) up to the saturation (100 % FC) and the rate of N_2O emission started to decline after the soil reaches the maximum water holding capacity (>100 % FC).

The rate of N_2O emission will be minimum in a higher FC condition (waterlogged or flooded soils); this is due to less aeration (restricted) and low emission (N_2O) leading to a complete denitrification process, thereby emitting N_2 gas (Dalal et al. 2003). The maximum level of N_2O emission

Table 1 Properties of soil used in these experiments (wastewater-irrigated soil)

Soil parameters	Properties
pH	6.3 ± 1.5
EC (μS/cm)	299.6 ± 1.9
Moisture (%)	11.8 ± 0.1
Total C (mg/kg)	10,523 ± 1.1
Total N (mg/kg)	1165.9 ± 1.4
Nitrate-N (mg/kg)	53.7 ± 0.6
Ammonia-N (mg/kg)	14.9 ± 0.02
Total P (mg/kg)	223.1 ± 0.02
Olsen-P (mg/kg)	73.6 ± 0.08
K (mg/kg)	2849.8 ± 0.01
Ca (mg/kg)	2266.7 ± 0.01
Mg (mg/kg)	709.2 ± 0.09
Na (mg/kg)	308.9 ± 0.05
Fe (mg/kg)	10.4 ± 0.06
Al (mg/kg)	10.7 ± 0.06
B (mg/kg)	4.2 ± 0.02
Zn (mg/kg)	0.4 ± 0.03
Mn (mg/kg)	17.4 ± 0.01

was recorded at 75 to 100 % FC in both N treatment levels (500 and 1000 mg/kg of urea-N). Overall, high N addition with high soil moisture (75 to 100 % FC) resulted in a significant amount of N losses which were about 30 mg/pot/day emitted as N_2O. This was nearly twofold higher than that of the other treatment level of 500 mg/kg of soil N as urea (Fig.3). Soil N_2O production was highly dependent on the oxygen (O_2) supply and water-filled pore space (WFPS) (Bhandral et al. 2007; Dalal et al. 2003). The high N_2O emission can be expressed in a situation like low O_2 partial pressure (<0.5 vol.%) and high WFPS (>60 %) (Saggar et al.2013). N_2O is primarily produced in soil by the activities of microorganisms during nitrification and denitrification processes. The ratio of N_2O production depends on oxygen supply or water-filled pore space, decomposable organic carbon, N and substrate supply (Dalal et al. 2003).

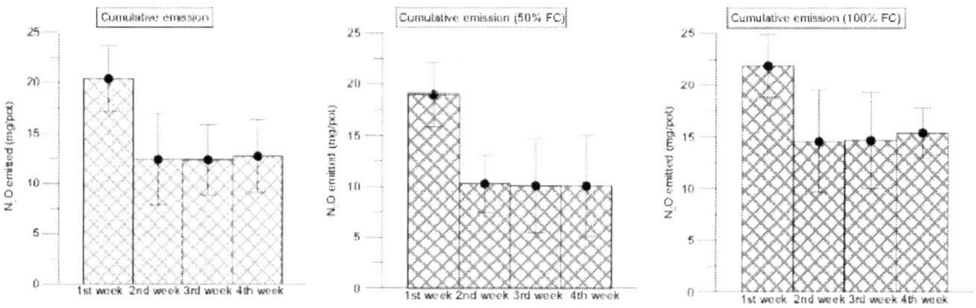

Fig. 3 Nitrous oxide emission from soil treated with urea with 500 and 1000 mg/kg—with varying moisture gradient (25, 50, 75, 100 and 120 % FC)

EFFECTS OF AWW IRRIGATION ON NITROUS OXIDE EMISSION

Effects of nutrient loading

Abattoir wastewater is the major source of N to the soil. Irrigation of nutrient-rich AWW increased the soil fertility and nutrient loss, especially N. The soil nutrient loss was significantly higher under AWW irrigation compared to that in non-irrigated soils. The N_2O emission rate significantly varied between the two levels of wastewater application. The distinct N_2O peak was observed during the first week after the application of AWW in all the crop species grown and non-cropping condition. The peak was at 22 mg/pot in 100 % FC and 18 mg/pot in 50 % FC in non-cropped conditions (average of the initial period). According to Bowwman (1996), up to 0.16 % of total N applied to the soil can be lost as N_2O emission within a day after fertiliser application (Dalal et al. 2003; Ruseel 1993). In comparison to all the treatments (including the AWW 50 % FC with and without a crop, AWW 100 % FC with a crop), the AWW treatment without a crop showed statistically significant N_2O emissions compared to the other treatment levels (Fig. 4). These increases were recorded from the first week after application of AWW. Similar results were found in Bhandral et al. (2007). They observed the peak emission of N_2O from the meat industry wastewater-irrigated soils in the first few days after treatment.

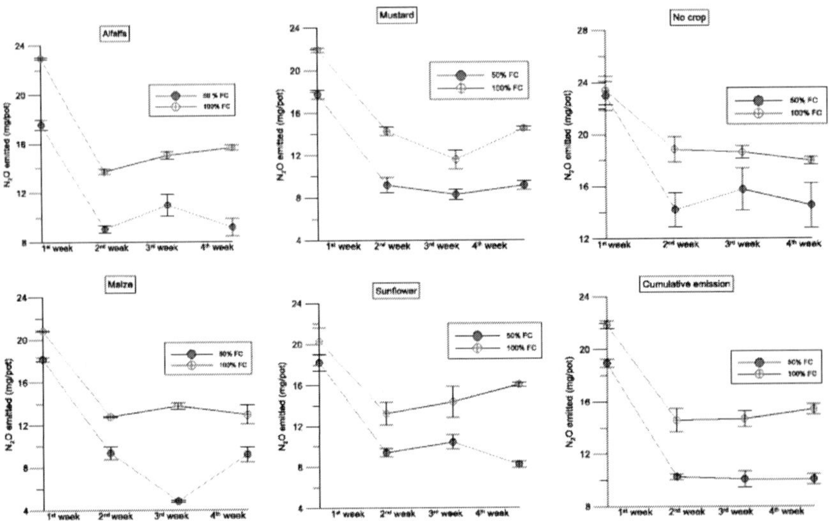

Fig. 4 Nitrous oxide emission from soil irrigated with abattoir wastewater with 50 and 100 % FC (moisture gradients)—measured at 4 weeks after planting (*ALF* alfalfa, *MU* mustard, *MZ*maize, *SF* sunflower, *NC* no crop)

The peak emission of N_2O from the AWW 100 % FC (no crop) was 22 % higher than that of the AWW 50 % FC (no crop) which is 53 % times higher than that of the cropped pot. A peak N_2O emission flux of 21.9 mg/pot/day was recorded under the non-cropping condition at the first week after planting; this indicates that there are greater chances for the loss of applied N if there is no crop to utilise the nutrients. According to Cardenas et al. (2010), high N input can increase the annual N_2O emission fluxes in a pasture under a range of N fertiliser inputs and their study concluded that higher emissions are possible under increased nutrient supply. A significant increase was recorded in consecutive weeks, but the differences were not much significant as compared to the first week. The overall emission data of 4 weeks suggest that there were highly significant effects of a high load of nutrient addition through wastewater in terms of N loss.

Effects of time

The N_2O emission was significant in 100 % FC at 2 weeks after the plantation in all the pots. The N_2O emission was high at the first 2 weeks after the planting and declined subsequently to reach the background levels within 4 weeks. Application of AWW irrigation has increased the soil N_2O emission rate in a short-duration study at the greenhouse for 4 weeks. The total emission in the first week was higher compared to that of the rest of the 3 weeks in both soils. The highest N_2O emission was recorded in the 100 % FC in the first week for all the four pots, which was about 21 to 22 mg/pot. The AWW irrigation affected N_2O emissions at both levels of irrigation (50 and 100 % FC). The N_2O emission increased initially (the first week after the treatment) and declined in the consecutive weeks of the experimental period (for example, the fourth week) (Fig. 5).

Fig. 5 The effects of abattoir wastewater (50 and 100 % FC) on cumulative N_2O emission of four plant species (alfalfa, mustard, maize and sunflower) measured at the fourth week after planting

Application of AWW (at 100 % FC) increased N_2O emission by 8 % in the first week and 43 % in the second week after application. A similar study by Singh et al. (2008) states that the application of urine resulted in a high amount of N_2O losses in their incubation studies. The authors found that the peak of N_2O emissions was within a week after application of dairy cow urine and observed a maximum of 11.2 mg of N_2O/kg of soil. Also, their studies showed that the maximum emission was reached before day 25 of the 40-day incubation study (Singh et al.2008). In the 4-week treatment period, N_2O emissions from 100 % FC (no crop) treatment in the first 2 weeks remained significantly higher than the N_2O emitted from the 50 % FC (no crop).

Effects of plant species used

There was a significant difference between AWW 50 % FC and 100 % FC irrigation in the overall N_2O emission of all the four crops used in this study. For example, in comparison with 50 % FC, 30, 23, 17 and 11 % (*H. annuus*, *S. alba*, *M. sativa* and *Z. mays*) of increases in N loss were recorded as N_2O in 100 % FC. N_2O emission was significantly ($p < 0.05$) higher in the application of 100 % FC of AWW to all the four crops grown. A maximum of 27.5 mg/pot was recorded in the 100 % FC condition in the second week of application with AWW. In general, N_2O emissions were significantly lower from all the crops irrigated with wastewater than those of the non-crop condition in both levels of treatment (50 and 100 % FC). The cumulative emissions recorded in the 50 % FC were 17.5, 17.7, 18.1 and 18.2 mg/pot for *H. annuus*, *S. alba*, *M. sativa* and *Z. mays*, respectively (non-cropped was 23 mg/pot). Similarly, the cumulative emissions recorded in the 100 % FC were 22, 21, 20 and 20 mg/pot from *H. annuus*, *S. alba*, *M. sativa* and *Z. mays*, respectively (non-cropped was 23.4 mg/pot) (Fig. 6). Pots with plants showed lower emissions than the pots without plants. Overall, the cumulative (average) N_2O emission was 13 % higher in high-field-moisture pots (100 % FC) than that in the moisture-deficit pots (50 % FC) (Table 2). The second highest N_2O emission was recorded on the second week after application of the AWW. A similar result was found in Bhandral et al. (2007), and they found that the initial increase in the N_2O emission after the application of meat industry wastewater declined progressively with time.

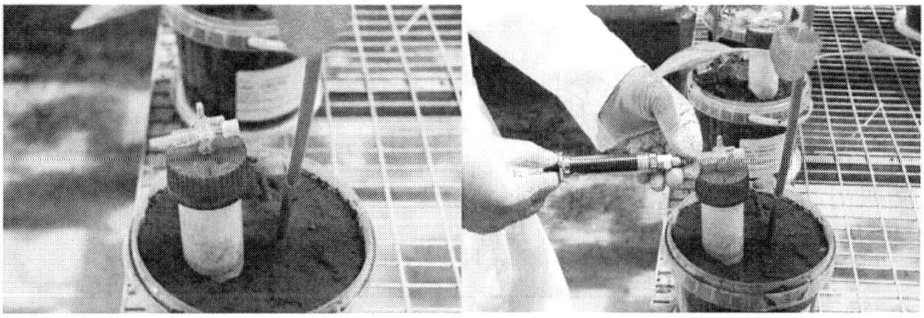

Fig. 6 Nitrous oxide emission from soil irrigated with abattoir wastewater with 100 % FC (moisture gradients)—measured at 4 weeks after planting

Table 2 Descriptive statistics: the effects of abattoir wastewater irrigation on N_2O emission: cumulative effects of 50 and 100 % FC ($n = 30$)

Properties	Range	Minimum	Maximum	Mean	Std. deviation	Variance	Skewness	Kurtosis
Nitrate-N (mg/pot)	621.2	291.4	912.6	489.8	193.1	37,294.2	0.9	–
Ammonia-N (mg/pot)	128.0	60.0	188.0	101.1	39.9	1588.0	0.9	–
N_2O week 1 (mg/pot)	12.8	14.7	27.5	20.4	3.4	11.4	0.5	−0.5
N_2O week 2 (mg/pot)	19.9	7.6	27.5	12.4	4.5	20.3	1.8	3.4
N_2O week 3 (mg/pot)	18.8	3.5	22.3	12.4	5.1	26.2	0.1	−1.0
N_2O week 4 (mg/pot)	18.6	0.6	19.2	12.7	4.7	21.9	−0.9	0.2
Cumulative emission (mg/pot)	55.8	33.0	88.8	57.9	13.9	193.9	0.1	−0.5

– equal to zero

Effects of soil moisture

Among the two levels of AWW irrigation treatment, N_2O emissions were highest for AWW 100 % FC (16.6 mg/pot), whereas the 50 % FC emitted 12.3 mg/pot (cumulative emission average). Bhandral et al. (2007) suggested that the highest N_2O emission is possible if soil is irrigated/applied with a high dose of urine, ammonium and urea.

The N_2O emission was higher (66.5 mg/pot) at 100 % FC ($p < 0.01$) than at 50 % FC (49 mg/pot) which may be attributed to higher soil moisture content in the former treatment (Table 3). According to a study by Bateman and Baggs (2005), N_2O emission was high under the elevated soil moisture condition. Soil N_2O emission was significantly affected by soil moisture and a nutrient input source (Weier et al. 1993). Similarly, Maag and Vinther (1996) noticed that soil with an optimum moisture condition and high N source can produce increased N_2O emissions. Results of decreased N_2O production at low moisture content were obtained by Dalal et al. (2003). A study by Russel et al. (1993) on AWW irrigation to land treatment in a pastureland suggests that, under favourable conditions (e.g. a soil temperature below 12 °C, soil pH 5.9 and abattoir effluent pH of 5), N loss can reach a maximum within a day. Soil moisture was one of the most important factors influencing N_2O emissions from tropical forest soils (Kiese and Butterbach-Bahl 2002). In moist soils, the reduction of NO_3^- proceeds in a series of steps, producing NO_2, nitric oxide (NO), nitrous oxide (N_2O) and N_2 gas. Denitrification results not only in the loss of a valuable plant nutrient but also in the release of N_2O (greenhouse gas), which is implicated in the destruction of atmospheric ozone (Bolan et al. 2004). In a flooded soil condition, denitrification has been considered to be a major pathway of N loss (Aulakh et al. 2001). Nitrifier denitrification can be a prime contributor to total N_2O production from the soil. Nitrifier denitrification contributed more to N_2O production than total conventional denitrification of NO_3^- at both 50 and 70 % WFPS.

Table 3 Effects of abattoir wastewater irrigation on N_2O emission ($n = 30$); a comparison of cumulative effects of two types of moisture gradient by the two-sample *t* test

Properties	50 % FC	100 % FC	Sig. diff.
N_2O emitted—week 1 (mg/pot)	19 ± 3.1	21.9 ± 3	n.s.
N_2O emitted—week 2 (mg/pot)	10.2 ± 2.8	14.6 ± 4.9	$p < 0.05$
N_2O emitted—week 3 (mg/pot)	10.1 ± 4.6	14.7 ± 4.7	n.s.
N_2O emitted—week 4 (mg/pot)	10 ± 4.9	15.4 ± 2.4	$p < 0.001$
Cumulative emission (mg/pot)	49.3 ± 12.5	66.5 ± 9.3	$p < 0.001$

n.s. not significant

Statistical significance

The overall emissions of N_2O were positively correlated with nitrate-N and ammonia-N for all the 4 weeks of treatment, but the N_2O emission was negatively correlated with plant DM yield. This means that high plant growth can minimise the N loss as N_2O emission as a result of elevated N uptake. Similar to soil moisture content, N_2O emission was positively correlated with ammonia-N, pH and Olsen P (Jha et al. 2012). In the current study, the soil was collected from the land treatment site (AWW irrigated) which was rich in soil nutrients such as nitrate-N, ammonia-N and Olsen P. Overall, the plots irrigated with AWW 100 % FC showed significantly higher N loss (N_2O) than the plots irrigated with 50 % FC for all four plant species used (Table4). The overall AWW irrigation (50 and 100 % FC) effects on the plant-induced denitrification of four plant species were interpreted using principal component analysis (PCA) for the individual parameters. The effects of the irrigation loading rate on N losses through N_2O were clearly explained using PCA. The wastewater loading rates ($n = 30$) of both treatments (AWW 100 and 50 % FC) showed distinction from their nutrient loss as N_2O. The AWW with 100 % FC irrigated pots (all the four plants) greatly varied from the abattoir 50 % FC irrigated sample reflecting the quantity of N loss and their frequency in different time periods. The factor loading float shows clearly the effects of wastewater irrigation on the soil properties.

Table 4 Effects of abattoir wastewater irrigation on N_2O emission ($n = 30$); a comparison of cumulative effects of two types of moisture gradient by Pearson correlation coefficient

Properties	Nitrate-N (mg/pot)	Ammonia-N (mg/pot)	N_2O week 1 (mg/pot)	N_2O week 2 (mg/pot)	N_2O week 3 (mg/pot)	N_2O week 4 (mg/pot)	Cumulative emission (mg/pot)	DM yield (mg/pot)
Nitrate-N (mg/pot)	1							
Ammonia-N (mg/pot)	0.509**	1						
N_2O week 1 (mg/pot)	0.146	0.017	1					
N_2O week 2 (mg/pot)	0.348	0.035	0.655**	1				
N_2O week 3 (mg/pot)	0.252	0.159	0.432*	0.255	1			
N_2O week 4 (mg/pot)	0.282	0.336	0.570**	0.620**	0.488**	1		
Cumulative emission (mg/pot)	0.335	0.187	0.804**	0.784**	0.719**	0.854**	1	
DM yield (mg/pot)	−0.076	−0.246	−0.391*	−0.316	−0.324	-0.285	−0.412*	1

*Correlation is significant at the 0.05 level; **correlation is significant at the 0.01 level

Similar to a high loading rate, factors 1 (nitrate-N) and 2 (ammonia-N) explained 51 and 21 % of variation, respectively. Therefore, nearly 50 % of the total N loss as determined by the nutrient load was supplied through wastewater addition. The properties such as ammonia-N and nitrate-N were largely influenced (factors 1 and 2) by the overall emission of N_2O. The PCA method showed that AWW irrigation was significantly different in the individual treatments in terms of soil N loss and DM yield in the current pot experiment. The highest N_2O emission was recorded at 100 % FC, due to the supply of a high rate of nitrate-N through AWW.

Effects of AWW irrigation on plant-induced denitrification—nutrient loss and utilisation

This study demonstrated the influence of AWW irrigation on the plant-induced denitrification and soil productivity. The effects of AWW irrigation on soil nutrient loss as N_2O are presented in Table 5. In the present study, the nutrient input included wastewater- and soil-derived nutrients; the output included plant uptake and loss through N_2O emission. Using the above information, the percentage of N loss through N_2O was calculated to study the effects of AWW irrigation on plant-induced denitrification, and soil fertility and productivity changes (Eq. 1).

Table 5 Effects of abattoir wastewater irrigation on N loss as N_2O emission (%) in two types of loading rate in four crops grown under greenhouse condition (n = 30)

Crops	N added through effluent (mg/pot)		N emitted (mg/pot)		% of N emitted		DM yield (mg/pot)	
	50 % FC	100 % FC	50 % FC	100 % FC	50 % FC	100 % FC	50 % FC	100 % FC
Alfalfa	973.2	2031.9	17.6	23.0	1.8	1.1	7333.3	9333.3
Mustard	973.2	2031.9	17.8	21.9	1.8	1.1	14,333.3	16,666.7
Maize	973.2	2031.9	18.2	20.8	1.9	1.0	24,666.7	32,666.7
Sunflower	973.2	2031.9	18.2	20.3	1.9	1.0	28,666.7	39,333.3
No crop	973.2	2031.9	23.0	23.4	2.4	1.2	0	0

$$\%\text{of added N emitted} = \frac{\text{N emitted (mg/pot)}}{\text{N added by AWW (mg/pot)}} \times 100 \qquad (1)$$

The results from this study showed that 50 % FC treated soil had higher N_2O emissions in terms of percentage losses (high differences in input and nutrient loss ratio) compared to the 100 % FC. Similarly, the non-cropped soils were very highly variable between the crops due to the absence of plants. In the current experiment, results showed that nearly 1.2 to 2.4 % of applied N was lost as N_2O. Moiser et al. (1996) showed that nearly 2.5 % of the total N applied to the soil was lost as N_2O from agriculture land. The relationship between N input and overall N_2O emission found in this study was positively correlated, similar to the previous research report by IPCC (1999) and studies by Dobbie et al. (1999).

DISCUSSION

The result from the greenhouse experiments clearly demonstrated that the addition of nutrients or AWW irrigation enhanced the nitrous oxide emission. The effect of nutrient (N) addition on the soil generally increases the soil N loss through N_2 and N_2O, if it is not utilised by green cover, and the excess N may find a pathway to enter the atmosphere or be leached to groundwater (Reichenau et al. 2016). The following reasons could be attributed to the high nitrous oxide emission from the wastewater-irrigated soils: (i) the rate of nutrient loading through abattoir wastewater irrigation/high soil fertility, (ii) the process of denitrification and (iii) the anaerobic condition created by the excessive addition of abattoir wastewater. The rates of nutrient addition on nitrous oxide emission in soil have been reported by many researchers; for example, Zhang et al. (2016) reported that increased nitrate or phosphate loading resulted in stronger influence on denitrification than single nutrient additions that stimulate denitrification rates in sandy sediments in aquatic/terrestrial transition zones.

The abattoir wastewater irrigation had significant impacts on the physiochemical properties of the soils. Nitrous oxide (N_2O) emissions can be significantly affected by the amounts of N available in soils, and also forms of nutrients, local climate and soil conditions are the determining factors of the effects (Peng et al. 2011). Increased N availability generally leads to increased N_2O emission (Davidson et al. 1996) because this stimulates the microbial processes of nitrification and denitrification, which produce the gas fluxes. The relationship between N input (500 and 1000 mg/kg) on cumulative N_2O emission found in this study was significant (N losses increase with the increasing application of N) (Fig. 3). The results have been consistent with earlier reported studies on N_2O emissions (Liu et al. 2005; van Groenigen et al. 2004; Willén et al. 2016) (Fig. 3). Similarly, Chmura et al. (2016) assessed the impact of nutrient additions on greenhouse gas fluxes using dark static chambers in a microtidal and a macrotidal marsh (Both were experimentally fertilised for 6 years). They found that N_2O fluxes are likely to vary with the source of pollutant nutrients, but emissions will be lower if N is not accompanied by an adequate supply of P.

Overall, the application of AWW caused significant N_2O loss. In our experiment, we compared the two treatments of nutrient loadings 50 % FC and 100 % FC and we found that there was a highly significant correlation between both treatments in terms of nitrous oxide emission ($p < 0.001$) (Fig. 4). A number of studies have reported the effect of split fertilisation or low-rate application which showed significant reduction in N_2O emissions (Yu et al. 2016). There was a higher percentage reduction

(28 %) in cumulative N_2O emissions under the split urea application compared with the single fertilisation, although these emissions were influenced by the N fertiliser rate and soil moisture (Yu et al. 2016). The results suggest that low-rate application of AWW is a potential strategy for reducing N_2O emissions in a wastewater-irrigated soil or in a land treatment site. Similarly, the results by Neto et al. (2016) showed that N fertiliser increased N_2O emissions from the soil, especially when urea was used. The emission factor for N fertiliser was 0.46 ± 0.33 %.

There was no difference between urea fertiliser and abattoir wastewater nutrient addition on nitrous oxide emission values. Values of N_2O fluxes were reported for each week and compared with cumulative emission. The application of AWW at the rate of 100 % FC reached a peak of 24 mg/pot emission in the first few weeks after planting (Fig. 5). A study by van der Weerden et al. (2016) recommended an N_2O emission inventory for New Zealand's agricultural soils and found that there was no difference between urea fertiliser in terms of N_2O emission due to the different origins and characteristics of these N sources. For example, in New Zealand's agricultural soils, N_2O emissions have means of 0.6 and 0.3 % for urea fertiliser and FDE, respectively (der Weerden et al. 2016).

According to a study report by Rowe et al. (2012), the effect of N deposition on mineralisable N stock was more apparent in more organic soils, whereas the effect on nitrate proportion was more apparent in more mineral soils. With the high proportions of nitrate (over 40 %) that responses also depend on soil C content and site temperature (Table 1). The proportions of mineral N and nitrate were both strongly influenced by the N deposition rate and by interactions with soil C content (Rowe et al. 2012). In soil, urea is rapidly hydrolysed to ammonium (NH_4^+) ions, a part of which may be lost as ammonia (NH_3^-) and subsequently as nitrous oxide (N_2O) (Singh et al. 2013). The rate of denitrification was higher in soils incubated at saturation than in soils incubated at FC. Brown et al. (2012) found that denitrification was the dominant microbial source of N_2O, and responded to increased soil water content and higher labile carbon availability. Elevated precipitation increased soil emissions of N_2O, especially in combination with added nitrogen and heat. The reduction to N_2 plus absorption by water primarily depends on soil properties, such as the availability of mineral N (substrate for nitrification and denitrification), soil oxygen and water content, soil temperature, pH and redox conditions, and the availability of labile organic C and N (Chapuis et al. 2007; Yan et al. 2012). Urease inhibitor increased the plant N uptake but did not result in a significant increase in herbage DM yields from urea fertiliser. Urease inhibitor was effective in reducing NH_3^- and N_2O emissions from both the urine and urea treatments, with the reduction in N_2O emissions varying with plant N uptake (Singh et al. 2013). However, the effect of changing soil moisture on DR and N_2O/N_2 ratio may vary with the type of soil, its nutrient status and the management practices followed on the farm (Jha et al. 2012).

At wastewater irrigation sites, nitrate is formed during the irrigation events. The soil air is rapidly displaced by the wastewater and the soil becomes saturated. Denitrification and nitrous oxide emission rates are at a maximum during this period. As the site drains, the number of anoxic sites decreases and background rates re-establish (Russell et al. 1993). Peng et al. (2011) observed that peak N_2O fluxes induced by N treatments were concentrated in short periods (2 to 3 weeks) after fertilisation in summer and in soil thawing periods in early spring. The weekly N_2O emission was calculated for each crop, and we found significant differences among the crops used with two levels of AWW irrigation. Overall, the N_2O fluxes were significantly higher in 100 % FC compared to 50 % FC (Fig. 6) probably due to a larger amount of AWW irrigation with nutrients (Table 1). The three N levels increased annual N_2O emissions significantly ($P < 0.05$) in the medium and high N loading treatments compared with the control. A similar pattern was reported by Russell et al. (1993) in which peak rates at the pasture sites were higher with primary-treated effluent (1–137 g N_2O-N ha/h) than with anaerobic effluent (1–62 g N_2O-N ha/h). This was attributed to the higher organic carbon concentration in primary-treated effluent and possibly soil temperature peak nitrous oxide emission rates increased with increasing surface soil temperature. A study by Bhandral et al. (2007) found that among the N sources, the highest emissions were measured with nitrate application, emissions being ten times more than those from other N sources for compacted soil. Also, they reported that the soil compaction caused a sevenfold increase in the N_2O flux; the total N_2O fluxes for the entire experimental period ranged from 2.62 to 61.74 kg N_2O-N/ha for the compacted soil and 1.12 to 4.37 kg N_2O-N/ha for the uncompacted soil.

The N_2O emissions were very low in pots with high DM yield as found in this study and shown in Table 5, possibly due to a high amount of nutrient uptake by the crops for growth and development. On the other hand, not-cropped conditions show a higher nutrient loss (high N_2O emission fluxes recorded in this study) (Table 5). The increased N_2O emission from the control pot was due to the absence of crop or nutrient utilisation by crops. A twofold increase in the N_2O emission rate was recorded for the non-cropped pot compared to the cropped pot. In this short-duration study, major peaks were observed on the first 2 weeks after the treatment including the cropped and non-cropped control; this might have continued to increase for a few weeks after the treatment if we continued to irrigate the field at the same irrigation rate (Bhandral et al. 2007). Soils are the main sources of the greenhouse gas N_2O. The N_2O emission at the soil surface is the result of production and consumption processes of agricultural systems (Chapuis et al. 2007).

In grazed pastures, loss of N occurs mainly through ammonia (NH_3^-) volatilisation, the release of gaseous N such as nitric oxide (NO) and nitrous oxide (N_2O) through biological denitrification, and nitrate (NO_3^-) leaching, which has both economic and environmental implications (Bolan et al. 2004). In the nearly water-saturated soil (90 % WFPS), N_2O production was, as expected, dominated by conventional denitrification of NO_3^- (Kool et al.2011). A similar result was reported by Weitz et al. (2001) and Zhou et al. (2008) that N_2O emission was insensitive to soil moisture, possibly resulting from lower soil nitrogen content (0.13–0.21 %). Differences were observed in nitrification–denitrification rates in flooding conditions because they control the availability of N and oxygen. Nitrification was controlled not only by oxygen but also by the amount of ammonium sulphate added and the high pH of the water; these two latter factors could increase the NH_3 concentration, with inhibition of the coupled nitrification–denitrification (Carrasco et al. 2004). The results of Bateman and Baggs (2005) indicate N_2O production during heterotrophic nitrification in our soil at 50 % WFPS and the possibility of aerobic denitrification. Thus, several processes may simultaneously produce N_2O in soil at 60 % WFPS and below. Nitrification was the main source of N_2O in soils at 35–60 % WFPS, indicating the significance of this process for global warming.

CONCLUSIONS

A nutrient-rich water source (e.g. AWW) can supply sufficient or a surplus amount of primary nutrients to the soil and plants. However, a significant amount of applied nutrient (N) is lost to air, soil and water through various processes by N_2O and ammonia (NH_3) emissions and nitrate leaching. These losses are considered environmental hazards due to the ill effects caused by N_2O as a potential greenhouse gas and nitrate in groundwater as a potential water pollutant. Hence, minimising and management of nitrogenous pollutants become more important in the current era of sustainable agriculture. The mitigation options to reduce N loss include the use of nitrogenous inhibitors (NI), growing bioenergy crops and adopting efficient farm budgeting (applying nutrients only when it is necessary). The emission of nitrous oxide was high at the first 2 weeks after planting and declined subsequently reaching the background levels within 4 weeks. Nitrous oxide emission was higher for 100 % FC than for 50 % FC. The peak nitrous oxide emission flux was recorded in the non-cropping condition during the first week after planting; this indicated that there are higher risks of applied nitrogen loss through nitrous oxide emission in the absence of plant uptake of nitrogen. Overall, nitrous oxide emission was 15 % higher in the non-cropped than cropped treatments. The conclusions that can be drawn from this study in relation to plant-induced denitrification are as follows: the N_2O emission rate increased with increasing soil moisture content (50 to 100 % FC), the N_2O emission rate increased with increasing N input through chemical fertiliser or wastewater irrigation, the N_2O emission rate decreased in the presence of plants and increased under the no-crop condition due to plant uptake and dry matter yield was found to be significantly higher in the 100 % FC than the 50 % treatment. The rate of N loss varied with the rate of AWW irrigation; nearly 1.2 to 2.4 % of applied N was lost as N_2O. The main conclusions that can be drawn from this study in relation to plant-induced denitrification are the following: the N_2O emission rate increased with increasing soil moisture content, the N_2O emission rate increased with increasing N input through wastewater irrigation and the N_2O emission rate decreased with the presence of plants which can be attributed to plant uptake.

ACKNOWLEDGMENTS

The first author would like to thank the University of South Australia for the UPS scholarships and CRC for Contamination Assessment and Remediation of the Environment for the CRC-CARE scholarships to undertake research on nutrient management in abattoir wastewater-discharged soil.

REFERENCES

1. Aulakh MS, Khera TS, Doran JW, Bronson KF (2001) Denitrification, N_2O and CO_2 fluxes in rice-wheat cropping system as affected by crop residues, fertilizer N and legume green manure. Biol Fertil Soils 34:375–389
2. Bateman E, Baggs E (2005) Contributions of nitrification and denitrification to N_2O emissions from soils at different water-filled pore space. Biol Fertil Soils 41:379–388
3. Baumert KA, Herzog T, Pershing J (2005) Navigating the numbers: greenhouse gas data and international climate policy. World Resour Inst

4. Bhandral R, Saggar S, Bolan N, Hedley M (2007) Transformation of nitrogen and nitrous oxide emission from grassland soils as affected by compaction. Soil Tillage Res 94:482–492

5. Bolan NS, Saggar S, Luo J, Bhandral R, Singh J (2004) Gaseous emissions of nitrogen from grazed pastures: processes, measurements and modelling, environmental implications, and mitigation. Adv Agron 84:37–120

6. Bouwman A (1996) Direct emission of nitrous oxide from agricultural soils. Nutr Cycl Agroecosyst 46:53–70

7. Brown JR et al. (2012) Effects of multiple global change treatments on soil N_2O fluxes. Biogeochemistry 109:85–100

8. Cardenas L, Thorman R, Ashlee N, Butler M, Chadwick D, Chambers B, Cuttle S, Donovan N, Kingston H, Lane S (2010) Quantifying annual N2O emission fluxes from grazed grassland under a range of inorganic fertiliser nitrogen inputs. Agric Ecosyst Environ 136:218–226

9. Carrasco D, Fernández-Valiente E, Ariosa Y, Quesada A (2004) Measurement of coupled nitrification-denitrification in paddy fields affected by Terrazole, a nitrification inhibitor. Biol Fertil Soils 39:186–192

10. Cassidy DP, Belia E (2005) Nitrogen and phosphorus removal from an abattoir wastewater in a SBR with aerobic granular sludge. Water Res 39:4817–4823.

11. Cerri CC, Maia SM, Galdos MV, Cerri CE, Feigl BJ, Bernoux M (2009) Brazilian greenhouse gas emissions: the importance of agriculture and livestock. Scientia Agricola 66(6):831–843

12. Chapuis-Lardy L, Wrage N, Metay A, JL CHOTTE, Bernoux M (2007) Soils, a sink for N_2O? A review. Glob Chang Biol 13:1–17

13. Chen M, Ma LQ (2001) Comparison of three aqua regia digestion methods for twenty Florida soils. Soil Sci Soc Am J 65:491–499

14. Chmura GL, Kellman L, van Ardenne L, Guntenspergen GR (2016) Greenhouse gas fluxes from salt marshes exposed to chronic nutrient enrichment. PLoS One 11:e0149937

15. Crutzen PJ, Mosier AR, Smith KA, Winiwarter W (2008) N2O release from agro-biofuel production negates global warming reduction by replacing fossil fuels. Atmos Chem Phys 8:389–395

16. Czepiel P, Crill P, Harriss R (1995) Nitrous oxide emissions from municipal wastewater treatment. Environ Sci Technol 29:2352–2356

17. Dalal RC, Wang W, Robertson GP, Parton WJ (2003) Nitrous oxide emission from Australian agricultural lands and mitigation options: a review. Soil Res 41:165–195

18. Davidson E, Matson PA, Brooks P (1996) Nitrous oxide emission controls and inorganic nitrogen dynamics in fertilized tropical agricultural soils. Soil Sci Soc Am J 60:1145–1152

19. Dobbie K, McTaggart I, Smith K (1999) Nitrous oxide emissions from intensive agricultural systems: variations between crops and seasons, key driving variables, and mean emission factors. J Geophys Res Atmos 104:26891–26899

20. Food and Agriculture Organisation (FAO) (2006) Livestock's long shadow—environmental issues and option www.fao.org

21. Garnaut R, Howes S, Jotzo F, Sheehan P (2008) Emissions in the Platinum Age: the implications of rapid development for climate-change mitigation. Oxf Rev Econ Policy 20;24(2):377–401

22. Gunnarsson I, Sigfússon B, Stefánsson A, Arnórsson S, Scott SW, Gunnlaugsson E (2011) Injection of H2S from Hellisheiði Power Plant, Iceland. In: Proceedings of the 36th Workshop on Geothermal Reservoir Engineering

23. Intergovernmental Panel on Climate Change (IPCC) (1999) Climate Change 2001 – The Scientific Basis. Contributions of working group I to the Third Assessment Report of the Intergovernmental Panel on Climate Change. Cambridge University Press, Cambridge, UK and New York, NY, USA, p 881

24. Intergovernmental Panel on Climate Change (IPCC) (2007). IPCC Secretariat Genevawww.ipcc.ch

25. Jha N, Saggar S, Tillman R, Giltrap D (2012) Changes in denitrification rate and N_2O/N_2 ratio with varying soil moisture conditions of New Zealand pasture soils. Advanced nutrient management: gains from the past-goals for the future. Fertilizer and Lime Research Centre Occasional Report

26. Kampschreur MJ, Tan NC, Kleerebezem R, Picioreanu C, Jetten MS, Loosdrecht MCV (2007) Effect of dynamic process conditions on nitrogen oxides emission from a nitrifying culture. Environ Sci Technol 42:429–435

27. Kampschreur MJ, Temmink H, Kleerebezem R, Jetten MS, van Loosdrecht MC (2009) Nitrous oxide emission during wastewater treatment. Water Res 43:4093–4103

28. Kiese R, Butterbach-Bahl K (2002) N_2O and CO_2 emissions from three different tropical forest sites in the wet tropics of Queensland, Australia. Soil Biol Biochem 34:975–987

29. Kool DM, Dolfing J, Wrage N, Van Groenigen JW (2011) Nitrifier denitrification as a distinct and significant source of nitrous oxide from soil. Soil Biol Biochem 43:174–178

30. Liu XJ, Mosier AR, Halvorson AD, Zhang FS (2005) Tillage and nitrogen application effects on nitrous and nitric oxide emissions from irrigated corn fields. Plant Soil 276:235–249

31. Longhurst R, Roberts A, O'Connor M (2000) Farm dairy effluent: a review of published data on chemical and physical characteristics in New Zealand. N Z J Agric Res 43:7–14

32. Luo J, Lindsey S, Xue J (2004) Irrigation of meat processing wastewater onto land. Agric Ecosyst Environ 103:123–148

33. Maag M, Vinther FP (1996) Nitrous oxide emission by nitrification and denitrification in different soil types and at different soil moisture contents and temperatures. Appl Soil Ecol 4:5–14

34. Matheyarasu R, Seshadri B, Bolan N, Naidu R (2012) Nutrient management in effluents derived from agricultural industries: an Australian perspective. WIT Press

35. Matheyarasu R, Seshadri B, Bolan NS, Naidu R (2014) Impacts of abattoir waste-water irrigation on soil fertility and productivity

36. McCarl BA, Schneider UA (2000) US agriculture's role in a greenhouse gas emission mitigation world: an economic perspective. Rev Agric Econ 22:134–159

37. Meinshausen M, Meinshausen N, Hare W, Raper SC, Frieler K, Knutti R, Frame DJ, Allen MR (2009) Greenhouse-gas emission targets for limiting global warming to 2 C. Nature 458(7242):1158–1162

38. Meat & Livestock Australia (MLA) (2010) Industry environmental sustainability review A.ENV.0086

39. Mosier A, Parton W, Valentine D, Ojima D, Schimel D, Delgado J (1996) CH4 and N_2O fluxes in the Colorado shortgrass steppe: 1. Impact of landscape and nitrogen addition. Glob Biogeochem Cycles 10:387–399

40. Oenema O, Wrage N, Velthof GL, van Groenigen JW, Dolfing J, Kuikman PJ (2005) Trends in global nitrous oxide emissions from animal production systems. Nutr Cycl Agroecosyst 72:51–65

41. Olsen SR, Cole CV, Watanabe FS, Dean LA (1954) Estimation of available phosphorus in soils by extraction with sodium bicarbonate. US Dep Agric Circ 939

42. Organisation for Economic Co-operation and Development (OECD) -FAO (2011) Agricultural Outlook 2011–2020 OECD-FAO:14

43. Peng Q, Qi Y, Dong Y, Xiao S, He Y (2011) Soil nitrous oxide emissions from a typical semiarid temperate steppe in inner Mongolia: effects of mineral nitrogen fertilizer levels and forms. Plant Soil 342:345–357

44. Preston BL, Jones RN, Scientific C (2006) Climate change impacts on Australia and the benefits of early action to reduce global greenhouse gas emissions. CSIRO Australia

45. Ravishankara A, Daniel JS, Portmann RW (2009) Nitrous oxide (N_2O): the dominant ozone-depleting substance emitted in the 21st century. Science 326:123–125

46. Rayment GE, Lyons DJ (2011) Soil chemical methods: Australasia vol 3. CSIRO publishing

47. Reichenau TG, Klar CW, Lenz-Wiedemann VI, Fiener P, Schneider K, (2016) Nitrate leaching, regional assessment of global change impacts. Springer, pp. 303–310

48. Rowe EC, Emmett BA, Frogbrook ZL, Robinson DA, Hughes S (2012) Nitrogen deposition and climate effects on soil nitrogen availability: influences of habitat type and soil characteristics. Sci Total Environ 434:62–70

49. Russell J, Cooper R, Lindsey S (1993) Soil denitrification rates at wastewater irrigation sites receiving primary-treated and anaerobically treated meat-processing effluent. Bioresour Technol 43:41–46

50. Saggar S, Andrew R, Tate K, Rodda N, Hedley C, Townsend J (2002) Measurements and modelling of nitrous oxide emissions from dairy pastures. In: Currie LD, Loganathan P (eds) Proceedings of the workshop on dairy farm soil management. Massey University, Palmerston North, pp. 201–214

51. Saggar S, Bolan N, Bhandral R, Hedley C, Luo J (2004) A review of emissions of methane, ammonia, and nitrous oxide from animal

excreta deposition and farm effluent application in grazed pastures. N Z J Agric Res 47:513–544

52. Saggar S et al. (2013) Denitrification and N_2O: N_2 production in temperate grasslands: processes, measurements, modelling and mitigating negative impacts. Sci Total Environ 465:173–195

53. Saggar S, Giltrap DL, Davison R, Gibson R, de Klein CA, Rollo M, Ettema P, Rys G (2015) Estimating direct N_2O emissions from sheep, beef, and deer grazed pastures in New Zealand hill country: accounting for the effect of land slope on the N_2O emission factors from urine and dung. Agric Ecosyst Environ 205:70–78

54. Schneider UA (2000) Agricultural sector analysis on greenhouse gas emission mitigation in the United States. Texas A&M University

55. Shine KP, Fuglestvedt JS, Hailemariam K, Stuber N (2005) Alternatives to the global warming potential for comparing climate impacts of emissions of greenhouse gases. Clim Chang 68:281–302

56. Singh J, Saggar S, Giltrap D, Bolan NS (2008) Decomposition of dicyandiamide (DCD) in three contrasting soils and its effect on nitrous oxide emission, soil respiratory activity, and microbial biomass—an incubation study. Soil Res 46:517–525

57. Singh J, Kunhikrishnan A, Bolan NS, Saggar S (2013) Impact of urease inhibitor on ammonia and nitrous oxide emissions from temperate pasture soil cores receiving urea fertilizer and cattle urine. Sci total Environ 465:56–63

58. Siqueira Neto M, Galdos MV, Feigl BJ, Cerri CE, Cerri CC (2016) Direct N2O emission factors for synthetic N-fertilizer and organic residues applied on sugarcane for bioethanol production in central-southern Brazil. GCB Bioenergy 8:269–280

59. Sparling G et al. (2006) Nutrient leaching and changes in soil characteristics of four contrasting soils irrigated with secondary-treated municipal wastewater for four years. Soil Res 44:107–116

60. Tsujimoto Y, Masuda J, Fukuyama J, Ito H (1994) N_2O emissions at solid waste disposal sites in Osaka City. Air Waste 44:1313–1314

61. van der Weerden T, Cox N, Luo J, Di H, Podolyan A, Phillips R, Saggar S, de Klein C, Ettema P, Rys G (2016) Refining the New Zealand nitrous oxide emission factor for urea fertiliser and farm dairy effluent. Agric Ecosyst Environ 222:133–137

62. van Groenigen JW, Kasper GJ, Velthof GL, van den Pol-van Dasselaar A, Kuikman PJ (2004) Nitrous oxide emissions from silage maize

fields under different mineral nitrogen fertilizer and slurry applications. Plant Soil 263:101–111

63. Weier KL, Doran JW, Power JF, Walters DT (1993) Denitrification and the dinitrogen/nitrous oxide ratio as affected by soil water, available carbon, and nitrate. Soil Sci Soc Am J 57(1):66–72

64. Weitz A, Linder E, Frolking S, Crill P, Keller M (2001) N_2O emissions from humid tropical agricultural soils: effects of soil moisture, texture and nitrogen availability. Soil Biol Biochem 33:1077–1093

65. Willén A, Jönsson H, Pell M, Rodhe L (2016) Emissions of nitrous oxide, methane and ammonia after field application of digested and dewatered sewage sludge with or without addition of urea. Waste Biomass Valorization 7:281–292

66. Yan C, Zhang H, Li B, Wang D, Zhao Y, Zheng Z (2012) Effects of influent C/N ratios on CO2 and CH4 emissions from vertical subsurface flow constructed wetlands treating synthetic municipal wastewater. J Hazard Mater 203:188–194

67. Yu Y, Zhao C, Jia H (2016) Ability of split urea applications to reduce nitrous oxide emissions: a laboratory incubation experiment. Appl Soil Ecol 100:75–80

68. Zhang L, Yao X, Tang C, Xu H, Jiang X, Zhang Y (2016) Influence of long-term inundation and nutrient addition on denitrification in sandy wetland sediments from Poyang Lake, a large shallow subtropical lake in China. Environ Pollut

69. Zhou YA, Pijuan M, Zeng RJ, Yuan Z (2008) Free nitrous acid inhibition on nitrous oxide reduction by a denitrifying-enhanced biological phosphorus removal sludge. Environ Sci Technol 42(22):8260–8265

10
CHAPTER

Adsorption Technique for the Removal of Organic Pollutants from Water and Wastewater

Mohamed Nageeb Rashed[1]

[1] *Aswan Faculty of Science, Aswan University, Aswan, Egypt*

1. INTRODUCTION

Organic pollution is the term used when large quantities of organic compounds. It originates from domestic sewage, urban run-off, industrial effluents and agriculture wastewater. sewage treatment plants and industry including food processing, pulp and paper making, agriculture and aquaculture. During the decomposition process of organic pollutants the dissolved oxygen in the receiving water may be consumed at a greater rate than it can be replenished, causing oxygen depletion and having severe consequences for the stream biota. Wastewater with organic pollutants contains large quantities of suspended solids which reduce the light available to photosynthetic organisms and, on settling out, alter the characteristics of the river bed, rendering it an unsuitable habitat for many invertebrates. Organic pollutants include pesticides, fertilizers, hydrocarbons, phenols, plasticizers, biphenyls, detergents, oils, greases, pharmaceuticals, proteins and carbohydrates [1-3].

Toxic organic pollutants cause several environmental problems to our environment. The most common organic pollutants named persistent organic pollutants (POPs). POPs are compounds of great concern due to their toxicity,persistence, long-range transport ability [4] and bioaccumulation in animals [5], travel long distances and persist in living organisms. POPs are carbon-based chemical compounds and mixtures (twelve pollutants) that include industrial chemicals such as polychlorinated biphenyls (PCBs),polychlorinated dibenzo-p-dioxins and dibenzofurans (PCDD/Fs), and some organochlorine pesticides (OCPs), such as hexachlorobenzene (HCB) or dichloro-diphenyl-trichloroethane (DDT), dibenzo-p-dioxins (dioxins) and dibenzo-p-furans (furans) [6]. PCDD/Fs are released to the environment as by-products of several processes, like waste incineration or metal production [7]. Many of these compounds have been or continue to be used in large quantities and due to their environmental persistence, have the ability to bioaccumulate and biomagnify [8].

Efficient techniques for the removal of highly toxic organic compounds from water have drawn significant interest. A number of methods such as coagulation, filtration with coagulation, precipitation, ozonation, adsorption, ion exchange, reverse osmosis and advanced oxidation processes have been used for the removal of organic pollutants from polluted water and wastewater. These methods have been found to be limited, since they often involve high capital and operational costs. On the other hand ion exchange and reverse osmosis are more attractive processes because the pollutant values can be recovered along with their removal from the effluents. Reverse osmosis, ion exchange and advanced oxidation processes do not seem to be economically feasible because of their relatively high investment and operational cost.

Among the possible techniques for water treatments, the adsorption process by solid adsorbents shows potential as one of the most efficient methods for the treatment and removal of organic contaminants in wastewater treatment. Adsorption has advantages over the other methods because of simple design and can involve low investment in term of both initial cost and land required. The adsorption process is widely used for treatment of industrial wastewater from organic and inorganic pollutants and meet the great attention from the researchers. In recent years, the search for low-cost adsorbents that have pollutant –binding capacities has intensified. Materials locally available such as natural materials, agricultural wastes and industrial wastes can be utilized as low-cost adsorbents. Activated carbon produced from these materials can be used as adsorbent for water and wastewater treatment [9].

2. ADSORPTION PHENOMENON

Adsorption is a surface phenomenon with common mechanism for organic and inorganic pollutants removal. When a solution containing absorbable solute comes into contact with a solid with a highly porous surface structure, liquid–solid intermolecular forces of attraction cause some of the solute molecules from the solution to be concentrated or deposited at the solid surface. The solute retained (on the solid surface) in adsorption processes is called adsorbate, whereas, the solid on which it is retained is called as an adsorbent. This surface accumulation of adsorbate on adsorbent is called adsorption. This creation of an adsorbed phase having a composition different from that of the bulk fluid phase forms the basis of separation by adsorption technology.

In a bulk material, all the bonding requirements (be they ionic, covalent, or metallic) of the constituent atoms of the material are filled by other atoms in the material. However, atoms on the surface of the adsorbent are not wholly surrounded by other adsorbent atoms and therefore can attract adsorbates. The exact nature of the bonding depends on the details of the species involved, but the adsorption process is generally classified as physicsorption (characteristic of weak Van Der Waals forces) or chemisorption (characteristic of covalent bonding). It may also occur due to electrostatic attraction.

As the adsorption progress, an equilibrium of adsorption of the solute between the solution and adsorbent is attained (where the adsorption of solute is from the bulk onto the adsorbent is minimum). The adsorption amount (qe, mmol g^{-1}) of the molecules at the equilibrium step was determined according to the following equation:

$$qe = V\left(Co\text{-}Ce\right) /M \quad (1)$$

where V is the solution volume (L); M is the mass of monolithic adsorbents (g); and Co and Ce are the initial and equilibrium adsorbate concentrations, respectively.

Other definition of adsorption is a mass transfer process by which a substance is transferred from the liquid phase to the surface of a solid, and becomes bound by physical and/or chemical interactions. Large surface area leads to high adsorption capacity and surface reactivity [10].

2.1. Adsorption Isotherms And Models

An adsorption isotherm is the presentation of the amount of solute adsorbed per unit weight of adsorbent as a function of the equilibrium concentration in the bulk solution at constant temperature. Langmuir and Freundlich adsorption isotherms are commonly used for the description of adsorption data.

The Langmuir equation is expressed as:

$$Ce / qe = 1 / bXm + Ce / Xm, \tag{2}$$

Where Ce is the equilibrium concentration of solute (mmol L^{-1}), qe is the amount of solute adsorbed per unit weight of adsorbent (mmol g^{-1} of clay), Xm is the adsorption capacity (mmol g^{-1}), or monolayer capacity, and b is a constant (L $mmol^{-1}$).

The Freundlich isotherm describes heterogeneous surface adsorption. The energy distribution for adsorptive sites (in Freundlich isotherm) follows an exponential type function which is close to the real situation. The rate of adsorption/desorption varies with the strength of the energy at the adsorptive sites. The Freundlich equation is expressed as:

$$\log qe = \log k + 1 / n \log Ce , \tag{3}$$

Where k (mmol g^{-1}) and $1/n$ are the constant characteristics of the system [11].

3. TYPES OF ADSORBENTS

Different types of adsorbents are classified into natural adsorbents and synthetic adsorbents. Natural adsorbents include charcoal, clays, clay minerals, zeolites, and ores. These natural materials, in many instances are relatively cheap, abundant in supply and have significant potential for modification and ultimately enhancement of their adsorption capabilities. Synthetic adsorbents are adsorbents prepared from Agricultural products and wastes, house hold wastes, Industrial wastes, sewage sludge and polymeric adsorbents. Each adsorbent has its own characteristics such as porosity, pore structure and nature of its adsorbing surfaces. Many waste materials used include fruit wastes, coconut shell, scrap tyres, bark and other tannin-rich materials, sawdust, rice husk, petroleum wastes, fertilizer wastes, fly ash, sugar industry wastes blast furnace slag, chitosan and seafood processing wastes, seaweed and algae, peat moss, clays, red mud, zeolites, sediment and soil, ore minerals etc.

Activated carbons as adsorbent for organic pollutants consists in their adsorption a complex process and there still exists considerable difficulty. The main cause of this difficulty results from the large number of variables involved. These include, for example, electrostatic, dispersive and chemical interactions, intrinsic properties of the solute (for example solubility and ionization constant), intrinsic properties of the adsorbent (such pore size distribution), solution properties (in particular, pH) and the temperature of the system [12].

Activated carbons (AC) (both granular activated carbon (GAC) and powdered activated carbons (PAC)) are common adsorbents used for the removal of undesirable odor, color, taste, and other organic and inorganic impurities from domestic and industrial waste water owing to their large surface area, micro porous structure nonpolar character and due to its economic viability.The major constituent of activated carbon is the carbon that accounts up to 95% of the mass weight In addition, active carbons contain other hetero atoms such as hydrogen, nitrogen, sulfur, and oxygen. These are derived from the source raw material or become associated with the carbon during activation and other preparation procedures [13-14]. Putra et al. [15] investigated the removal of Amoxicillin (antibiotic) from pharmaceutical effluents using bentonite and activated carbon as adsorbents. The study was carried out at several pH values. Langmuir and Freundlich models were then employed to correlate the equilibria data on which both models fitted the data equally well. While chemisorption is the dominant adsorption mechanism on the bentonite, both physicosorption and chemisorption played an important role for adsorption onto activated carbon.

Adsorption of methane on granular activated carbon (GAC) was studied. The results showed that with decreasing temperature or increasing methane uptake by GAC the adsorption efficacy decreased. Interactions between the methane molecules and the surface of carbon increase the density of adsorbed methane in respect to the density of compressed gas. The effect that the porosity and the surface chemistry of the activated carbons have on the adsorption of two VOC (benzene and toluene) at low concentration (200 ppm) was also studied. The results show that the volume of narrow micropores (size <0.7 nm) seems to govern the adsorption of VOC at low concentration, specially for benzene adsorption. AC with low content in oxygen surface groups has the best adsorption capacities. Among the AC tested, those prepared by chemical activation with hydroxides exhibit the higher adsorption capacities for VOC. The adsorption capacities achieved are higher than those previously shown in the literature for these conditions, especially for toluene. Adsorption capacities as high as 34 g benzene/100 g AC or 64 g toluene/100 g AC have been achieved [16].

4. ADSORPTION OF DYES

Adsorption techniques are used as high quality treatment processes for the removal of dissolved organic pollutants, such as dyes, from industrial wastewater. Dyes consider as type of organic pollutants. The textile, pulp and paper industries are reported to utilize large quantities of a number of dyes, these pollutant may be found in wastewaters of many industries generating considerable amounts of colored wastewaters, toxic and even carcinogenic, posing serious hazard to aquatic living organisms. Dyes represent one of the problematic groups; they are emitted into wastewater from various industrial branches, mainly from the dye manufacturing and textile finishing and also from food coloring, cosmetics, paper and carpet industries. It is well known that the dye effluents from dyestuff manufacturing and textile industries, may exhibit toxic effects on microbial populations and can be toxic and/or carcinogenic to mammalian animal. Most dyes used in textile industries are stable to light and are not biologically degradable. Furthermore, they are resistant to aerobic digestion. [17].

On searching for economical and available starting materials; different low cost adsorbents were used for the removal of dyes. Activated rice husk was used as cheap adsorbent for color removal from wastewater [18]. Hamdaoui [19] reported that the maximum adsorption of basic dye, methylene blue, onto cedar sawdust and crushed brick was 60 and 40 mg L^{-1}, respectively. Wood-shaving bottom ash (WBA) was used for the removal of Red Reactive 141 (RR141), and azo reactive dyes. WBA/H_2O and WBA/H_2SO_4 adsorbents were made by treating WBA with water and 0.1 M H_2SO_4, respectively; to increase adsorption capacity. The effects of different parameters on adsorption (effect of contact time, initial pH of solution, dissolved metals and elution) were studied.The maximum dye adsorption capacities of WBA/H_2O and WBA/H_2SO_4 obtained from a Langmuir model at 30 °C were 24.3, 29.9, and 41.5 mg l^{-1}, respectively. In addition, WBA/H_2O and WBA/H_2SO_4 could reduce colour and high chemical oxygen demand (COD) of real textile wastewater [20]. Beer brewery waste has been shown to be a low-cost adsorbent for the removal of methylene blue dye from the aqueous solution. The results of preliminary adsorption kinetics showed that the diatomite waste could be directly used as a potential adsorbent for removal of methylene blue on the basis of its adsorption–biosorption mechanisms [21].

Sewage sludge was applied for the preparation of activated carbon adsorbent. Activated carbon adsorbent prepared from sewage sludge has being identified as a potentially attractive material for wastewater. Research studies has been conducted to demonstrate the uses of treated sewage sludge for the removal of dyes from wastewater and polluted water

[22-27]. Otero et al. [27] produced activated carbon by chemically activation and pyrolysis of sewage sludge. The properties of this type of material was studied by liquid-phase adsorption using crystal violet, indigo carmine and phenol as adsorbates. Three prepared activated carbon,of different particle sizes, were used ASS-g1 (particle diameter<0.12 mm), ASS-g2 (0.12<particle diameter<0.5 mm) and PSS-g2(0.12<particle diameter<0.5 mm). Crystal violet dye adsorption has been higher (Qmax 263.2 mg/g using AAS, 270 mg/g using ASS and 184 mg/g using PPS) than indigo carminc (Qmax 60.04 mg/g using AAS, 54.8 mg/g using ASS and 30.8 mg/g using PPS). They proposed that activated carbons made from sewage sludge show promise for the removal of organic pollutants from aqueous streams.

TABLE 1. Selected adsorbents used for dyes removal from polluted water

Dye	Adsorbent	Adsorption capacity	Reference
Reactive Blue 2	activated carbon	0.27,mmol/g	[31]
Reactive Red 4,	activated carbon	0.24 mmol/g	[31]
Reactive Yellow 2	activated carbon	0.11 mmol/g	[31]
Everzol Black B	Sepiolite	120.5 g/kg	[32]
Everzol Red 3BS	Sepiolite	108.8 g/kg	[33]
Everzol Red 3BS	Zeolite	111.1 g/kg	[32]
Everzol Black B	Zeolite	60.6 g/kg	[32]
Orange-G	bagasse fly ash	1.245 g/kg	[33]
Methyl Violet	bagasse fly ash	3.712 g/kg	[33]
Acid Blue 113	amino-functionalized nanoporous silica SBA-3	769 g/kg	[34]
Acid Red 114	amino-functionalized nanoporous silica SBA-3	1000 g/kg	[34]
Acid Green 28	amino-functionalized nanoporous silica SBA-3	333 g/kg	[34]

Acid Yellow 127	amino-functionalized nanoporous silica SBA-3	1250 g/kg	[34]
Acid Orange 67	amino-functionalized nanoporous silica SBA-3	2500 g/kg	[34]
Acid Blue 25	waste tea activated carbon	203.34 mg/g	[35]
Methylene Blue	bituminous coal-based activated carbon	580 mg/g	[36]
Methylene Blue	coal-based activated carbon coal-based activated carbon(KOH washed)	(252 mg/g 234.0 mg/g	[37]
Methylene Blue	activated carbon from Cotton stalk-based	180.0 mg/g	[38]
Methylene Blue	activated carbon from Posidonia oceanica (L.) dead leaves:	285.7 mg/g	[39]
Methylene Blue	Salix psammophila activated carbon	225.89 mg/g	[40]
Methylene Blue	activated carbon from flamboyant pods (Delonix regia)	890mg/g	[41]
Methylene Blue	activated carbon from from Oil palm wood-based	90.9 mg/g	[42]
Methylene Blue	activated carbon from Oil palm shell-based	243.9 mg/g	[43]

The utilization of fly ash in removing dyes from textile wastewater was investigated. A calcium-rich fly ash had been used as adsorbent for the removal of Congo red dye under different conditions. It was observed that the maximum adsorption obtained was between 93% and 98%. under the studies conditions [28]. Wang et al. [29] had reported the use of treated and non-treated fly ash for the removal of methylene blue and basic dye from a wastewater solution. The adsorption capacity for acid treated fly ash was found to be 2.4 x 10-5 mol/g, while non-treated fly ash showed an adsorption capacity of 1.4 x 10-5 mol/g. Wang et al. [30] in another investigation also found that the porous unburned carbon in the fly ash was responsible for the adsorption of the dye, and not the fly ash itself. Table (1) shows selected adsorbent used for dyes removal from polluted water

5. ADSORPTION OF PHENOLS

Since 1860, phenol has been in production, with its basic use as an antiseptic. During late 19th century and thereafter the use of phenol has been further extended to the synthesis of dyes, aspirin, plastics, pharmaceuticals, petrochemical and pesticide chemical industries. In fact, by 2001, the global phenol production has reached an impressive 7.8 million tons [44].

Among the different organic pollutants in wastewater, phenols are considered as priority pollutants since they are harmful to plants, animals and human, even at low concentrations. The major sources of phenolic are steel mills, petroleum refineries, pharmaceuticals,petrochemical, coke oven plants, paints,coal gas, synthetic resins, plywood industries and mine discharge. The wastewater with the highest concentration of phenol (>1000 mg/L) is typically generated from coke processing. Phonolic compounds are also emanated from resin plants with a concentration range of 12–300 mg/L. Environmental Protection Agency (EPA) has set a limit of 0.1 mg/L of phenol in wastewater. The World Health Organization (WHO) is stricter on phenol regulation. It sets a 0.001 mg/L as the limit of phenol concentration in potable water.

Adsorption of phenolic compounds from aqueous solutions by activated carbon is one of the most investigated of all liquid-phase applications of carbon adsorbents [45]. Several adsorbents were used treatment wastewater and removal of phenols. The adsorption isotherms for mono-, di-, and trichlorophenols from aqueous solutions on wood-based and lignite-based carbons were investigated. The adsorptive capacity for 2,4-DCP was found to be 502 mg/g and Freundlich model gave a best fit the experimental data [46]. Zogorski et al. [47] studied the kinetics of adsorption of phenols on GAC. They observed that 60% to 80% of the adsorption occurs within the first hour of contact followed by a very slow approach to the final maximum equilibrium concentration.

In another study, the extent of adsorption of 2,4-dichlorophenol was found to be a function of pH. The presence of surface functional groups also affected the adsorption of phenols onto activated carbon. The presence of dissolved oxygen on activated carbon increased the adsorptive capacity for phenolic compounds This increase in adsorptive capacity was attributed to the oligomerization of the compounds through oxidative coupling reactions [48].

Hamdaouia et al. [49] studied and modeled the adsorption equilibrium isotherms of five phenolic compounds from aqueous solutions onto GAC. The five compounds selected were Phenol (Ph), 2-chlorophenol (2-CP), 4-chlorophenol (4-CP), 2,4-dichlorophenol (DCP), and 2,4,6-trichlorophenol

(TCP). They also observed that the interaction of phenolic compounds with activated carbon surface occurred in localized monolayer adsorption type, i.e. adsorbed molecules are adsorbed at definite, localized sites. Uptake of phenols increased in the order Ph < 2-CP < 4- CP < DCP < TCP, which correlated well with respective increase in molecular weight, cross-sectional area, and hydrophobicity and decrease in solubility and pKa.

Sawdust, a very low cost adsorbent was used, after carbonization, for the removal of phenol from industrial waste waters. The equilibrium adsorption level was determined as a function of the solution pH, temperature, contact time, adsorbent dose and the initial concentration. The adsorption maximum for phenol using sawdust was 10.29 mg/L [50].

Adsorbents, carbonaceous materials, activated carbon (AC), bagasse ash (BA) and wood charcoal (WC), were used for removal of phenol from water [51].The results showed the removal efficiencies for phenol–AC, phenol–WC and phenol–BA, approximately 98%, 90% and 90%, respectively. Removal efficiency of phenol slightly increased when the pH of adsorption system decreased. Yapar and Yilmar [52] reported the adsorptive capacity of some clays and natural zeolite materials found in Turkey for the removal of phenol. They found that calcined hydrotalcite was the best among the studied adsorbents in which adsorbed 52% of phenol from a solution of 1000 mg/L phenol at the adsorbent/phenol ratio of 1:100 while the others adsorbed only 8% of phenol. Also, silica gel, activated alumina, AC, fitrasorb 400 and Hisir 1000 adsorbent were examined as adsorbents for the removal of phenol from aqueous solution. They found that Hisir 1000 was the best among the tested materials [53]. Das and Patnaik [54] utilized blast furnace flue dust (BFD) and slag to investigate phenol adsorption through batch experiment.

Bromophenols (2-bromophenol, 4-bromophenol and 2, 4- dibromophenol) considered as one of toxic organic phenol. Industrial wastes was used as low cost adsorbent for the removal of these pollutants. The results show the maximum adsorption on carbonaceous adsorbent prepared from fertilizer industry waste 40.7, 170.4 and 190.2 mg g^{-1} for 4-bromophenol 2-bromophenol and 2, 4-dibromophenol, respectively. As compared to carbonaceous adsorbent, the other three adsorbents (viz., blast furnace sludge, dust, and slag) adsorb bromophenols to a much smaller extent [55]. Table (2) represented the adsorption efficiencies of different adsorbents for the removal of phenols.

TABLE 2. Adsorption capacities of different adsorbents for the removal of phenols.

Organic Pollutants	**Adsorbent**	**Adsorption Maxmium**	**Ref.**
Phenol	Porous Clay	14.5 mg/g	[56]
2,5-dichlorophenol	Porous Clay	45.5 mg/g	[56]
3,4-dichlorophenol	Porous Clay	48.7 mg/g	[56]
3,5-dichlorophenol	(cetyl-pyridinium-Al PILC),	97.2 mg/g	[57]
phenol	Hemidesmus Indicus Carbon(HIC)	370 ppm	[58]
phenol	Commercial Activated Carbon(CAC)	294 ppm	[58]
phenol	NORIT Granular Activated Carbon (NAC 1240)	74.07 mg/g	[59]
phenol	NORIT Granular Activated Carbon 1010	166.6 mg/g	[59]
phenol	Active carbon	257 mg/g	[60]

Organic Pollutants	Adsorbent	Adsorption Maxmium	Ref.
phenol	Mesoporous carbon CMK-3-100°C	347 mg/g	[60]
phenol	Mesoporous carbon CMK-3-130 °C	428 mg/g	[60]
phenol	Mesoporous carbon CMK-3-150 °C	473 mg/g	[61]
phenol	Leaf litter of *Shorea roubsta*	76%	[60]
phenol	activated phosphate rock (1M HNO3)	83.34 mg/g	[62]
phenol	Natural clay	15 mg/g	[63]
2,4-dichlorophenol	activated carbon derived from oil palm empty fruit bunch (EFB)	232.56 mg/g	[64]

6. ADSORPTION OF PESTICIDES AND HERBICIDES

Pesticides and herbicides, intentionally released into the environment, are ubiquitous in aquatic systems; they are often detected at low levels and commonly occur in the form of complex mixtures [64-65]. Leaching of chemical fertilizers and pesticides, applied to agricultural and forest land, is one of the main reasons for organic pollution in several water streams. Pesticides and herbicides are harmful to life because of their toxicity, carcinogenicity and mutagenicity [66]. Therefore toxicity of pesticides and their degradation products is making these chemical substances a potential hazard by contaminating the environment. They have raised serious concerns about aquatic ecosystem and human health because of the long-term accumulation of their single and/or combined toxicological effects [67]. The contamination of ground water,surface water and soils, by pesticides and herbicides are currently a significant concern, and this because of increasing use of pesticides in agriculture, and domestic activities [68].

Among newly developed pesticides, organophosphorous pesticides are most commonly used. This class of chemicals is divided into several forms; however the two most common forms are phosphates and phosphorothionates. Methyl parathion (O,O-dimethyl O-4-nitrophenyl phosphorothioate) is a class I insecticide. Once methyl parathion introduced into the environment from spraying on crops, droplets of methyl parathion in the air fall on soil, plants or water. While most of the methyl parathion will stay in the areas where it is applied, some can move to areas away from where it was applied by rain, fog and wind [69].

Modified polymer adsorbents were prepared for the removal of organic pollutants from water and wastewater. Adsorption of organic pollutants using cyclodextrin-based polymer (CDPs) as adsorbent, is an efficient technique with the advantages of specific affinity, low cost and simple design [65, 70-71]. Cyclodextrin polymers (CDPs) can be synthesized using cyclodextrin (CD) as complex molecule and polyfunctional substance (e.g., epichlorohydrin (EPI)) as cross-linking agent.Though a number of CDPs with various structures and properties have been developed [72,73], it is still ambiguous how CDP properties affect adsorption affinity toward organic contaminants, particularly mixed pollutants. Liu et al. [66] illustrated the cross-linked structure of cyclodextrin polymer and related adsorption mechanisms in Scheme (1).

Other modified polymer adsorbent was used in herbicides treatment, porous polymeric adsorbents were used for the adsorption of herbicides (alachlor, amitrole, trifluralin and prometryn) from liquid solution. Two adsorbent resins were investigated, the highly hydrophobic Amberlite XAD-4 (polystyrene–divinylbenzene copolymer) and the functionalized more hydrophilic XAD-7 (nonionic aliphatic acrylic polymer). The adsorption were sussccufuly at pH 6.5 [74].

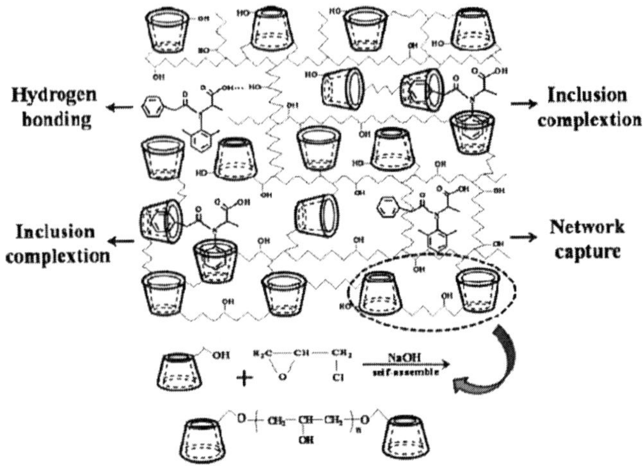

SCHEME 1. Cross-linked structure of cyclodextrin polymer and related adsorption mechanisms, from Liu et al. [66].

Other adsorbent used widely for the removal of pesticides is activated carbon. Activated carbons (ACs) prepared from agricultural and industrial wastes were used for the removal of pesticides from polluted water. Activated carbons produced from agricultural residues (olive kernel, corn cobs, rapeseed stalks and soya stalks) via physical steam activation were tested for the removal of Bromopropylate (BP) pesticide from water. The results show maximum adsorption capacity(q_m) of the pesticide on adsorbents (ACs) prepared from corn cob, olive kernel, soya stalks and rapeseed stalks at values 7.9x 10^{-2},12.3 x 10^{-2}, 11.6 x 10^{-2} and 18.9 x 10^{-2} mmol/L, respectively. The BP removal from water achieved in this study was 90–100% for all ACs [75].

Ayranci and Hoda [76] studied the adsorption of pesticides,pesticides ametryn [2- (ethylamino)-4-isopropylamino-6-methyl-thio-s-triazine], aldicarb (2-methyl-2-(methylthio) propionaldehyde o-methylcarbamoyloxime], diuron [N-(3,4-dichlorophenyl)- N,N-dimethyl urea] and dinoseb [2-(sec-butyl)- 4,6-dinitrophenol], from aqueous solution onto high specific area activated carbon-cloth adsorbent (ACC) and found that the maximum adsorption capacity of ACC for ametryn, diuron, dinose and aldicarb were 354.61,421.58, 301.84 and 213.06 mg/g, respectively.

Djilani et al. [77] developed new activated carbon adsorbents from lignocellulosic wastes of vegetable origin (coffee grounds (CG), melon seeds (MS) and orange peels (OP)). The adsorption efficiency of these new adsorbents was tested with organic pollutants: o-nitrophenol and pnitrotoluene. The elimination ratio obtained with new adsorbents was in the range from 70% to 90%. The time necessary to attain the adsorption equilibrium was between 75 and 135 min.

The ability of MgAl layered double hydroxides (LDHs) and their calcined products to adsorb besticides contaminants, 2,4-dinitrophenol (DNP) and 2-methyl-4,6-dinitrophenol (DNOC) from water was assessed. Adsorption tests were conducted on LDHs with variable Mg/Al ratios (and variable layer charge), pH values, contact times and initial pesticide concentrations to identify the optimum conditions for the intended purpose. All adsorbents except the carbonate-containing hydrotalcite possessed a very high adsorption capacity for both contaminants. As noted above, the adsorption of the pesticides on the calcined LDHs was only 25–40% [78]

Activated carbon prepared from banana stalk by potassium hydroxide (KOH) and carbon dioxide (CO_2) activation (BSAC) was explored for its ability to remove the pesticides, 2,4-dichlorophenoxyacetic acid (2,4-D) and bentazon. The percent removal efficiency of 2,4-D decreased from 98.4 to 85.4% as the 2,4-D initial concentration increased from 50 to 300 mg L^{-1}. For bentazon, the percent removal efficiency decreased from 96.5 to 61.6% as the bentazon initial concentration increased from 25 to 250 mg L^{-1}. Therefore, the adsorption of 2,4-D and bentazon by BSAC has strong dependence on the initial concentration of the pesticides. Maximum adsorption capacity (qc) for 2,4 dichlorophenoxyacetic acid (2,4-D) were 168.03 mg g^{-1}, and for bentazon 100.95 mg g^{-1} [79]

In the other study for using activated carbon as adsorbent, magnetic and graphitic carbon nanostructures was used for the removal of a pesticide (2,4-dichlorophenoxyacetic acid) from aqueous solution.The magnetic and graphitic carbon nanostructures as adsorbents were prepared from two different biomasses, cotton and filter paper. The resultant adsorbents were characterized with TEM and N2 adsorption–desorption methods. The adsorption capacities for the prepared adsorbents from filter paper and cotton are about 77 and 33 mg/g, respectively [80].

Chemically and thermally treated watermelon peels (TWMP) have been utilized for the removal of methyl parathion (MP) pesticide from water. The effect of process variables such as pH of solution, shaking speed, shaking time, adsorbent dose, concentration of solution and temperature have been optimized. Maximum adsorption (99±1%) was achieved for (0.38–3.80)×10−4 mol dm^{-3} of MP solution, using 0.1 g of adsorbent in 20 ml of solution for 60 min agitation time at pH 6. The developed adsorption method has been employed to surface water samples with percent removal 99%±1 [81].

7. ADSORPTION OF OTHER ORGANIC POLLUTANTS

Other organic pollutants were found as a pollutants in water and wastewater, this include pharmaceutical effluents, surfactants, organic solvents, phthalates, hydrocarbons, esters, alcohols,volatile, semi-volatile and non-volatile chlorinated organic pollutants. Activated carbons, calys and clay minerals are used widely for the removal of organic pollutants.

7.1. Adsorption On Activated Carbon

Adsorption on activated carbon is currently the most frequently used technology for removing organic pollutants from aqueous industrial sludge, surface waters and drinking water. Methyl tert-butyl ether (MTBE) is an organic pollutants used mainly as a fuel component in fuel if gasoline engine and also as a solvent. The adsorption of methyl tert-butyl ether (MTBE) by granular activated carbon was investigated, the maximum adsorption capacity of MTBE on granular activated carbon was 204.1 mg/g. Results illustrate that granular activated carbon is an effective adsorbent for methyl tert-butyl ether and also provide specific guidance into adsorption of methyl tert-butyl ether on granular activated carbon in contaminated groundwater [82].

A novel triolein-embedded activated carbon composite adsorbent was developed. Results suggested that the novel composite adsorbent was composed of the supporting activated carbon and the surrounding triolein-embedded cellulose acetate membrane. The adsorbent was stable in water, for no triolein leakage was detected after soaking the adsorbent for five weeks. The adsorbent had good adsorption capability to dieldrin, which was indicated by a residual dieldrin concentration of 0.204 μgL^{-1}. The removal efficiency of the composite adsorbent was higher than the traditional activated carbon adsorbent [83]. Essa et al. [84] studied the potential of chemical activated date pits as an adsorbent. The date pits were impregnated with 70% phosphoric acid followed by thermal treatment between 300 to 700°C. The effects of activation temperature and acid concentration on pore surface area development were studied. Samples prepared at 500°C showed a specific area of 1319 m^2/g and total pore volume of 0.785 cm^3/g. Aqueous phenol adsorption trends using the local activated carbon sample were compared to a commercial sample (Filtrasorb-400).

Five commercially available types of activated carbon (GAC 1240, GCN 1240, RB 1, pK 1-3, ROW 0.8 SUPRA)are prepared and used to remove organic chlorinated compounds from wastewater of a chemical plant. The various types of activated carbon were tested on the basis of Freundlich adsorption isotherms for 14 pure organic chlorinated compounds, of molecular weight ranging from that of dichloromethane (MW ¼ 84.93 g mol^{-1}) to hexachlorobenzene (MW ¼ 284.78 g mol^{-1}). The best adsorbent (GAC 1240 granulated activated carbon) was selected and used in a laboratory fixed bed column to assess its removal efficiency with respect to the tested organic chlorinated compounds. Removal efficiency was always higher than 90% (Table 3) [85]

TABLE 3. Removal efficiency (%) of chlorinated compounds from wastewater by five commercially available types of activated carbon, from Pavonia et al [85]

Substance	Precenage of adsorption efficiency (%)
Dichloromethane	98.3
Trichloromethane	98.8
1,1,1- Trichloromethane	99.0
Carbon tetrachloride	99.0
1,2-Dichloroethane	82.8
Trichloroethylene	94.7
1,1,2- Trichloroethane	86.3
Tetrachloroethylene	91.6
1,1,1,2- Tetrachloroethane	87.3
Trans 1,4-dichloro-2-butene	94.2
1,2,4-Trichlorobenzene	99.2
1,2,3-Trichlorobenzene	90.5
Hexachloro-1,3-butadiene	99.4
Hexachlorobenzene	95.1

Antibiotic considered as one of the pharmacological organic pollutants. Choi et al. [86] investigated the treatment of seven tetracycline classes of antibiotic (TAs) from raw waters (synthetic and river) using coagulation and granular activated carbon (GAC) filtration. Their results ended to that both coagulation and GAC filtration were effective for the removal of TAs, and the removal efficiency depended on the type of TAs. GAC filtration was relatively more effective for removal of tetracycline (TC), doxycycline-hyclate (DXC), and chlortetracycline-HCl (CTC), which were difficult to be removed by coagulation. Putra et al. [87] investigated the removal of Amoxicillin (antibiotic) from pharmaceutical effluents using bentonite and activated carbon as adsorbents. The study was carried out at several pH values. Langmuir and Freundlich models were then employed to correlate the equilibria data on which both models fitted the data equally well. While chemisorption is the dominant adsorption mechanism on the bentonite, both physisorption and chemisorption played an important role for adsorption onto activated carbon. Ruiz et al. [88] studied the removal of paracetamol (anti analgesic drug) from aqueous solutions using chemically modified activated carbons. The effect of the chemical nature of the activated carbon material such as carbon surface chemistry and composition on the removal of paracetamol was studied. The surface heterogeneity of the carbon affected the rate of paracetamol removal. They found that after oxidation the wettability of the carbon was enhanced, which favored the transfer of paracetamol molecules to the carbon pores. At the same time the overall adsorption rate and removal efficiency are reduced in the oxidized carbon due to the competitive effect of water molecules. Tris (2-chloroethyl) phosphate (TCEP), iopromide, naproxen, carbamazepine, and caffeine drugs were quite frequently observed in both surface waters and effluents from waste water treatment plants. The elimination of these chemicals during drinking water and wastewater treatment processes at full- and pilot-scale also was investigated. Conventional drinking water treatment methods such as flocculation and filtration were relatively inefficient for contaminant removal, while efficient removal (99%) was achieved by granular activated carbon (GAC).

Liu et al. [89] studied the removal effects of organic pollutants in drinking water (42 species organic pollutants in 11 categories) by activated carbon, haydite and quartz sand with the method of solid-phase extraction (SPE). The removal rates of total peak area of organic pollutants by activated carbon, haydite and quartz were 70.35%, 29.68% and 37.36%. Among all, activated carbon showed the best removal effect to most organic pollutants contents, and quartz sand to species. So if activated carbon - quartz sand combined processes were adopted, organic pollutants species and total peak area could be reduced simultaneously. The removal rates for phthalates, hydrocarbons, esters and alcohols were 62.62%, 75.83%, 72.52% and 62.99%, respectively. The adsorption capability of activated carbon for dimethyl phthalate and di-n-butyl phthalate, two priority pollutants in water, were preferable. The removal rates reached 93.27% and 57.02%. For haydite, there were 26 kinds of organic pollutants in 10 categories in corresponding treated water. The total peak area was removed by 29.68%. The removal effects for amines, alkanes and phenols were satisfactory and the removal rates were 68.71%, 49.97% and 41.19%, respectively. But in the case of phthalates, esters and aldehydes the effects were not obviously, the same as dimethyl phthalate and di-nbutyl phthalate. The species of organic pollutants were reduced from 36 to 20 by quartz sand. Most notably, xylene, dimethyl phthalate and di-n-butyl phthalate in raw water were removed efficiently. Xylene and di-n-butyl phthalate in tap water were removed absolutely, and dimethyl phthalate was removed by 59.59%.

Cellulose acetate (CA) embedded with triolein (CA-triolein), was prepared as adsorbent for the removal of persistent organic pollutants (POPs) from micro-polluted aqueous solution. The comparison of CA-triolein, CA and granular activated carbon (GAC) for dieldrin removal was investigated. Results showed that CA-triolein absorbent gave a lowest residual concentration after 24 h although GAC had high removal rate in the first 4 h adsorption. Then the removal efficiency of mixed POPs (e.g. aldrin, dieldrin, endrin and heptachlor epoxide), absorption isotherm, absorbent regeneration and initial column experiments of CA-triolein were studied in detail. The linear absorption isotherm and the independent absorption in binary isotherm indicated that the selected POPs are mainly absorbed onto CA-triolein absorbent by a partition mechanism. Thermodynamic calculations showed that the absorption was spontaneous, with a high affinity and the absorption was an endothermic reaction. Rinsing with hexane the CA-triolein absorbent can be regenerated after absorption of POPs. No significant decrease in the dieldrin removal efficiency was observed even when the absorption–regeneration process was repeated for five times. The results of initial column experiments showed that the CA-triolein absorbent did not reach the breakthrough point at a breakthrough empty-bed volume (BV) of 3200 when the influent concentration was 1–1.5 l g/L and the empty-bed contact time (EBCT) was 20 min [90]

Activated coke (AC) was studied to adsorb organic pollutants from coking wastewater. The study initially focused on the sorption kinetics and equilibrium sorption isotherms of AC for the removal of chemical oxygen demand (COD) from coking wastewater. The results showed that when the dose of AC was 200 g L^{-1}, 91.6% of COD and 90% of color could be removed after 6 h of agitation at 40°C. The kinetics of adsorption of COD from coking wastewater onto AC was fit to the pseudo-second order model. The adsorption of COD onto AC was enhanced with an increase of temperature, indicating that the adsorption process would be a chemical adsorption rather than a physical one [91].

Bottom ash, a kind of waste material generated from thermal coal-fired power plants, is generally used in road bases and building materials. Bottom ash was used to remove the organic pollutants in coking wastewater and papermaking wastewater. Particular attention was paid on the effect of bottom ash particle size and dosage on the removal of chemical oxygen demand (COD). The results show that the COD removal efficiencies increase with decreasing particle sizes of bottom ash, and the COD removal efficiency for coking wastewater is much higher than that for papermaking wastewater due to its high percentage of particle organic carbon (POC). Different trends of COD removal efficiency with bottom ash dosage are also observed for coking and papermaking wastewaters because of their various POC concentrations [92].

7.2. Adsorption On Clays And Clay Minerals Adsorbents

Several adsorbents were used for the removal of these pollutants. One of an effective and low cost adsorbents is calys and clay minerals. Natural clay minerals due to their high surface area and molecular sieve structure are very effective sorbents for organic contaminants of cationic or polar in character.Natural and modified clay minerals and zeolites are good candidates for improving activated carbon(AC) performance, because they have large surface areas for retention of pollutants [93].

Adsorbent prepared from organoclays and activated carbon were shown to remove a variety of organic contaminants [94-95]. Montmorillonite was applied as adsorbent for the removal of cationic surfactants, while the hydrophilic surface of montmorillonite was modified and used as adsorbent [96]. Calcined hydrocalcites was prepared and used to remove organic anionic pesticides [97] from polluted water. Vesicle–clay complexes in which positively charged vesicles composed of didodecyldimethy-lammonium bromide (DDAB) were adsorbed on montmorillonite and removed efficiently anionic (sulfentrazone, imazaquin) and neutral (alachlor, atrazine) pollutants from water. These complexes (0.5% w:w) removed 92–100% of sulfentrazone, imazaquin and alachlor,and 60% of atrazine from a solution containing 10 mg/L. A synergistic effect on the adsorption of atrazine was observed when all pollutants were present simultaneously (30 mg/L each), its percentage of removal being 85.5. Column filters (18 cm) filled with a mixture of quartz sand and vesicle–clay (100:1, w:w) were tested. For the passage of 1 L (25 pore volumes) of a solution including all the pollutants at 10 mg/L each, removal was complete for sulfentrazone and imazaquin, 94% for alachlor and 53.1% for atrazine, whereas removal was significantly less efficient when using activated carbon. A similar advantage of the vesicle–clay filter was observed for the capacities of removal. Table (4) show the removal efficiencies of organic pollutants from water by DDAB-Clay complex adsorbent [98]

Data in parentheses correspond to the procedure when dried complex was added to the solution, whereas the other case correspond to the removal after incubation with DDAB followed by clay addition.

Data in brackets are the predicted removal by Langmuir equation (binding coeffients)

Zhao et al. [99] prepared mesoporous silica materials and used it for adsorption of organic pollutants in water. Mesoporous silica materials is performed using self-assembling micellar aggregates of two surfactants: cetylpyridinium bromide (CPB) and cetyltrimethylammonium bromide (CTAB). The retention properties have been studied of these two kinds mesoporous silicas towards environmental pollutants (mono-, di-, tri-

TABLE 4. Removal of organic pollutants from water by DDAB-Clay complex adsorbent [a,b] (From Undabeytiaa et al. [98])

% removal	Initial herbicide conc. (mg/L)	DDAB Conc.(Mm)	Pollutants
37.3±0.3	8	3	Atrazine
(57.4± 0.7)	8	5	
56.6±1.0			
(59.8±0.1)			
[59.0]			
94.9±0.8	10	3	Alachlor
(85.6±1.0)	10	6	
95.5±0.1			
(87.2±0.2)			
[92.7]			
73.6±0.4	10	3	Imazaquin
(92.0±0.3)	10	6	
75.52±0.3			
(92.3±1.0)			
[95.8]			
98.0±1.0	10	3	Sulfentrazone
(99.7±0.1)	10	6	
99.7±0.1			
(99.7±0.1)			
[99.9]			

chloroacetic acid, toluene, naphthalene and methyl orange). The effect of the composition (presence and absence of surfactants, different kinds of surfactants) on the sorption performance has been considered. They found that materials show excellent retention performance toward chloroacetic acids, toluene, naphthalene and methyl orange. The materials without surfactants does not show, if any, affinity for ionic and non-ionic analytes.

The applicability of mesoporous aluminosilica monoliths with three-dimensional structures and aluminum contents with 19≤Si/Al≥1 was studied as effective adsorbents of organic molecules from an aqueous solution. Mesocage cubic Pm3n aluminosilica monoliths were successfully fabricated using a simple, reproducible, and direct synthesis (scheme 2).

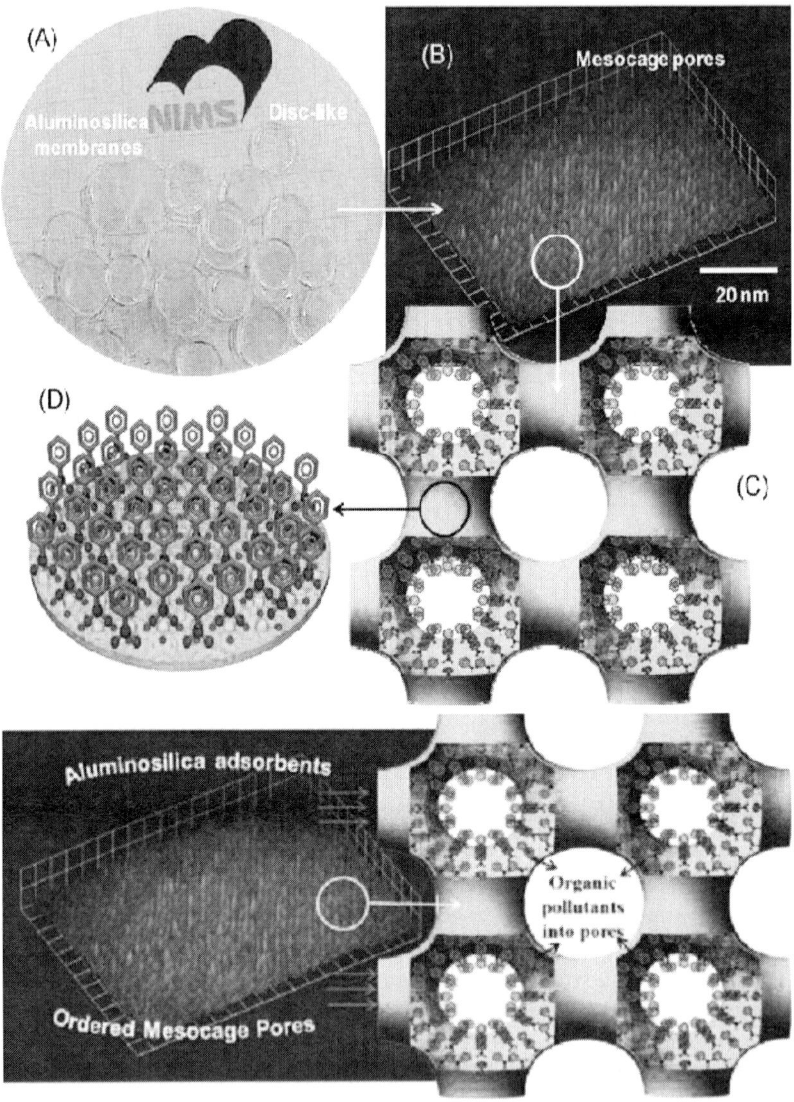

SCHEME 2. Aluminosilica monoliths with a disc-like shape (A) and mesocage pores (B) as adsorbents (C) of organic compounds (I–IV) inside the mesocage cavity and onto pore surfaces of 3D cubic Pm3n structures (D). Note that 3D TEM image (B) was recorded with aluminosilica monoliths with a Si/Al ratio of 4 (From El-Safty et al. [100]).

The acidity of the monoliths significantly increased with increasing amounts of aluminum species in the silica pore framework walls. The batch adsorption of the organic pollutants onto (10 g/L) aluminosilica monoliths was performed in an aqueous solution at various temperatures. These adsorbents exhibit efficient removal of organic pollutants (e.g., aniline, pchloroaniline,o-aminophenol, and p-nitroaniline) of up to 90% within a short period (in the order of minutes). In terms of proximity adsorption, the functional acid sites and the condensed and rigid monoliths with tunable periodic scaffolds of the cubic mesocages are useful in providing easy-to-use removal assays for organic compounds and reusable adsorbents without any mesostructural damage, even under chemical treatment for a number of repeated cycles [100].

Modified clays were used as adsorbents for the removal of organic pollutants from wastewater. Two pillared clays are synthesized by intercalation of solutions of aluminium and zirconium and evaluated as adsorbents for the removal of Orange II and Methylene Blue from aqueous solutions. The contact time to attain equilibrium for maximum adsorption was found to be 300 min. Both clays were found to have the same adsorption capacity when Orange II was used as adsorbent, whereas the adsorption capacity of Zr-PILC was higher (27 mg/g) than that of Al-PILC (21 mg/g)for Methylene Blue. The adsorption kinetics of dyes has been studied in terms of pseudo-first- and -second-order kinetics, and the Freundlich, Langmuir and Sips isotherm models have also been applied to the equilibrium adsorption data. The addition of NaCl has been found to increase the adsorption capacities of the two pillared clays for Orange II Pillared InterLayered Clays (PILCs) are porous materials that can be obtained by the intercalation of soils, thereby creating high value added materials from natural solids [101]. Also, pillared clay adsorbent was used for the removal of benzo(a)pyrene and chlorophenols [102], chlorinated phenols from aqueous solution by surfactant-modified pillared clays [103] and herbicide Diuron on pillared clays [104].

8. CONCLUSION

Organic pollutants in the ecosystem, especially persistent organic pollutants (POPs),are of the most important environmental problems in the world. The literature reviewed revealed that there has been a high increase in production and utilization of organic pollutants in last few decades resulting in a big threat of pollution. Efficient techniques for the removal of highly toxic organic compounds from water and wastewater have drawn significant interest. Adsorption is recognized as an effective and low cost technique for the removal of organic pollutants from water and wastewater, and produce high-quality treated effluent. This chapter highlighted the removal of organic pollutants using adsorption technique with different kinds of natural and synthetic adsorbents.

Many researches have given considerable attention aimed at establishing to the removal efficiency of organic pollutants by adsorption technique. To decrease treatment costs, attempts have been made to find inexpensive alternative activated carbon (AC), from waste materials of industrial, domestic and agricultural activities. Also, clays and natural clay minerals, due to their high surface area and molecular sieve structure, are very effective adsorbents for organic contaminants. The chapter focus, reviews and evaluates literature dedicated on the adsorption phenomenon, different types of natural and synthetic adsorbents, adsorption of dyes, phenols, pesticides and other organic pollutants. Finally it ended with recent researches of organic pollutants adsorption on activated carbons, clays and clay minerals.

ABBREVIATIONS

AC Activated carbon, PCP Pentachlorophenol

CAC Commercial Activated carbon, PCDD/Fs polychlorinated dibenzo-p-dioxins and dibenzofurans

PAC powdered activated carbons, DDT dichloro-diphenyl-trichloroethane

ACC Activated carbon cloth adsorbent, DNOC 2,4-dinitrophenol (DNP) and 2-methyl-4,6-dinitrophenol

WC Wood charcoal, HCB Hexachlorobenzene

GAC Granular activated carbon, CD Cyclodextrin

HIC Hemidesmus Indicus carbon, BP Bromopropylate

POPs Persistent organic pollutants, LDHs MgAl layered double hydroxides

BET Buruner, Emmett, and Teller, MTBE Methyl tert-butyl ether

TCP 2,4,6-trichlorophenol, MP Methyl parathion

DCP Dichlorophenol, TCEP Tris (2-chloroethyl) phosphate

BA Bagasse ash, TC Tetracycline

BFD Blast furnace flue dust, DXC Doxycycline-hyclate

TWMP Treated watermelon peels, CTC Chlortetracycline-HCl

PPC Potato peels charcoal, CA Cellulose acetate

WBA Wood-shaving bottom ash, CA-triolein Cellulose acetate (CA) embedded with triolein

CDPs Cyclodextrin-based polymer, EBC Empty-bed contact time

CG Coffee grounds, DDAB Ddidodecyldimethy- lammonium bromide

MS Melon seeds, CPB Cetylpyridinium bromide

OP Orange peels, CTAB Cetyltrimethylammonium bromide

MB Methylene blue, EPA Environmental Protection Agency

RR141 Red Reactive, 141 WHO TheWorld Health Organisation

POPs Persistent organic pollutants, PhAcS Pharmaceutically active substances

OCPs Organochlorine pesticides, PCBs polychlorinated biphenyls

TMP Trimethoprim, PCP Pentachlorophenol

PhAcS Pharmaceutically active substances, PCDD/Fs polychlorinated dibenzo-p-dioxins and dibenzofurans

PCBs polychlorinated biphenyls, DDT Dichloro-diphenyl-trichloroethane

COD Chemical oxygen demand, MW Molecular Weight

REFERENCES

1. Ali, I., Mohd. Asim, Tabrez A. Khan. Low cost adsorbents for the removal of organic pollutants from wastewater. Journal of Environmental Management 2012;113,170-183.
2. Ali, I., Aboul-Enein, H.Y., Chiral. Pollutants: Distribution, Toxicity and Analysis by Chromatography and Capillary Electrophoresis. John Wiley & Sons, Chichester, 2004.
3. Damià, B. Emerging Organic Pollutants in Waste Waters and Sludge. Springer, Berlin, 2005.
4. Harrad, S., Persistent Organic Pollutants Enviromental Behaivour and Pathways for Human Exposure. Kluwer Academic Publishers, Norwell,2001.

5. Burkhard, L.P., Lukazewycz, M.T. Toxicity equivalency values for polychlorinated biphenyl mixtures. Environ. Toxicol. Chem. 2008, 27, 529–534.

6. Clive, T. Persistent Organic Pollutants: Are we close to a solution? Published by the Canadian Arctic Resources Committee,Volume 26, Number 1, Fall/Winter.2000.

7. Fiedler, H..Persistent Organic Pollutants. Springer, New York, 2003.

8. Ritter, L. ; Solomon, K.R.; Forget, J. Persistent organic pollutants. An assessment report Canadian Network of Toxicology Centres 620 Gordon Street Guelphon Canada, 2000.

9. Crini G. Recent developments in polysaccharide-based materials used as adsorbents in wastewater treatment,.Prog. Polym Sci. 2005,30, 38–70,

10. Kurniawan, T. A. and Lo, W.H.. Removal of refractory compounds from stabilized landfill leachate using an integrated H2O2 oxidation and granular activated carbon (GAC) adsorption treatment. Water Research,2009, 43, 4079-4091.

11. Aly, Osman M and Faust, Samuel D. Adsorption processes for water treatment. Butterworth Publisher (Boston) 509 p, 1987.

12. Moreno-Castilla C. Adsorption of organic molecules from aqueous solutions on carbon materials. Carbon 2004, 42,83–94.

13. Amokrane, A., Comel, C. and Veron, J. Landfill leachate pretreatment by coagulation-flocculation. Water Research,1997, 31, 2775-2782.

14. [14] Surmacz-Gorska, J. Degradation of Organic Compounds in Municipal Landfill Leachate. Lublin, Publishers of Environmental Engineering Committee of Polish Academy of Sciences,2001.

15. Putraa, E. K., Pranowo, R., Sunarsob, J., Indraswati, N. and Ismadji, S. Performance of activated carbon and bentonite for adsorption of amoxicillin from wastewater: Mechanisms, isotherms and kinetics water research. 2009, 43, 2419 - 2430.

16. Lillo-Rodenas M.A., D. Cazorla-Amoros, A. Linares-Solano. Behaviour of activated carbons with different pore size distributions and surface oxygen groups for benzeneand toluene adsorption at low concentrations. Carbon 2005, 43,1758–1767.

17. Ardejani F. Doulati, Kh. Badii, N. Yousefi Limaee, N.M. Mahmoodi, M. Arami, S.Z. Shafaei, A.R. Mirhabibi.Numerical modelling and laboratory studies on the removal of Direct Red 23 and Direct Red 80 dyes from textile effluents using orange peel, a low-cost adsorbent. Dyes and Pigments 2007,73, 178-185.

18. Gupta Vinod K., Alok Mittal, Rajeev Jain, Megha Mathur, Shalini S. Adsorption of Safranin-T from wastewater using waste materials- activated carbon and activated rice husks Journal of Colloid and Interface Science,2006,303, 80–86.

19. Hamdaoui Oualid. Batch study of liquid-phase adsorption of methylene blue using cedar sawdust and crushed brick Journal of Hazardous Materials 2006,B135,264–273.

20. Leechart P., Woranan Nakbanpote b,1, Paitip Thiravetyan. Application of 'waste' wood-shaving bottom ash for adsorption of azo reactive dye Journal of Environmental Management 2009,90,912–920

21. Tsai Wen-Tien, Hsin-Chieh Hsub, Ting-Yi Su b, Keng-Yu Lin b, Chien-Ming Lin. Removal of basic dye (methylene blue) from wastewaters utilizing beer brewery waste. Journal of Hazardous Materials 2007,154 (1-3)73-78.

22. Rashed. M.N.. Acid Dye Removal from Industrial Wastewater by Adsorption on Treated Sewage Sludge, Int. J. Environment and Waste Management, 2011, 7, Nos. 1/2,175-191.

23. Rozada F., M. Otero, A. Moran, A.I. Garcıa. Adsorption of heavy metals onto sewage sludge-derived materials. Bioresource Technology 2009,99, 6332–6338.

24. Rio, S., Catherine, F.B., Laurence, L.C., Philippe, C. and Pierre, L.C. Experimental design methodology for the preparation of carbonaceous sorbents from sewage sludge by chemical activation- application to air and water treatments. Chemosphere 2005,58, 423–437.

25. Martin, M.J., Artola, A., Balaguer, M.D. and Rigola, M..Towards waste minimization in WWTP: activated carbon from biological sludge and its application in liquid phase adsorption, J. Chem. Technol. Biotechnol., 2002,77, 825–833.

26. Gulnaz, O., Kayaa, A., Matyar, F. and Arikan, B. Sorption of basic dyes from aqueous solution by activated sludge', Journal of Hazardous Materials B, 2004, 108, 183–188.

27. Otero M., F. Rozada, L.F. Calvo, A.I. Garcıa, A. Moran Adsorption on adsorbents from sewage sludge Elimination of organic water pollutants using adsorbents obtained from sewage sludge Dyes and Pigments 57,2003, 55–65.

28. Acemioglu, 2004Acemioglu,B. Adsorption of Congo red from aqueous solution onto calcium rich fly ash. Journal of Colloid Interface Science, 2004, 274, 371-379.

29. Wang, S., Boyjoo, Y., Choueib, A. and Zhu, J. Utilization of fly ash as low cost adsorbents for dye removal. Chemeca, 2004,26-29.

30. Wang, S., Boyjoo, Y., and Zhu, J. Sonochemical treatment of fly ash for dye removal from wastewater. Journals of Hazardous Materials, 2005,Vol. B126, 91-95.

31. Al-Degs Yahya S. a, Musa I. El-Barghouthi a, Amjad H. El-Sheikh a, Gavin M. Walker. Effect of solution pH, ionic strength, and temperature on adsorption behavior of reactive dyes on activated carbon, Dyes and Pigments 2008,77, 16-23.

32. Orhan Ozdemira, Bulent Armaganb, Mustafa Turanb, Mehmet S. C, elikc. Comparison of the adsorption characteristics of azo-reactive dyes on mezoporous minerals,Dyes and Pigments 2004, 62, 49–60.

33. Indra D. Mall, Vimal C. Srivastava, Nitin K. Agarwal. Removal of Orange-G and Methyl Violet dyes by adsorption onto bagasse fly ashdkinetic study and equilibrium isotherm analyses, Dyes and Pigments 2006,69,210-223

34. Mansoor Anbia, Samira Salehi. Removal of acid dyes from aqueous media by adsorption onto amino-functionalized nanoporous silica SBA-3, Dyes and Pigments 2012, 94,1-9

35. Auta M., B.H. Hameed. Preparation of waste tea activated carbon using potassium acetate as an activating agent for adsorption of Acid Blue 25 dye, Chemical Engineering Journal 2011, 171, 502–509

36. Emad N. El Qada, Stephen J. Allen, Gavin M. Walker. Adsorption of Methylene Blue onto activated carbon produced from steam activated bituminous coal: A study of equilibrium adsorption isotherm. Chemical Engineering Journal 2006, 124, 103–110

37. G. Gong, Q. Xie, Y. Zheng, S. Ye, Y. Chen. Regulation of pore size distribution in coal-based activated carbon, New Carbon Mater. 2009, 24, 141–146

38. B.S. Girgis, E. Smith, M.M. Louis, A.N.A. El-Hendawy. Pilot production of activated carbon from cotton stalks using H3PO4, J. Anal. Appl. Pyrol. 2009,86, 180–184

39. Mehmet Ulas Durala, Levent Cavasa,b, Sergios K. Papageorgiouc, Fotis K. Katsarosc. Methylene blue adsorption on activated carbon prepared from Posidonia oceanica (L.) dead leaves: Kinetics and equilibrium studies. Chemical Engineering Journal 2011, 168, 77–85

40. Yongze Bao, Guilan Zhang. Study of Adsorption Characteristics of Methylene Blue onto Activated Carbon Made by Salix Psammophila. Energy Procidia 2012,16, 1141 – 1146

41. Alexandro M.M. Vargas, Andre L. Cazetta, Marcos H. Kunita, Tais L. Silva, Vitor C. Almeida. Adsorption of methylene blue on activated carbon produced from flamboyant pods (Delonix regia): Study of adsorption isotherms and kinetic models. Chemical Engineering Journal 2011, 168,722–730

42. Ahmad A.L., M.M. Loh, J.A. Aziz. Preparation and characterization of activated carbon from oil palm wood and its evaluation on methylene blue adsorption, Dyes Pigments 2007, 75, 263–272.

43. Tan I.A.W., A.L. Ahmad, B.H. Hameed. Enhancement of basic dye adsorption uptake from aqueous solutions using chemically modified oil palm shell activated carbon, Colloids Surf. A 2008,318, 88–96.

44. Das V.Srihari– Ashutosh. Adsorption Of Phenol From Aqueous Media By An Agro-Waste (Hemidesmus Indicus) Based Activated Carbon.Applied Ecology And Environmental Research 2011,7(1): 13-23.

45. Radovic, L. R., Moreno-Castilla, C. and Rivera-Utrilla, J. Carbon materials as adsorbents in aqueous solutions. In: Radovic, L.R. (Ed.). Chemistry and Physics of Carbon, A Series of Advances, 2001,27, 227-405.

46. Dobbs R. A. and Cohen, J. M.. Carbon adsorption isotherms for toxic organics. In: EPA-600/8-80-023, Municipal Environmental Research Laboratory, Office of Research and Development, Cincinnati, Ohio 1980.

47. Zogorski, J. S., Faust, S. D. and Haas Jr., J. H. The kinetics of adsorption of phenols by granular activated carbon. J. Colloid Interface Sci.,1976, 55, 329-341.

48. Snoeyink, V. L., McCreary, J. J. and Murin, C. J. Carbon adsorption of trace organic compounds. Cincinnati, Ohio, Municipal Research Lab Office of Research and Development, U.S. Environmental Protection Agency,1977.

49. Hamdaoui, O. and Naffrechoux, E. Modeling of adsorption isotherms of phenol and chlorophenols onto granular activated carbon Part I. Two-parameter models and equations allowing determination of thermodynamic parameters. Journal of Hazardous Materials,2007, 147, 381-394.

50. Larous. S1, Meniai A-H1 M.F.F. Sze, V.K.C. Lee, G. McKay.The use of sawdust as by product adsorbent of organic pollutant from wastewater: adsorption of phenol. Desalination 2008,218, 323–333

51. Mukherjee Somnath, Sunil Kumarb, Amal K. Misra, Maohong Fan.Removal of phenols from water environment by activated carbon, bagasse ash and wood charcoal. Chemical Engineering Journal 2007,129, 133–142.

52. Yapar S., M. Yilmar. Removal of phenol by using montmorillonite, clinoptilolite and hydrotalcite, Adsorption 2004, 10, 287–298.

53. Roostari N., F.H. Tezel. Removal of phenol from aqueous solution by adsorption, J. Environ. Manage. 2004,70, 157–164.

54. Das C.P., L.N. Patnaik. Removal of phenol by industrial solid waste, Practice Period. Hazard. Toxic Radioactive Waste Manage. ASCE 9,2005, (2) 135–140.

55. Bhatnagar Amit. Removal of bromophenols from water using industrial wastes as low cost adsorbents. Journal of Hazardous Materials 2007,B139, 93–102.

56. Sofía Arellano-Cárdenas,Tzayhrí Gallardo-Velázquez, Guillermo Osorio-Revilla, Ma. del Socorro López-Cortéz1 and Brenda Gómez-Perea. Adsorption of Phenol and Dichlorophenols from Aqueous Solutions by Porous Clay Heterostructure (PCH) J. Mex. Chem. Soc. 2005, 49(3), 287-291

57. Matthes, W., Kahr, G. Sorption of organic compounds by Al- and Zr-hydroxy-inter-calated and pillared bentonite. Clays and Clay Minerals 2000,48, 593–602.

58. V.Srihari, Ashutosh Das.Adsorption Of Phenol From Aqueous Media By An Agro-Waste (Hemidesmus Indicus) Based Activated Carbon. Applied Ecology And Environmental Research 2009, 7(1): 13-23.

59. Hawaiah Imam Maarof, Bassim H. Hameed, Abdul Latif Ahmad. Adsorption Isotherms for phenols onto Activated Carbon. AJChE,2004,4(1)-70-76

60. Enamul Haque, Nazmul Abedin Khan, Siddulu Naidu Talapaneni,Ajayan Vinu, Jonggeon Jegal and Sung Hwa Jhung. Adsorption of Phenol on Mesoporous Carbon CMK-3: Effect of Textural Properties, Bull. Korean Chem. Soc. 2010, Vol. 31, No. 6

61. Mishra S., Bahattachrya J. Potential of leaf litter for phenol adsorption-A Kinetic Study. Indian J.Chem Techn. 2006,13, 298-301

62. Atef S. Alzaydien and Waleed Manasreh. Equilibrium, kinetic and thermodynamic studies on the adsorption of phenol onto activated phosphate rock. International Journal of Physical Sciences 2009, 4 (4), 172-181,

63. Djebbar M. 1, F. Djafri1, M. Bouchekara and A. Djafri. Adsorption of phenol on natural clay.African Journal of Pure and Applied Chemistry 2012, 6(2), 15-25, 15.

64. Shaarani F.W., B.H. Hameed.Batch adsorption of 2,4-dichlorophenol onto activated carbon derived from agricultural waste, Desalination, 2010,255(1–3)159–164

65. Gilliom, R.J. Pesticides in U.S. streams and groundwater. Environmental Science & Technology 2007,41 (10), 3408-3414.

66. Liu Huihui, Xiyun Cai, Yu Wanga, Jingwen C. Adsorption mechanism-based screening of cyclodextrin polymers for adsorption and separation of pesticides from water.Water research 2011,4 5, 3 4 9 9 -3 5 1 1

67. International Agency for Research on Cancer (IARC), Overall evaluations of carcinogenicity: An updating of IARC monographs volumes 1 to 42, Supplement 7, WHO, Lyon, France, 1987.

68. Cui, N., Zhang, X., Xie, Q.,Wang, S.,Chen, J.,Huang, L., Qiao, X., Li, X., Cai, X. Toxicity profile of labile preservative bronopol in water: the role of more persistent and toxic transformation products. Environmental Pollution 2011,159 (2), 609-615.

69. Philip H.H., in: E.M. Michalenko,W.F. Jarvis, D.K. Basu,G.W. Sage,W.M. Meyland, J.A. Beauman, D.A. Gray (Eds.), Handbook of Environmental Fate and Exposure Data for Organic Chemicals, III, Lewis, Chelsea, MI, 1991.

70. Memon G. Zuhra, M.I. Bhanger, Mubeena Akhtar, Farah N. Talpur, Jamil R. Memon. Adsorption of methyl parathion pesticide from

water using watermelon peels as a low cost adsorbent.Chemical Engineering Journal, 2008,138, 616–621

71. Allabashi, R., Arkas, M., Hormann, G., Tsiourvas, D. Removal of some organic pollutants in water employing ceramic membranes impregnated with cross-linked silylated dendritic and cyclodextrin polymers. Water Research 2007,41 (2), 476-486

72. Romo, A., Penas, F.J., Isasi, J.R., Garcia-Zubiri, H.X., Gonzalez-Gaitano, G. Extraction of phenols from aqueous solutions by beta-cyclodextrin polymers. Comparison of sorptive capacities with other sorbents. Reactive & Functional Polymers 2008, 68 (1), 406-413.

73. Crini, G., Morcellet, M. Synthesis and applications of adsorbents containing cyclodextrins. Journal of Separation Science 2002, 25 (13), 789-813.

74. G. Kyriakopoulosa, D. Douliaa, E. Anagnostopoulos. Adsorption of pesticides on porous polymeric adsorbents.Chemical Engineering Science 2005,60, 1177 – 1186

75. Ioannidou Ourania A., Anastasia A. Zabaniotou, George G. Stavropoulos, Md. Azharul Islam, Triantafyllos A. Albanis. Preparation of activated carbons from agricultural residues for pesticide adsorption. Chemosphere 2010,80,1328–1336

76. Ayranci Erol, Numan Hoda. Adsorption kinetics and isotherms of pesticides onto activated carbon-cloth. Chemosphere 2005,60, 1600–1607

77. Djilani Chahrazed, Rachida Zaghdoudib, Ali Modarressid, Marek Rogalskid, Fayc¸ al Djazia, Abdelaziz Lallame. Chemical Engineering Journal 2012,189– 190, 203– 212

78. Chaara D., I. Pavlovic, F. Bruna, M.A. Ulibarri, K. Draoui, C. Barrig. Removal of nitrophenol pesticides from aqueous solutions by layered double hydroxides and their calcined products. Applied Clay Science 2010,50, 292–298

79. SalmanJ.M., V.O. Njokua, B.H. Hameeda. Adsorption of pesticides from aqueous solution onto banana stalk activated carbon. Chemical Engineering Journal 2011,174, 41– 48

80. Maryam Khoshnood, Saeid Azizian.Adsorption of 2,4-dichlorophenoxyacetic acid pesticide by graphitic carbon nanostructures prepared from biomasses.Journal of Industrial and Engineering Chemistry.2012,18(5) 1796-1800

81. Memon G. Zuhra, M.I. Bhanger, Mubeena Akhtar, Farah N. Talpur, Jamil R. Memon. Adsorption of methyl parathion pesticide from water using watermelon peels as a low cost adsorbent..Chemical Engineering Journal 2008,138, 616–621

82. Chen D. Z.,J. X. Zhang,J. M. Chen. Adsorption of methyl tert-butyl ether using granular activated carbon: Equilibrium and kinetic analysis Int. J. Environ. Sci. Tech.,2010, 7 (2), 235-242.

83. Jia RU, LIU IIuijuan, QU Jiuhui, WANG Aimin, DAI Ruihua.Characterization and adsorption behavior of a novel triolein-embedded activated carbon composite adsorbent.Chinese Science Bulletin 2005,50 (23)2788—2790.

84. Essa, M.H.;Al-Zahrani M.A.; Nesaratnam S.A. Step towards national reliance using locally produced activated carbon from dates. Third Saudi Technical Conference and Exhibition (STCEX-3), Riyadh, Saudi Arabia, 291-297,2004.

85. Pavonia B., D. Drusiana, A. Giacomettia, M. Zanetteb.Assessment of organic chlorinated compound removal from aqueous matrices by adsorption on activated carbon. Water Research 2006,40, 3571 – 3579.

86. Choi, K. J., Kima, S. G. and Kim, S. H. Removal of antibiotics by coagulation and granular activated carbon filtration. Journal of Hazardous Materials, 2008,151(1), 38-43.

87. Putraa, E. K., Pranowo, R., Sunarsob, J., Indraswati, N. and Ismadji, S. Performance of activated carbon and bentonite for adsorption of amoxicillin from wastewater: Mechanisms, isotherms and kinetics. Water research. 2009,43, 2419 – 2430.

88. Ruiz, B., Cabrita, I., Mestre, A. S., Parra, J. B., PiresJ., Carvalho, A. P. and Ania, C. O. Surface heterogeneity effects of activated carbons on the kinetics of paracetamol removal from aqueous solution. Applied Surface Science, 2010,256, 5171-5175.

89. Liu Hanchao, Feng Suping, Du Xiaolin, Zhang Nannan, Liu Yongli. Comparison of three sorbents for organic pollutant removal in drinking water. Energy Procedia 2012,18, 905 – 914.

90. Huijuan Liu, Jia Ru, Jiuhui Qu, Ruihua Dai, Zijian Wang, Chun Hu.Removal of persistent organic pollutants from micro-polluted drinking water by triolein embedded absorbent. Bioresource Technology 2009,100,2995–3002

91. Mo he Zhang, Quan lin Zhao, Xue Baib, Zheng fang Ye. Adsorption of organic pollutants from coking wastewater by activated coke Colloids and Surfaces A: Physicochem. Eng. Aspects 2010,362, 140–146

92. Wei-ling Sun, Yan-zhi Qu, Qing Yu, Jin-Ren Ni. Adsorption of organic pollutants from coking and papermaking wastewaters by bottom ash. Journal of Hazardous Materials 2008,154, 595–601

93. Watanabe, Y., Yamada, H., Tanaka, J., Komatsu, Y., Moriyoshi, Y. Environmental purification materials: removal of ammonium and phosphate ions in water system. Trans. Mater. Res. Soc. Jpn. 2004, 29, 2309–2312.

94. Beall, G.W. The use of organo-clays in water treatment. Appl. Clay Sci.2003, 24, 11–20.

95. Ake, C.L., Wiles, M.C., Huebner, H.J., McDonald, T.J., Cosgriff, D.,Richardson, M.B., Donelly, K.C., Phillips, T.D. Porous organoclay composite for the sorption of polycyclic aromatic hydrocarbons and pentachlorophenol from groundwater. Chemosphere 2003,51, 835–844.

96. Borisover, M., Graber, E.R., Bercovich, F., Gerstl, Z. Suitability of dye–clay complexes for removal of non-ionic organic compounds from aqueous solutions. Chemosphere 2001,44, 1033–1040.

97. Li, F., Wang, Y., Yang, Q., Evans, D.G., Forano, C., Duan, X. Study on adsorption of glyphosate (N-phosphonomethyl glycine) pesticide on MgAl-layered double hydroxides in aqueous solution. J. Hazard. Mater. 2005, 125, 89–95.

98. Tomas Undabeytiaa,, Shlomo Nirb, Trinidad Sanchez-Verdejoa, Jaime Villaverdea, Celia Maquedaa, Esmeralda Morilloa. A clay–vesicle system for water purification from organic pollutants.Water Research 2008,42, 1211 – 1219.

99. Zhao Y.X., M.Y. Ding, D.P. Chen. Adsorption properties of mesoporous silicas for organic pollutants in water. Analytica Chimica Acta 2005,542,193–198.

100. Sherif A. El-Safty, Ahmed Shahata, Mohamed Ismaela. Mesoporous aluminosilica monoliths for the adsorptive removal of small organic pollutants. Journal of Hazardous Materials 2012,201– 202, 23– 32.

101. Gil A., F.C.C. Assis, S. Albeniz, S.A. Korili Removal of dyes from wastewaters by adsorption on pillared clays. Chemical Engineering Journal 2011,168,1032–1040.

102. Srinivasan K.R., H.S. Fogler. Use of inorgano-organo-clays in the removal of priority pollutants from industrial wastewaters: adsorption of benzo(a)pyrene and chlorophenols from aqueous solutions, Clays Clay Miner.1990, 38, 287–293.

103. Michot L.J., T.J. Pinnavaia. Adsorption of chlorinated phenols from aqueous solution by surfactant-modified pillared clays, Clays Clay Miner. 1991,39, 634–641.

104. Bouras O., J..C. Bollinger, M. Baudu, H. Khalaf. Adsorption of diuron and its degradation products from aqueous solution by surfactant-modified pillared clays, Appl. Clay Sci. 2007,37, 240–250.

Index

Index